ADVANCES IN MATERIALS RESEARCH 13

ADVANCES IN MATERIALS RESEARCH

Series Editor-in-Chief: Y. Kawazoe

Series Editors: M. Hasegawa   A. Inoue   N. Kobayashi   T. Sakurai   L. Wille

The series Advances in Materials Research reports in a systematic and comprehensive way on the latest progress in basic materials sciences. It contains both theoretically and experimentally oriented texts written by leading experts in the field. Advances in Materials Research is a continuation of the series Research Institute of Tohoku University (RITU).

Please view available titles in *Advances in Materials Research* on series homepage http://www.springer.com/series/3940

Atsushi Muramatsu
Tokuji Miyashita

Editors

Nanohybridization of Organic-Inorganic Materials

With 162 Figures

Professor Dr. Atsushi Muramatsu
Professor Dr. Tokuji Miyashita
Tohoku University, Institute for Multidisciplinary Research for Advance Materials
Katahira 2-1-1, Aoba-ku, Sendai 980-8577, Japan
E-mail: mura@tagen.tohoku.ac.jp, miya@tagen.tohoku.ac.jp

Series Editor-in-Chief:

Professor Yoshiyuki Kawazoe
Institute for Materials Research, Tohoku University
2-1-1 Katahira, Aoba-ku, Sendai 980-8577, Japan

Series Editors:

Professor Masayuki Hasegawa
Professor Akihisa Inoue
Professor Norio Kobayashi
Professor Toshio Sakurai
Institute for Materials Research, Tohoku University
2-1-1 Katahira, Aoba-ku, Sendai 980-8577, Japan

Professor Luc Wille
Department of Physics, Florida Atlantic University
777 Glades Road, Boca Raton, FL 33431, USA

Advances in Materials Research   ISSN 1435-1889
ISBN 978-3-540-92232-2   e-ISBN 978-3-540-92233-9
DOI 10.1007/978-3-540-92233-9
Springer Heidelberg Dordrecht London New York

Library of Congress Control Number: 2009929028

© Springer-Verlag Berlin Heidelberg 2009
This work is subject to copyright. All rights are reserved, whether the whole or part of the material is
concerned, specifically the rights of translation, reprinting, reuse of illustrations, recitation, broadcasting,
reproduction on microfilm or in any other way, and storage in data banks. Duplication of this publication or
parts thereof is permitted only under the provisions of the German Copyright Law of September 9, 1965, in its
current version, and permission for use must always be obtained from Springer-Verlag. Violations are liable
to prosecution under the German Copyright Law.
The use of general descriptive names, registered names, trademarks, etc. in this publication does not imply,
even in the absence of a specific statement, that such names are exempt from the relevant protective laws and
regulations and therefore free for general use.

Cover design: SPi Publisher Services

Printed on acid-free paper

Springer is part of Springer Science+Business Media (www.springer.com)

Preface of Series by the Editor-in-Chief

This book entitled "Advanced Materials Characterization for Corrosion Products formed on the Steel Surface" is published as the seventh volume in the series of "Advances in Materials Research" edited by Professors Yoshio Waseda and Shigeru Suzuki. The book is composed of 11 chapters, the contents of which altogether try to solve the currently important basic problems in corrosion and also to help the steel-making industries to realize substantially better process engineering. It will contribute to the society world widely "for preventing environmental degradation" which the editors stressed as the subtitle of the book.

The book contains recent high-level research results starting from the fundamental mechanism of corrosion to various measurement tools which are used for characterization to reveal the structures of corrosion products both statically and dynamically.

As the series editor, I would like to thank Dr. Claus Ascheron of Springer-Verlag, who always has interest in and kindly takes care of our research activity to encourage publication of this series of books.

Sendai   *Yoshiyuki Kawazoe*
December 2005

Preface

Synthesis and application of nanoparticles have numerously been reported by researchers belonging to many fields, such as materials science, chemistry, physics, etc. Nanoparticles themselves are well known to give fascinating characters in their application. However, the interest in their improvement and promotion is now turning to the hybridization of organic and/or inorganic nanomaterials. Although nanolevel hybridization is an outstandingly novel and original technique, it involves many difficulties to attain desired natures in industrial application.

On the other hand, not only the combinations of various materials but also their methods surely result in a wonderful variety of hybridized characteristics. However, an integrated and well-classified book on them is not found in any fields of science, since it seems very difficult to abridge them from the interdisciplinary point of view. In this regard, only the researchers in Hybrid Nano-material Research Center (HyNaM), Institute of Multidisciplinary Research for Advanced Materials, Tohoku University, and their colleague, can provide an overall review of such a research field. So, this book will be focused on the synthesis, characterization, and process of nanohybrid materials, including nanoparticles and ultrathin film, and gives a comprehensive survey of achievements by HyNaM.

This book represents the fundamental aspects of nanohybrid materials in synthesis procedure, characterization, and processes with some selected examples, from both the basic science and the applied engineering points of view. This is the first major compilation of new advances which covers current status and topics on new synthetic information of nanohybrid materials composed of organic and/or inorganic ones in nanometer level in one volume. The contributors were selected for their long and continuing expertise in their respective fields. We would like to express our gratitude to all those involved in manuscript preparation and publishing. In particular, we should mention the significant contributions of Dr. J. Kano, for his assistance in preparing the electronic typesetting of this volume.

Sendai

May 2009

Atsushi Muramatsu

Tokuji Miyashita

Contents

Part I Nanohybridization of Nanoparticles

---

1 Nanohybridized Synthesis of Metal Nanoparticles and Their Organization

Kensuke Naka and Yoshiki Chujo 3

1.1 Introduction 3

1.2 Design of Nanoparticle-Based Building Blocks 4

    1.2.1 Stabilized by Thiol Ligand 5

    1.2.2 Stabilized by Amine Ligands 7

    1.2.3 Stabilized by Bipyridyl Ligands 8

    1.2.4 Stabilized by Phosphine Ligands 8

    1.2.5 Controlling Numbers of Functional Group on Nanoparticle Building Blocks 9

1.3 Well-Defined Nanoparticle Hybrids with Linear Polymers 10

    1.3.1 Polymer-Grafted Metal Nanoparticles 10

    1.3.2 π-Conjugated Polymer Metal Nanoparticle Hybrids 12

1.4 Organization of Metal Nanoparticles 13

    1.4.1 Overview of Metal Nanoparticle Organizations 13

    1.4.2 Organization of Metal Nanoparticles by Dendritic Molecules 15

    1.4.3 Self-Organized Nanocomposites 19

1.5 Organic–Metal 1D Nanostructures 25

1.6 Flocculation of Metal Nanoparticles by Stimuli 27

    1.6.1 Photoresponsive Aggregation 27

    1.6.2 Metal Nanoparticles Modified with an Ionic Liquid Moiety 28

1.7 Conclusion and Outlook...................................... 32

References 33

X   Contents

2 Organic–Inorganic Hybrid Liquid Crystals: Innovation Toward "Suprahybrid Material"

Kiyoshi Kanie and Atsushi Muramatsu 41

2.1 Materials Innovation Toward "Suprahybrid Material" by the Utilization of Organic–Inorganic Hybridization 41

2.2 Organic Liquid Crystals and Lyotropic Liquid–Crystalline Inorganic Fine Particles 42

2.3 Development of Organic–Inorganic Hybrid Liquid Crystals........ 44

    2.3.1 Hybridization of Calamitic Liquid–Crystalline Amines with Monodispersed TiO_2 Nanoparticles [4] 44

    2.3.2 Hybridization of Calamitic Liquid–Crystalline Phosphates with Monodispersed α-Fe_2O_3 Fine Particles [5] 48

2.4 Summary and Prospect 51

References 52

3 Polymer-Assisted Composites of Trimetallic Nanoparticles with a Three-Layered Core–Shell Structure for Catalyses

Naoki Toshima 55

3.1 Introduction 55

3.2 Fabrication of a Core/Shell Structure 56

3.3 Fabrication of Pd/Ag/Rh Trimetallic Nanoparticles............. 59

3.4 Catalytic Properties of Pd/Ag/Rh Trimetallic Nanoparticles 64

3.5 Au/Pt/Rh Trimetallic Nanoparticles 68

3.6 Pt/Pd/Rh Trimetallic Nanoparticles 74

3.7 Concluding Remarks.. 76

References 78

4 Fabrication of Organic Nanocrystals and Novel Nanohybrid Materials

Tsunenobu Onodera, Hitoshi Kasai, Hidetoshi Oikawa, and Hachiro Nakanishi 81

4.1 Introduction 81

4.2 PDA Nanocrystals... 83

    4.2.1 Fabrication Technique: Reprecipitation Method........... 83

4.3 Optical Properties of PDA Nanocrystals....................... 87

4.4 Hybridized PDA Nanocrystals 88

    4.4.1 Silver-Deposited PDA Nanocrystals Fabricated Using Surfactants as Binder........................... 88

    4.4.2 Silver-Deposited PDA Nanocrystals Produced by Visible-Light-Driven Photocatalytic Reduction......... 94

4.5 Summary and Future Scopes................................. 98

References 99

Contents   XI

Part II Nanohybridized Thin Films

5 Polymer Nanoassemblies and Their Nanohybridization with Metallic Nanoparticles

Masaya Mitsuishi, Jun Matsui, and Tokuji Miyashita 103
5.1 Introduction 103
5.2 Hybrid Nanoassemblies via Polymer Nanosheets 106
5.3 Spectroscopic Properties of Hybrid Assemblies 111
5.4 Hybrid Polymer Nanosheet Fabricated from Core–Shell
     Nanoparticles 114
    5.4.1 Synthesis of Magnetic Nanoparticle Covered
     with Poly(N-Alkylmethacrylamide)s 115
    5.4.2 Monolayer Behavior of p(alkylMA)s-Coated Iron Oxide
     Nanoparticles at the Air–Water Interface 116
    5.4.3 Surface Morphology of Nanoparticle Monolayer at
     Different Deposition Pressures 119
    5.4.4 Multilayer Formation 120
5.5 Conclusion 121
References 122

6 Single Molecular Film for Recognizing Biological Molecular Interaction: DNA–Protein Interaction and Enzyme Reaction

Kazue Kurihara..125
6.1 Introduction 125
6.2 Surface Forces Measurement 125
6.3 Interaction Between Nucleic Acid Bases 126
6.4 Surface Immobilization of Protein Modified with His-Tag 127
    6.4.1 pH Dependence of DSIDA Interactions 129
    6.4.2 Interactions Between DSIDA Monolayers
     in the Presence of Cu^{2+} 130
    6.4.3 AFM Imaging of Immobilized His-Tag Sigma A Protein ...130
6.5 Specific Interactions Between Enzyme–Substrate Complexes:
     Heptaprenyl Diphosphate Synthase 130
    6.5.1 Interaction Between Subunit I and Subunit II
     upon Approach 131
    6.5.2 Interaction Between Subunit I and Subunit II
     upon Separation 132
    6.5.3 Selectivity in Substrate–Enzyme Complexation 134
6.6 Sequence-Dependent Interaction Between a Transcription Factor
     and DNA...136
6.7 Conclusion 136
References 136

XII   Contents

Part III Nanohybridized Fine Porous Materials

---

7 Organic–Inorganic Hybrid Mesoporous Silica
Satoru Fujita, Mahendra P. Kapoor, and Shinji Inagaki 141
7.1   Introduction 141
7.2   Postsynthetic Grafting Methods 142
7.3   Direct Synthesis from Monosilylated Organic Precursors.........143
7.4   Direct Synthesis from Polysilylated Organic Precursors..........146
    7.4.1   Synthesis of PMOs from 100% Polysilylated Organic
        Precursors 147
    7.4.2   Synthesis of PMOs by Cocondensation 152
7.5   PMOs with Crystal-Like Pore Walls 157
    7.5.1   Synthesis of Various PMOs with Crystal-Like Pore Walls ..158
    7.5.2   Synthesis of PMO with Crystal-Like Pore Walls Obtained
        from Allyl Precursors 161
    7.5.3   Functionalization of PMO with Crystal-Like Pore Walls ...163
7.6   Summary..164
References 165

8 Organic–Inorganic Hybrid Zeolites
Containing Organic Frameworks
Katsutoshi Yamamoto and Takashi Tatsumi 171
8.1   Introduction 171
8.2   Synthesis and Physical Properties of ZOL 172
8.3   Crystallization Scheme 180
8.4   Proof of Organic Framework 183
8.5   Applications and Prospects 188
References 189

Part IV Characterization and Process of Nanohybridized Materials

---

9 Characterization of Metal Proteins
Masaki Unno and Masao Ikeda-Saito 193
9.1   Introduction 193
9.2   Proteins and Metals 193
9.3   Metals in Biology...195
    9.3.1   Iron (Fe) 196
    9.3.2   Copper (Cu) 197
    9.3.3   Zinc (Zn) 197
    9.3.4   Calcium (Ca) 198
9.4   Special Metal Cofactors 199
    9.4.1   Iron–Sulfur Cofactor 199
    9.4.2   Hemes..201

Contents   XIII

9.5   Functions of Metalloproteins 202
        9.5.1 O_2 Transport 202
        9.5.2   Electron Transfer 202
        9.5.3   Structural Roles for Metal Ions 203
        9.5.4   Function of Metalloenzymes............................. 203
9.6   Physical Methods in Characterizing Metalloproteins 204
        9.6.1   X-Ray Crystallography 204
        9.6.2   Optical Absorption Spectroscopy 207
        9.6.3   X-Ray Absorption Spectroscopy 209
        9.6.4   Vibrational Spectroscopy 210
        9.6.5   Electron Paramagnetic Resonance Spectroscopy 211
        9.6.6   Nuclear Magnetic Resonance Spectroscopy 212
        9.6.7   Mössbauer Spectroscopy 213
General References 215
References 216

10 Electron Microscopy Characterization
of Hybrid Metallic Nanomaterials
Daisuke Shindo and Zentaro Akase 219
10.1   Introduction 219
10.2   Principles of Electron Microscopy............................ 219
        10.2.1   Transmission Electron Microscopy 219
        10.2.2   High-Angle Annular Dark-Field STEM 223
        10.2.3   Analytical Electron Microscopy....................... 224
        10.2.4   Electron Holography 227
10.3   Specimen Preparation....................................... 230
        10.3.1   Ultramicrotomy 230
        10.3.2   Ion Milling.. 231
        10.3.3   Focused Ion Beam 232
10.4   Applications 233
        10.4.1   Carbon–Metal System (Carbon-Encapsulated Metal
                  Nanoparticles)...................................... 233
        10.4.2   Oxide–Oxide System ($YBa_2Cu_3O_y$ High-T_c
                  Superconductor)..................................... 236
        10.4.3   Metal-Based Compound System
                  (Nd–Fe–B Nanocomposite Magnet) 238
        10.4.4   Metal–Oxide System (Co-CoO Recording Tape) 239
        10.4.5   Polymer–Metal System (Silver–Epoxy Paste) 242
References 245

11 Supercritical Hydrothermal Synthesis
of Organic–Inorganic Hybrid Nanoparticles
T. Adschiri and K. Byrappa 247
11.1   Introduction 247
11.2   Supercritical Hydrothermal Technique 249
11.3   Apparatus 251

XIV Contents

11.4 Basic Principles of Supercritical Hydrothermal Synthesis 253
11.5 Supercritical Hydrothermal Synthesis of Metal Oxides. 258
11.6 Supercritical Hydrothermal Synthesis of Hybrid Organic–Inorganic
     Nanoparticles 262
     11.6.1 Organic–Inorganic Hybrid Nanoparticles 262
     11.6.2 Supercritical Hydrothermal Organic–Inorganic Hybrid
            Nanoparticles 263
     11.6.3 Mechanism of Formation of Organic–Inorganic Hybrid
            Nanoparticles 265
     11.6.4 Experimental Study of the Supercritical Hydrothermal
            Synthesis of Organic–Inorganic Hybrid Nanoparticles with
            Size and Shape Control 268
     11.6.5 Dispersibility of Organic–Inorganic Hybrid Nanoparticles . . 271
     11.6.6 Evaluation Techniques for Organic Ligand Molecules
            in Organic–Inorganic Hybrid Nanoparticles 273
11.7 Self-Assembly of Hybrid Organic–Inorganic Nanoparticle 274
11.8 Conclusion 276
References 277

Index 281

List of Contributors

Tadafumi Adschiri
WPI Research Center
Advanced Institute for Materials
Research
Tohoku University, Katahira
Aoba-ku, Sendai 980-8577
Japan

Zentaro Akase
Institute of Multidisciplinary
Research for Advanced Materials
Tohoku University, Katahira
Aoba-ku, Sendai 980-8577
Japan

K. Byrappa
University of Mysore, P.B. No. 21
Manasagangotri P.O., Mysore 570006
India

Yoshiki Chujo
Department of Polymer Chemistry
Graduate School of Engineering
Kyoto University, Katsura
Nishikyo-ku, Kyoto 615-8510
Japan

Satoru Fujita
Toyota Central R&D Labs., Inc.
Nagakute, Aichi 480-1192
Japan

Masao Ikeda-Saito
Hybrid Nano-material Research
Center, Institute of Multidisciplinary
Research for Advanced Materials
Tohoku University, Katahira
Aoba-ku, Sendai 980-8577
Japan

Shinji Inagaki
Toyota Central R&D Labs., Inc.
Nagakute, Aichi 480-1192
Japan

Kiyoshi Kanie
Hybrid Nano-material Research
Center, Institute of Multidisciplinary
Research for Advanced Materials
Tohoku University, Katahira
Aoba-ku, Sendai 980-8577
Japan

Mahendra P. Kapoor
Toyota Central R&D Labs., Inc.
Nagakute, Aichi 480-1192
Japan

Hitoshi Kasai
Institute of Multidisciplinary
Research for Advanced Materials
Tohoku University, Katahira
Aoba-ku, Sendai 980-8577
Japan

XVI   List of Contributors

Kazue Kurihara
Hybrid Nano-material Research
Center, Institute of Multidisciplinary
Research for Advanced Materials
Tohoku University, Katahira
Aoba-ku, Sendai 980-8577
Japan

Jun Matsui
Hybrid Nano-material Research
Center, Institute of Multidisciplinary
Research for Advanced Materials
Tohoku University, Katahira
Aoba-ku, Sendai 980-8577
Japan

Masaya Mitsuishi
Hybrid Nano-material Research
Center, Institute of Multidisciplinary
Research for Advanced Materials
Tohoku University, Katahira
Aoba-ku, Sendai 980-8577
Japan

Tokuji Miyashita
Hybrid Nano-material Research
Center, Institute of Multidisciplinary
Research for Advanced Materials
Tohoku University, Katahira
Aoba-ku, Sendai 980-8577
Japan

Atsushi Muramatsu
Hybrid Nano-material Research
Center, Institute of Multidisciplinary
Research for Advanced Materials
Tohoku University, Katahira
Aoba-ku, Sendai 980-8577
Japan

Kensuke Naka
Department of Chemistry and
Materials Technology Graduate
School of Technological Science
Kyoto Institute of Technology
Matsugasaki, Sakyo-ku, Kyoto
606-8585
Japan

Hachiro Nakanishi
Institute of Multidisciplinary
Research for Advanced Materials
Tohoku University, Katahira
Aoba-ku, Sendai 980-8577
Japan

Hidetoshi Oikawa
Institute of Multidisciplinary
Research for Advanced Materials
Tohoku University, Katahira
Aoba-ku, Sendai 980-8577
Japan

Tsunenobu Onodera
Institute of Multidisciplinary
Research for Advanced Materials
Tohoku University, Katahira
Aoba-ku, Sendai 980-8577
Japan

Daisuke Shindo
Institute of Multidisciplinary
Research for Advanced Materials
Tohoku University, Katahira
Aoba-ku, Sendai 980-8577
Japan

Takashi Tatsumi
Catalytic Chemistry Division,
Chemical Resources Laboratory,
Tokyo Institute of Technology,
4259-R1-9 Nagatsuta, Midori-ku
Yokohama 226-8503
Japan

Naoki Toshima
Department of Applied Chemistry,
Faculty of Engineering, and
Advanced Materials Institute,
Tokyo University of Science
Yamaguchi, SanyoOnoda-shi
Yamaguchi 756-0884
Japan

Masaki Unno
Hybrid Nano-material Research
Center
Institute of Multidisciplinary
Research for Advanced Materials
Tohoku University, Katahira
Aoba-ku, Sendai 980-8577
Japan

Katsutoshi Yamamoto
Department of Chemical
and Environmental Engineering
Faculty of Environmental
Engineering
The University of Kitakyushu
1-1 Hibikino, Wakamatsu-ku
Kitakyushu 808-0135
Japan

Part I

Nanohybridization of Nanoparticles

1

Nanohybridized Synthesis of Metal Nanoparticles and Their Organization

Kensuke Naka and Yoshiki Chujo

1.1 Introduction

Metal nanoparticles have various unusual chemical and physical properties compared with those of metal atoms or bulk metal due to the quantum size effect and their large superficial area, which make them attractive for applications such as optics, electronics, catalysis, and biology [1, 2]. The catalytic properties of metal nanoparticles have generated great interest over the past decade. Among various metal nanoparticles, gold nanoparticles have tremendously high molar absorptivity in the visible region. Particle aggregation results in further color changes of gold nanoparticles solutions due to mutually induced dipoles that depend on interparticle distance and aggregate size. This phenomenon can be applied to various sensing systems [3–7]. The assembly of gold, silver, or copper nanoparticle monolayers is one of the ideal substrate for surfaced-enhanced Raman scattering (SERS) [8, 9].

Bare metal nanoparticles are prepared by employing physical methods such as mechanic subdivision of metallic aggregates and evaporation of a metal in a vacuum by resistive heating or laser ablation. Chemical methods such as the reduction of metal salts in solution are the most convenient ways to control the size of the particles and modified the surface chemical composition. To exploit nanoparticle properties for future device fabrication, self-organization of nanoparticles in a controlled manner is required. To construct such devices, new fabrication techniques must be developed with suitable metal nanoparticle-based building blocks. Several patterns for self-assemblies of the metal nanoparticles are schematically illustrated in Fig. 1.1.

A number of outstanding reviews on the synthesis and assembly of metal nanoparticle-based building blocks have appeared [2, 10]. This chapter highlights recent fabrication techniques of hybrid metal nanoparticles by controlled self-organization of the metal nanoparticle-based building blocks. Design of nanoparticle hybrids will be first focused on using building blocks for further assemblies. Several recent efforts have been concentrated on a system that

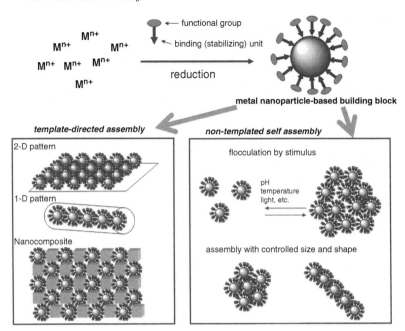

Fig. 1.1. Preparation of metal nanoparticle-based building blocks and their utilization for bottom-up nanofabrication

involves metal nanoparticles and dendritic molecules such as dendrimers and cubic silsesquioxanes. The organization of metal nanoparticles in superstructures of desired shape and morphology by using the dendritic molecules will be the main topic of this chapter. Several examples will also be described for the metal nanoparticles with various kinds of stimuli-responsive property, which involves aggregation or flocculation of metal nanoparticles in solution.

1.2 Design of Nanoparticle-Based Building Blocks

To exploit nanoparticle properties for future device fabrication by a "bottom-up method," fundamental and key challenge is the design of the nanoparticle-based building blocks. From a synthetic chemical viewpoint, these metal nanoparticles required to be functionalized with a wide variety of organic moieties using simple chemical process. The most important requirement for this purpose is that repeatedly isolated and redissolved in common solvents, and handle and characterize as usual molecules. The metal nanoparticles can be stabilized by solvents or ions (Fig. 1.2). Although they are eventually useful for catalysis, they tend to irreversibly aggregate over time or when removed from the solvent. To prevent the agglomeration, metal nanoparticles are protected by polymers which have coordination properties for the

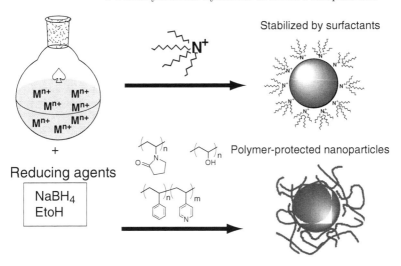

Fig. 1.2. Preparation of metal nanoparticles by chemical reduction in the presence of cationic surfactants (**a**) and polymers (**b**)

metal surfaces that are usually prepared as colloidal forms by reduction of metal ions in the presence of polymeric stabilizers such as poly(vinyl alcohol), poly(vinylpyrrolidone), and poly(vinyl ether) (Fig. 1.2) [11, 12]. Although these polymer-stabilized metal nanoparticles are stable in solution and easily prepared, requirement of large amount of polymers to stabilized metal nanoparticles inhibits close-packed assembly of the metal cores. Close-packed assembly of the metal nanoparticles is expected to produce complex electronic and functions based on quantum mechanical coupling of conduction electrons localized in each nanoparticle [13, 14]. Stabilization of metal nanoparticles by ligands as capping agents can enable further manipulation, control solubility characterization, and facilitate their analysis. Examples of ligand-stabilized metal nanoparticles are overviewed in the following sections.

1.2.1 Stabilized by Thiol Ligand

Mercapto groups (RSH) have been used as stabilizers of metal nanoparticles, especially for gold, in recent years, since Brust and coworkers [15] reported the preparation of ligand-stabilized gold nanoparticles by protecting the nanoparticles with a self-assembled monolayer of dodecanethiolate (Fig. 1.3). In this method (the Brust–Schiffrin method), $AuCl_4^-$ is transferred from aqueous phase to toluene using tetraoctylammonium bromide as a phase-transfer reagent and then reduced by $NaBH_4$ with alkanethiols, yielding nanoparticles having average core diameters in the range of 2–8 nm. The size of the resulting gold nanoparticles decreases with increasing thiol/$HAuCl_4$ reaction molar ratio. The crude product is modestly polydisperse, but can be separated into rather monodisperse samples by fractional

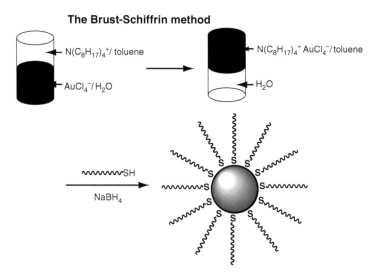

Fig. 1.3. Preparation of gold nanoparticles protected with monolayer of alkanethiols by two-phase method (the Brust–Schiffrin method)

precipitation. These monolayer-protected gold nanoparticles are repeatedly isolated and redissolved in common organic solvents such as toluene, hexane, and dichloromethane without irreversible aggregation or fusion. They are easy to handle and characterize just as stable organic compounds do. Monolayer-protected silver and palladium nanoparticles can also be prepared by the same protocol [14, 16].

Based on this system, efficient strategies to functionalize the metal nanoparticles have been developed. Murray et al. [17, 18] reported that surface functionalization can be achieved by simple place-exchange reactions of the alkanethiol monolayer-protected metal nanoparticles with ω-functionalized alkanethiolates (Fig. 1.4). The rate and equilibrium stoichiometry are controlled by factors that include the reaction feed mole ratio as the equation shown in Fig. 1.4. Alternatively, ω-functionalized alkanethiolates and dialkyl disulfides can be directly employed instead of alkanethiols as the same protocol in a single phase system [1].

The alkanethiolate monolayer-protected nanoparticles are insoluble in water, and while the place-exchanged nanoparticles bearing polar ω-functionalities dissolve in several polar solvents, none has proven to be water soluble. Murray et al. [19] prepared water-soluble gold nanoparticles stabilized by a water-soluble tiopronine which can be repeatedly isolated and redissolved. Sodium (3-mercaptopropionate) (MPA)-stabilized gold nanoparticles were successfully prepared by citrate reduction of $HAuCl_4$ in the presence of MPA. Simultaneous addition of citrate and MPA is essential to obtain their stable dispersions. The size of the particles can be controlled by the ratio of MPA/gold [20].

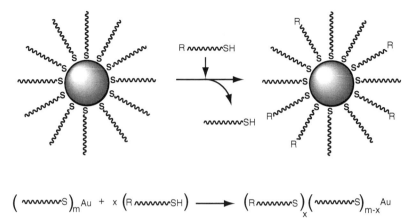

Fig. 1.4. General scheme for the place-exchange reaction between alkanethiol monolayer-protected gold nanoparticles and various functional thiols. In the bottom scheme, x and m are the number of new and original ligands, respectively

1.2.2 Stabilized by Amine Ligands

The Brust method of nanoparticle synthesis can be applied to generate amine-stabilized nanoparticles by simply substituting an amine for the thiol ligands (Fig. 1.5). Leff et al. [21] demonstrated that 2.5–7.0 nm average diameter gold nanoparticles can be stabilized by n-alkylamines. Jana and Peng [22] reported the synthesis of monodispersed gold, silver, copper, and platinum nanoparticles in a single organic phase. Although amines form only weakly bound and chemically unstable monolayers on bulk gold surfaces, the amine-capped nanoparticles are nearly as stable as their thiol-capped counterparts. Hiramatsu and Osterloh [23] reported a method for a large-scale synthesis of organoamine-protected gold and silver nanoparticles in 6–21 nm for Au and 8–32 nm for Ag size ranges and with polydispersities as low as 6.9%. The organoamine-protected gold nanoparticles of variable sizes formed by refluxing a solution of tetrachloroauric acid and oleylamine in toluene over the course of 120 min. The reducing equivalents in the reaction are provided by the amine, which can undergo metal ion-induced oxidation to nitriles. The weakly absorbed oleylamine on the nanoparticles can be readily displaced with aliphatic thiols by adding a solution of the oleylamine-ligated gold nanoparticles in toluene to a boiling solution of 5–10 equivalents (based on gold) of the thiol in the same solvent. Thiol-capped silver nanoparticles are obtained analogously at room temperature in chloroform. Aslam et al. [24] reported the synthesis of water-soluble gold nanoparticles with core diameters of 9.5–75 nm via reducing tetrachloroauric acid by oleylamine in water.

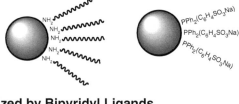

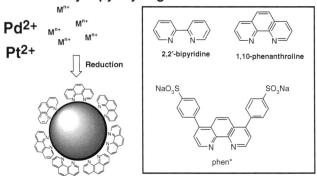

Fig. 1.5. Examples for ligand-stabilized metal nanoparticles prepared by reduction of metal ions in the presence of amine, phosphine, and bipyridyl ligands

1.2.3 Stabilized by Bipyridyl Ligands

Ligand-stabilized metal nanoparticles with bipyridyl derivatives were reported as a simple method to protect colloidal particles with skin of ligand molecules which are not removed by isolation and are completely dry (Fig. 1.5) [25–29]. Palladium(II) acetate can be reduced in acetic acid solution by 1 atom hydrogen at room temperature with 2,2'-bipyridine or 1,10-phenanthroline, followed by O_2 treatment [26]. Water-soluble, air-stable platinum nanoparticles were synthesized by stirring an acetic acid solution of platinum(II) acetate with 1,10-phenanthroline-4,7-bis(benzene-4-sulfonate) disodium salt (phen*) [28].

1.2.4 Stabilized by Phosphine Ligands

Phosphine-stabilized gold nanoparticles, originally formulated as Au_{55} $(PPh_3)_{12}Cl_6$ by Schmid et al. [30] have been widely studied as models for metallic catalysts and precursors to other functionalized nanoparticle building blocks possessing well-defined metallic cores. Stable gold colloids stabilized by phosphines became available if $HAuCl_4$ was reduced by trisodium citrate and treated with an excess of $PPh_2(C_6H_4SO_3Na-m)$ or better $P(C_6H_4SO_3Na-m)_3$, which can be isolated by concentration of dilute solutions and addition of ethanol [25]. Since the phosphine is easily oxidized in air, the synthesis has to be carried out in an inert atmosphere. Palladium and platinum colloids can be prepared by analogous procedures. Hutchison et al. [31] reported a convenient synthesis of 1.5 nm average diameter triphenylphosphine-stabilized gold nanoparticles. The phosphine-stabilized gold nanoparticles undergo rapid exchange of capping ligand phosphine with dissociated and added phosphine in dichloromethane solvent at room temperature [32]. Ligand exchange reactions of triphenylphosphine-stabilized nanoparticles with ω-functionalized thiols provides a versatile approach to functionalized, 1.5 nm gold nanoparticles from a single precursor [33–35].

1.2.5 Controlling Numbers of Functional Group on Nanoparticle Building Blocks

To fabricate more complex assemblies comprising nanoparticle building blocks, controlling the number of functional groups on a metal nanoparticle surface is required. Two groups reported synthesis of monofunctionalized gold nanoparticles by using peptide synthesis protocols (solid-phase reaction) [36,37]. The key point of this method is the low-density packing of functional groups present in many solid-phase supports. Each functional group on a common polystyrene Wang resin bead possesses a rough volume of at least ca. 9 nm^3 when suspended in DMF, and thus a nanoparticle with a diameter smaller or around 2 nm can be loaded on the solid phase through a single bond per particle. Direct evidence of monofunctionalization is revealed by dimerization of the isolated gold nanoparticles, which were treated by a slow addition of ethylenediamine as a bridging linker in the presence of a condensation reagent. TEM images demonstrated that the dimer species are dominant and that 55–66% of particles on the TEM grid are found to undergo dimerization.

Dimers are of special interest because of their application as substrates in surface-enhanced Raman spectroscopy (SERS). Although the above approach is limited to very small particles, Shumaker-Parry et al. [38] synthesized gold nanoparticle dimers by a solid-phase approach using a simple coupling reaction of asymmetrically functionalized particles. Although a single functional group was not formed on a metal nanoparticle surface here, this method can be used to generate dimers with a wide size range and containing two nanoparticles with different sizes.

With a single functional group attached to the surface, such nanoparticles can be treated and used as molecular nanobuilding blocks to react with other chemicals to form nanomaterials with all the nanoparticle building blocks linked together by covalent bonding. Huo et al. [39] reported synthesis of a "nanonecklace" from monofunctionalized gold nanoparticles and polylysine by using an activated reagent in a solution. Goodson III et al. examined the nature of the electromagnetic coupling and its influence on nonlinear properties of Au-necklace particles with the aid of time-resolved spectroscopy and they found the presence of strong electromagnetic coupling between the neighboring particles [40].

Stellacci et al. [41,42] showed that mixtures of thiolated molecules formed ordered alternating phase when assembled on surfaces of nanoparticles. These

10   K. Naka and Y. Chujo

types of domains will profoundly demarcate the two diametrically opposed singularities at the particle poles. In the case of a self-assembled ligand shell, the polar singularities manifest themselves as defect points, i.e., sites at which the ligands must assume a nonequilibrium tilt angle. Ligands at the poles should be the first molecules to be replaced in the place-exchange reactions. Gold nanoparticles coated with a binary mixture of 1-nonanethiol and 4-methylbenzenethiol were prepared. To place-exchange at the polar defects, the gold nanoparticles were dissolved in a solution containing 40 molar equivalent of 11-mercaptoundecanoic acid (MUA) activated by N-hydroxysuccinimide [43]. A two-phase polymerization reaction was performed by combining a toluene solution containing the MUA-functionalized nanoparticles with a water phase containing divalent 1,6-diaminohexane. A TEM image of a precipitate formed at the water-toluene interface showed a large population of linear chains of nanoparticles. The chains were soluble in dichloromethane and showed a film-forming property.

1.3 Well-Defined Nanoparticle Hybrids with Linear Polymers

1.3.1 Polymer-Grafted Metal Nanoparticles

Construction of well-defined polymers containing metal nanoparticles is one of the attractive targets for polymer and material chemistry and would become nanoparticle-based building blocks for further fabrication toward nanodevices. To connect a polymer end with a surface of metal nanoparticle, covalent and coordination bonds are considered to be the most suitable links as a similar concept as the numerous reports on the metal nanoparticles functionalized with the low molecular weight ligands. Attachment of metal nanoparticles to synthetic polymers adds film-forming properties to the metal nanoparticles and also provides the opportunity for microphase separation between the metal nanoparticles and the polymer matrix. The pioneer work in this field has been explored by Mirkin et al., who introduced a strategy for covalent attachment of DNA strands to gold nanoparticles [6].

A polymer-grafted metal nanoparticle was produced by reduction of metal ion with a bipyridyl-terminated polymer (Fig. 1.6) [44]. As described in Sect. 1.2.3, the bipyridyl ligands stabilized palladium and platinum nanoparticles were simply prepared and the ligand molecules were not removed even after isolation and drying. The palladium nanoparticles were synthesized by stirring an acetic acid solution of palladium(II) acetate and bipyridyl-terminated poly(oxyethylene) (bpy-POE) (MW = 2,000) (molar ratio 8:1) under 1 atm hydrogen at room temperature. After isolation, the polymer-grafted metal nanoparticles became soluble in various solvents such as CH_2Cl_2, $CHCl_3$, MeOH, acetone, and water. The solubility of the resulting nanoparticles was the same as that of poly(oxyethylene). These solutions were stable

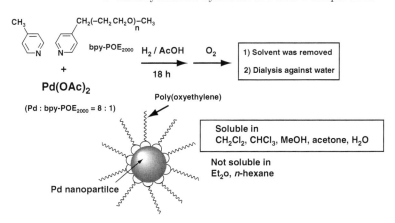

Fig. 1.6. Polymer-grafted palladium nanoparticles prepared by reduction of metal ions with H₂ in the presence of a bipyridyl-terminated poly(oxyethylene)

for more than half a year at room temperature under air. Although the formation of palladium nanoparticles was reported when an acetic acid solution of palladium(II) acetate was stirred with a large excess amount of poly(ethylene oxide) (MW = 900,000) under 1 atm hydrogen at room temperature, extremely larger amounts of the polymer (molar ratio 1:10) were required and the precipitation of the metal occurs after a comparably short time (1–2 days) [45].

Murray and coworkers [46] reported that a thiolated polymer, α-methoxy-ω-mercapto-poly(ethylene glycol) (PEG-SH, MW = 5,000), was used to produce polymeric monolayer-protected gold nanoparticles. The PEG-SH ligand was also selected because of the dissolution of LiClO₄ electrolyte. Thus, the resulting nanoparticles can be applied for new polymer electrolyte media, a semisolid having an ionically conductive nanophase around a metallic core. Thiol-functionalized poly(acrylamides) were also attached on the surfaces of metal nanoparticles [47, 48].

Gold nanoparticles grafted with hydrophobic homopolymer chains such as thiol-capped polystyrenes were prepared [49]. Kramer and Pine [50] reported about gold nanoparticles coated with a mixture of two different polymeric thiols, which created an amphiphilic shell as suggested by the observed accumulation of particles at the interface separating the domains of polystyrene and poly(2-vinylpyridine). Zubarev et al. [51] reported an efficient method to produce amphiphilic gold nanoparticles with an equal number of hydrophobic and hydrophilic arms. They used a V-shaped polybutadiene–poly(ethylene glycol) amphiphile containing a functional group at its junction point.

A more promising approach for the preparation of covalently attached polymer is given by the use of immobilized initiators for the in situ generation of the grafted polymer, which is the so-called a "grafting-to" method. This method can be applied for a variety of monomers utilizing radical, cationic, and anionic polymerization. To obtain a homogeneous grafted polymer, first

12 K. Naka and Y. Chujo

the surface grafting density has to be uniform, second the polydispersity index should be near 1, and finally all chains should be linear. These requirements are achieved by living polymerization. Fukuda and coworkers [52] reported synthesis of gold nanoparticles coated with well-defined, high-density polymer brushes by surface-initiated living radical polymerization. The "grafting-to" method applied for poly(methyl methacrylate) shells by surface-confined living radical polymerization [53], a thermoresponsive polymer by surface-induced reversible-addition-fragmentation chain-transfer polymerization [54], and ring-opening polymerization of lactones [55]. Palladium nanoparticles with surface initiators for ring-opening polymerization of 2-methyl-2-oxazoline were prepared by alcohol reduction method of palladium(II) acetate in the presence of a bipyridyl ligand [56].

1.3.2 π-Conjugated Polymer Metal Nanoparticle Hybrids

Composites of metal nanoparticles and π-conjugated polymers are useful for several applications [57]. Incorporation of metal nanoparticles enhances conductivity of the polymers [58]. The electronic structure of the polymer chain strongly influences the characteristic of embedded metal nanoparticles [59,60]. These composites have potential as catalysts since the π-conjugated polymer might provide a potentially efficient route for shuttling of electronic charge to the catalytic centers [61, 62]. If π-conjugated polymers have strong electron-donating properties, reduction of metal ions occurs via the electron transfer from the π-conjugated polymers to the metal ions, leading to the formation of metal nanoparticles. Polymers having reducing as well as stabilizing abilities would provide "clean" materials because the additional reducing agent would not be necessary. Huang and coworkers have reported about the simultaneous in situ reduction of metal ions, palladium(II) and gold(III), to their elemental forms, which, however, were not dispersed in most common solvents [63–65].

π-Conjugated polymer-protected gold, palladium, and platinum nanoparticles of narrow size distribution in stable colloidal form were prepared via reduction of each metal salt by a π-conjugated poly(dithiafulvene) (PDF) having electron donating properties (Fig. 1.7) [66,67] A series of π-conjugated PDFs have been prepared by cycloaddition polymerization of aldothioketenes and their alkanethiol tautomers, which were derived from aromatic diynes [68]. Reduction of metal ions by the π-conjugated polymer forms metal nanoparticles and the resulting oxidized polymer protected the metal nanoparticles. The resulting DMSO solution of the polymer-protected gold nanoparticles was stable without precipitation for more than a month at room temperature under air. Due to the effective expansion of π-conjugation of PDF by charge transfer, the absorption spectrum of the oxidized PDF showed a red-shift compared with that of the neutral PDF. The palladium nanoparticle dispersed π-conjugated PDF exhibited an anodic shift of oxidation potential for the dithiafulvene (DF) unit compared with that in the neutral PDF.

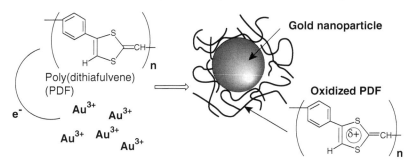

Fig. 1.7. Formation of π-conjugated poly(dithiafulvene) (PDF)-protected metal nanoparticles via reduction of metal ions by the π-conjugated electron-donating PDF

The oxidized polymer with delocalized positive charges provided both steric and electrostatic stabilization, protecting the metals as stable colloidal forms.

Polymers having pendant reducing groups were used to prepare metal nanoparticles, in which the additional step of introducing a reducing agent would not be necessary. Henpenius et al. [69] prepared gold nanoparticles inside polystyrene–oligothiophene–polystyrene triblock micelles in toluene without additional reducing agents. Gold nanoparticles were produced in aqueous solutions by a polyelectrolyte that possesses pendant terthiophene derivatives as the reducing group for HAuCl$_4$ [70]. Polysilane shell-crosslinked micelles, where the polysilane core is surrounded by a partially crosslinked shell of poly(methacrylic acid), can be used as the template for the synthesis of metal nanoparticles [71].

1.4 Organization of Metal Nanoparticles

1.4.1 Overview of Metal Nanoparticle Organizations

The realization of technologically useful nanoparticle-based materials depends not only on the quality of the nanoparticles (e.g., size and shape) but also on their spatial orientation and arrangement. The building and patterning of the metal nanoparticles into organized structures is a potential route to chemical, optical, magnetic, and electronic devices with useful properties [72–74]. Fabrication of nanoparticles into one-, two-, and three-dimensional structures is an attractive, challenging target for developing bottom-up nanofabricate techniques, since the collective properties of the resulting structures are expected to be different from those of the corresponding isolated nanoparticles [75, 76]. The development of practical strategies for the assembly of metal nanoparticles into order structure is thus an area of considerable current interest. The organization of metal nanoparticles in superstructures of desired shape and morphology is a challenging research area. Various strategies such as solvent evaporation [77, 78], electrostatic attraction [79], hydrogen bonding [80–82], DNA-driven assembly [5], and crosslinking induced by organic molecules [83] have been developed to form nanoparticle assemblies. Rotello et al. [84] have shown many excellent fabrication techniques to utilize polymers for controlled assemblies of metal nanoparticles.

Especially, the assembly of metal nanoparticles on solid supports has attracted substantial research efforts as a consequence of their unique electronic and optical properties [85–88]. For example, colloidal gold and silver nanoparticles are excellent building blocks for surface-enhanced Raman scattering-active substrates [89,90]. Different techniques, including Langmuir–Blodgett technique and electrophoretic deposition technique [91–94] have been used to obtain two-dimensional assembling of metal nanoparticles. One of the efficient methodologies for the organization involves the utilization of electrostatic interaction between a substrate and metal nanoparticles [95–98]. Schmitt et al. reported layered nanocomposite films prepared from colloidal metal nanoparticles through electrostatic interaction. Most successful routes required the surface modification of a solid substrate by positively charged polyelectrolytes because of the negatively charged metal nanoparticles.

The organization of metal nanoparticles in 1D assembly has met with limited success compared with 2D and 3D assemblies. Most of the successful methods required appropriate templates. Schmid and coworkers [99, 100] used ordered channels of porous alumina as a template to obtain linear arrangements of gold nanoparticles. The utilization of structured carbon and alumina substrate as a template also has been reported [101–103]. Teranishi and coworkers [101] reported the fabrication of one-dimensional arrangement of size-controlled gold nanoparticles in combination with a nanoscale ridge-and-valley structured substrate. Biological macromolecules have been used to build defined inorganic nanostructures. Among the biological macromolecules, DNA is one of the most interesting templates because of its diameter of only 2 nm and the micrometer-long distribution of well-defined sequence of DNA bases. Several papers reported the fabrication of one-dimensional arrangement templated by DNA [104–111]. Other biological materials such as peptide, virus, lipid, and biopolymers were also used [112–115]. The most important advantage of using biological materials is their single molecular weights, which provide controlled length of one-dimensional arrays. Nonbiological templates such as carbon nanotubes and polycation molecules templated self-assembly of gold nanoparticles also have been reported [116–118].

A scanning probe lithography technique, "dip-pen" nanolithography, has demonstrated the ability to pattern the metal nanoparticles with sub-100 nm resolution on substrates [119–121]. To draw a familiar analogy, the AFM tip acts as a "pen," the organic or inorganic molecules act as "ink," and the surface acts as a "paper" for nanostructures to be "drawn" on.

1.4.2 Organization of Metal Nanoparticles by Dendritic Molecules

This section will highlight the organization of metal nanoparticles by using dendritic molecules with 3D, 2D, and 1D arrangements. Dendrimers are monodisperse macromolecules with a regular and highly branched 3D architecture. The starburst structures are disk-like shapes in the early generations, whereas the surface branch cell becomes substantially more rigid and the structures are spheres [122]. Dendrimers have been attracting much attention as useful stabilizers of metal nanoparticles in solution, since the first successful report by Crooks and his coworkers in 1998 [123]. Most of these studies deal with the reduction in a solution where dendrimers are molecularly dissolved, essentially isolated from one another in solution, and metal ions are encapsulated in the interior space of the single dendrimer. The ions are reduced to metal atoms which self-assemble into a metal nanoparticle within the single dendrimer. Colloidal forms of gold nanoparticles in the 2–3 nm size regime were prepared by in situ reduction of $HAuCl_4$ in the presence of amine-terminated poly(amidoamine) (PAMAM) dendrimers [124–127]. The dendrimers operate as a very effective protective colloid for the preparation of gold particles since only a very small amount of the dendrimers is required to obtain nanometer size of gold particles compared to other linear polymers. Transmission electron microscopy (TEM) and dynamic light scattering (DLS) data suggested that the dendrimers adsorbed on the gold nanoparticles as a monolayer [127] The driving force for the interaction of the metal nanoparticles with the dendrimers is an association of gold with the primary amine terminal groups and the interior secondary and tertiary amines (especially for early generations). The resulting dendrimer/gold nanocomposites can be isolated from alcohol/water solutions by precipitation with tetrahydrofuran (THF) [126]. The required concentration is large in the case of weak interaction between platinum nanoparticles and dendrimers with amino groups [125].

Cubic silsesquioxanes are also regarded as the dendritic molecules for self-organization of metal nanoparticles. Octa(3-aminopropyl)octasilsesquioxane octahydrochloride (OAPOSS) is regarded as a structure equivalent of the G1.0 PAMAM dendrimer. In contrast to the PAMAM dendrimer, OAPOSS has an inner cubic rigid inorganic core with the size of 0.5 nm containing silicon and oxygen, which offers uniform interparticle spacing, good thermal and mechanical properties, and solvent resistance. Since the cubic silica core is rigid, the eight organic functional groups of OAPOSS are appended to the vertexes of the cube via spacer linkage [128,129]. Although the structure of the G1.0 PAMAM dendrimer is a disk-like flexible random shape, the functional groups of OAPOSS should become more rigid and the structure is spherical [130]. In contrast to the synthesis of the dendrimers, OAPOSS is simply prepared and isolated as precipitates by hydrolysis of aminopropyltriethoxysilane in an aqueous acidic methanol [131,132].

The OAPOSS-protected gold nanoparticles were prepared through reduction of $HAuCl_4$ in the presence of OAPOSS by $NaBH_4$ [133]. The size of the resulting gold nanoparticles increased with the increase in the HAuCl$_4$/OAPOSS molar ratio. The OAPOSS-protected gold nanoparticles with a diameter of 80 nm can be prepared at the HAuCl$_4$/OAPOSS molar ratio of 20. The OAPOSS-protected gold nanoparticles were assembled effectively on a slide glass [132]. The glass was immersed in the aqueous solution of the OAPOSS-protected gold nanoparticles. After the immersion, the glass was washed with H$_2$O several times to remove extra gold nanoparticles. An absorbance feature of the glass at 537 nm indicated the adsorption of the OAPOSS-protected gold nanoparticles. The absorbance continually increased with increasing the immersion time. The surface plasmon band was gradually redshifted from 525 to 566 nm. This is a consequence of overlap of the dipole resonances between neighboring gold nanoparticles on the glass substrate. That is, as the particle coverage increases, interparticle spacing becomes small compared to the incident wavelength. These results indicate that the gold nanoparticles have a positive charge and were immobilized densely on the glass substrate.

One of the efficient methodologies for the organization of 3D arrangement involves the utilization of chemical crosslinking by organic ligands. However, the combination of the metal nanoparticles with bifunctional linking molecules having rigid or flexible structures usually results in the formation of insoluble and uncontrollable aggregates in solutions. Unlike a bifunctional linker, stronger bonding of the dendritic molecules to the inorganic surfaces is expected due to a chelate or cluster effect. Rotello and coworkers [134] employed PAMAM dendrimers of different generations to both assemble gold nanoparticles and control the separation distance between them. In this system, the gold nanoparticles were functionalized with carboxylic acid terminal groups. Salt-bridge formation between the dendrimer surface-amine groups and the nanoparticle peripheral carboxylic acid groups led to electrostatic self-assembly between the dendrimer and nanoparticle component.

Chujo and Naka [135] reported an assembling of metal nanoparticles into spherical aggregates as colloidal forms in solution via self-organized spherical templates in a solution by using dendritic molecules such as OAPOSS- or amine-terminated PAMAM dendrimers (Fig. 1.8). When a methanol solution of OAPOSS and palladium(II) acetate was stirred at room temperature, the solution immediately became turbid, which suggested aggregate formation. The turbid solution gradually turned from yellow to black with an increase in the reaction time, indicating the reduction of the palladium ions. One drop of the turbid solution containing the obtained product was placed on a copper grid and allowed to evaporate the solvent under atmospheric pressure at room temperature. TEM showed that the spherical aggregates with a mean diameter of 70 nm were obtained (Fig. 1.8). These spherical aggregates were clearly composed of dark spots. These dark spots were individual palladium nanoparticles. Every nanoparticles, with a size of 4.0 ± 0.8 nm, appeared as a discrete entity in the nanosphere. Scanning electron microscopy (SEM) data also indicated the formation of spherical aggregates in size of 80 ± 20 nm.

1 Nanohybridized Synthesis of Metal Nanoparticles 17

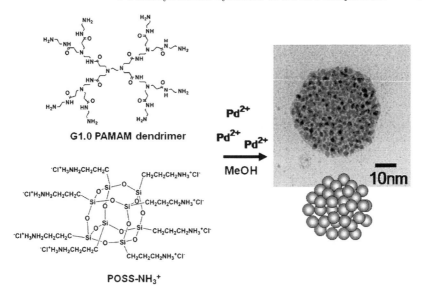

Fig. 1.8. Formation of spherical aggregates composed of palladium nanoparticles and the dendritic molecules (G1.0 PAMAM dendrimers or POSS–NH$_3^+$) in methanol

The resulting black colloidal solution was stable and neither precipitated nor flocculated over a period of several months.

The highly ordered spherical aggregates were composed of palladium nanoparticles, in which the dendritic molecules acted as crosslinkers and stabilizers for palladium nanoparticles. Palladium nanoparticles are attractive materials for catalysis [136, 137] and hydrogen storage [137]. Building the nanoparticles into hierarchical structures with stable manner should be required for these applications. TEM observations suggested that the density of palladium nanoparticles in the aggregates using OAPOSS as a template is higher than those using the amine-terminated G1 PAMAM dendrimer. From a TM-AFM image, the shapes of the aggregates using OAPOSS and the amine-terminated G1 PAMAM dendrimer were an oval and a spherical form on the plate, respectively. Increasing rigidity of the core of the dendritic molecules increased stability of the spherical form in the dry state.

The reaction and reaction-induced self-assembling process were explored in situ by means of a combined time-resolved method of small angle neutron scattering and small-angle X-ray scattering [138]. The dendrimer molecules and palladium(II) acetate first self-assemble themselves rapidly into spherical aggregates and thereafter their size was kept almost constant. Inside the aggregates, which serve as a template for the reaction, the reduction of palladium(II) acetate to palladium(0) and their self-assembly into palladium nanoparticles proceeds gradually with time over 12 h. The formation of the spherical aggregates was also supported by TEM.

Effect of different generation of the PAMAM dendrimers for the structure of the aggregates was observed. Using the G0 PAMAM dendrimer instead of the G1.0 PAMAM dendrimer under the same condition resulted in the formation of uniform, spherical aggregates with a diameter of 50 nm composed of palladium nanoparticles. On the other hand, the G2.0 PAMAM dendrimer decreased the size of the spherical aggregates. Although the exact reason for these phenomena is not entirely clear, we speculate that these results would be due to the difference of the functionality of the PAMAM dendrimers structure. That is, in the case of higher the PAMAM dendrimers generation, stronger interaction between colloids would be expected due to increase of the functional groups.

Wire-like aggregates of palladium nanoparticles were self-organized in an acidic methanol solution by using the G1.0 PAMAM dendrimer (Fig. 1.9) [139]. The combination of the G1.0 PAMAM dendrimer and metal ions in solution formed wire-like colloids, which would act as spatially constrained template for controlled synthesis of the metal nanoparticles. It is worth pointing out

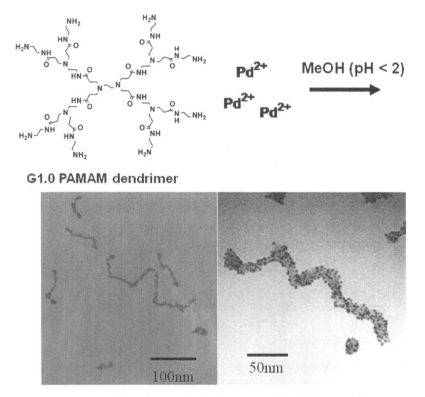

Fig. 1.9. TEM image (**a**) and magnified TEM image (**b**) of the wire-like aggregates of palladium nanoparticle with the G1 PAMAM dendrimers prepared in the acidic methanol solution. Scale bars of (**a**) and (**b**) are 500 and 100 nm, respectively

that the wire-like aggregates are formed spontaneously without any external force and templates. When the acidic methanol solution of the G1.0 PAMAM dendrimer and palladium(II) acetate was stirred at room temperature, the solution immediately became turbid. This suggests microscopic aggregate formation. The solutions gradually turned from yellow to black with increasing reaction time, indicating the reduction of palladium ions and the formation of palladium nanoparticles. One drop of the solution containing the obtained product was placed on a copper grid and allowed to evaporate the solvent under atmospheric pressure at room temperature. The TEM investigation showed the wire-like aggregates (Fig. 1.9). The length of most of the wire-like aggregates was more than $0.5\,\mu m$. These wire-like aggregates were clearly composed of dark spots. These dark spots were individual palladium nanoparticles. Every nanoparticles with a size of $4.0\pm0.8\,nm$, appeared as a discrete entity in the aggregates. A SEM investigation also indicated that the obtained products were wire-like structure. DLS showed that the average hydrodynamic radius was around $300\,nm$ indicating the presence of the aggregates in the solution.

1.4.3 Self-Organized Nanocomposites

A major challenge for nanoparticle-based nanocomposites is the preparation of thermally and mechanically stable nanocomposites with high content of metal nanoparticles and preventing both phase separation and aggregation of the metal nanoparticles in the host matrices. Several studies have been done on the self-organization of metal nanoparticles with monofunctionalized silsesquioxanes at a molecular level [100, 140]. Polyhedral octasilsesquioxane derivatives have strong tendency to crystallize. Rotello et al. [140] reported the formation of spherical aggregates with uniform internal spacing using diaminopyridine-functionalized polyhedral octasilsesquioxane and thymine-functionalized gold nanoparticles. Initially, there was the three points hydrogen bonding of the diaminopyridine unit with complementary thymine unit followed by subsequent aggregation and crystallization of the silsesquioxane moieties. Multifunctionalized cubic silsesquioxanes such as octa(3-aminopropyl)octasilsesquioxane (OAPOSS) were recently used to synthesize nanocomposites of metal nanoparticles. These types of cubic silsesquioxane provide alternative assemblies of functionalized metal nanoparticles. Rotello and coworkers [141] reported that electrostatic interaction between the carboxylic acid groups coated on gold nanoparticles and the ammonium groups on OAPOSS resulted into well-ordered nanocomposites featuring uniform interparticle spacings. Samples of electrostatically coupled nanoparticles were prepared by dropping solutions of carboxylic acid-functionalized gold nanoparticles in THF/MeOH into a solution of OAPOSS in $MeOH/H_2O$.

The use of OAPOSS has additional important advantages. First, amide bonds were formed between the reaction ion couples well defined in nanocomposites during subsequent chemical reaction to generate nanocomposites with improved chemical properties. Second, calcination of the nanocomposites of

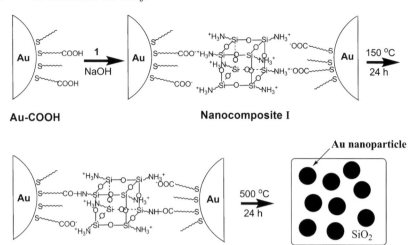

Fig. 1.10. Self-organized nanocomposites of functionalized gold nanoparticles with OAPOSS

the gold nanoparticles with OAPOSS would form silica–gold nanoparticle nanocomposites without the fuse of the gold nanoparticles, since the inorganic core of OAPOSS should be maintained by calcination on the contrast of organic linkers. Naka and Chujo [142] recently reported a self-organized nanocomposites of functionalized gold nanoparticles with OAPOSS via electrostatic interaction between the carboxylate anions and ammonium cations (Fig. 1.10). Subsequent chemical reaction between the reactive ion couples well defined in the nanocomposites generated amide bonds between the two components.

When a DMSO solution of OAPOSS and the carboxylic acid-functionalized gold nanoparticles (Au–COOH) was neutralized by an aqueous NaOH solution, the dark red solution faded completely and the gold nanoparticles precipitated out from the solution to form nanocomposite I (Fig. 1.10), which consisted of 11.0 wt% OAPOSS and 89.0 wt% Au–COO$^-$, estimated by thermogravimetric analysis (TGA) and elemental analysis of N. Even a half amount of OAPOSS against Au–COOH was added into a DMSO solution of Au–COOH, the TGA results indicated that the resulting nanocomposite consisted of 11.3 wt% OAPOSS and 88.7 wt% Au–COO$^-$, which was of the same composition as that of nanocomposite I. These results made it clear that the formed nanocomposites were not random aggregation between OAPOSS and Au–COO$^-$. The TEM image of nanocomposite I showed the existence of gold nanoparticles with a size of 6.0 ± 2.4 nm, which means no fuse of the gold nanoparticles occurred during the precipitation.

More stable nanocomposite II was produced after the nanocomposites I underwent a subsequent chemical reaction by heating at 150°C for 24 h (Fig. 1.10). An absorption peak corresponding to –CONH– bonds at

1,650 cm^{-1} appeared, indicating that amide bonds were formed between the well-defined reactive ion couples after the reaction. The average size of the gold nanoparticles in nanocomposite II was 6.4 ± 2.2 nm. Further heating of nanocomposite II at 500°C for 24 h resulted nanocomposite III in which the organic moiety of OAPOSS totally decomposed, and the inorganic cores of OAPOSS were maintained (Fig. 1.10) [143]. Generally, such a high heating temperature results in partial melting and interconnection between the gold nanoparticles [144]. The TEM image of nanocomposite III showed the existence of the gold nanoparticles with a size of 7.0 ± 2.5 nm. This suggests that no significant increase in particle size of the gold nanoparticles occurred even after the calcination at 500°C due to the rigid inorganic cores of OAPOSS.

The controlled organization of metal nanoparticles into multilayer films and porous nanocomposites has received intense attention in recent years for their uses in analytical and material chemistries, mainly because of their unique chemical and physical properties, which are different from those of bulk metals [2,8,9]. Self-assembled multilayer films of the metal nanoparticles have become a popular target in nanoscale material synthesis in the last few years [145] because they are complementary alternative to traditional preparation methods of metal films such as evaporation and plating. Self-assembled ultrathin multilayer films have been intensively investigated in recent years, since Decher et al. [146,147] introduced the method for preparing multilayer ultrathin films by the consecutive deposition of oppositely charged polyelectrolytes from dilute aqueous solution onto charged substrates. It was reported that a self-assembled multilayer poly(octadecylsiloxane) provided a nanostructured matrix for metal nanoparticle formation [148]. Silver nanoparticles stabilized by negatively charged polystyrene microspheres were transported into layer-by-layer (LBL) film structures via self-assembly between the microspheres and poly(ethyleneimine) [149] The work of Natan and coworkers [150] is representative of a bifunctional crosslinker-directed stepwise construction of conductive gold and silver colloidal multilayers. Controllable and reversible self-assembled multilayer films of gold nanoparticles were also efficient based on ligand/metal ion/ligand linkers [151]. Vapor-sensitive multilayer films were obtained via a LBL self-assembly of gold nanoparticles and dendrimers [152].

Naka and Chujo [153] reported that the LBL self-assembled multilayer films were successfully prepared via the adsorption of a facile approach of immersing the negatively charged glass substrates into OAPOSS and Au–COO$^-$ solutions alternatively. Electrostatic interaction between the –COO$^-$ on Au–COO$^-$ and the –NH$_3^+$ of OAPOSS was the driving force of the LBL self-assembly in a manner similar to the one reported by Caruso and coworkers, in which they created nanocomposite siliceous thin films by the LBL self-assembly in alternation with octa(3-aminopropyl)octasilsesquioxane and poly(styrene-4-sulfonate) to form thin films on planar and spherical supports [21]. A linear increase of the surface plasmon resonance of Au–COO$^-$ with the deposited bilayers indicates that each deposition cycle adds a virtually constant amount of Au–COO$^-$ on the film in each dipping cycle.

A high content and dense coverage of the gold nanoparticles provide the LBL multilayer films with bulk gold appearance and relatively high conductivity.

Films' colors shifted from dark red to yellow. The surface plasmon resonance band responsible for the dark red color exhibited by the gold nanoparticles arises from interband transition between the highly polarizable Au 5d band and the unoccupied states of the conduction band [13]. The intensity of the plasmon band is related to the quantity of Au–COO$^-$ deposited on the substrate. The darkening of the dark red color reveals increasing content of the gold nanoparticles from 1-, 3-, 5-, to 10-bilayer films. After 15 exposures to the 1/Au–COO$^-$ dipping cycle, the 15-bilayer film had a final appearance similar to bulk gold in color and reflectivity, suggestive of bulk gold behavior. The same phenomena were observed by Natan and coworkers [150].

Recent years have focused much attention on self-assembly of the metal nanoparticles to generate porous nanostructures. The porous metal nanostructures have potential application in the areas of advanced catalysis, electronics, optics, separations, and sensors [154]. A facile method was proposed to incorporate preformed gold nanoparticles within ordered macroporous materials [155]. Porous structures with metal nanoparticles were also produced by cyclodextrin-assisted incorporation of metal nanoparticles into porous silica [156]. Silica and gold nanoparticles were cooperatively assembled with lysine–cysteine diblock copolypeptides into hollow spheres [157], as the copolypeptide provided preferential attachment sites for silica and the gold nanoparticles.

Porous nanocomposites were prepared by precipitation of OAPOSS modified polystyrene (PS), latex particles, and Au–COO$^-$ followed by removal of the PS particles via solvent extraction (Fig. 1.11). The OAPOSS-modified

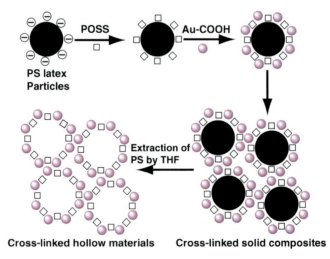

Fig. 1.11. Experimental procedures for preparation of porous nanocomposites by precipitation

PS latex particles precipitated out from the Au–COO$^-$ solution after being crosslinked by Au–COO$^-$. The size of the gold nanoparticles was 3.0 nm estimated from the XRD pattern of the porous nanocomposites. After the porous nanocomposites underwent a subsequent chemical reaction by heating at 150°C for 24 h, an absorption peak corresponding to –CONH– bonds at 1,650 cm^{-1} appeared in FT-IR spectra, indicating that amide bonds were formed between the well-defined reactive ion couples after the reaction, similar to the formation of nanocomposite II as shown in Fig. 1.10. The estimated size of the gold nanoparticles in the porous nanocomposites was also ≈3.0 nm, calculated from the XRD pattern of the porous nanocomposites after the reaction. It was concluded that no size growth of the gold nanoparticles occurred during the chemical reaction. This strategy associated with the LBL self-assembly technique was applied here to prepare porous nanocomposites from the two components. Nanocomposite films with exactly spherical pores were produced by the LBL self-assembly of the OAPOSS-coated PS latex particles and Au–COO$^-$ on the glass substrate. Figure 1.12 shows SEM images of the films after four dipping cycles before and after removal of the PS latex particles by solvent extraction or calcination at 500°C. The pore size and shape were intact after removal of the templates.

Physical and chemical crosslinking is an effective method of building and patterning nanostructures consisting of two or more types of nanoparticles with at least one type of metal nanoparticle. Incorporation of more than a

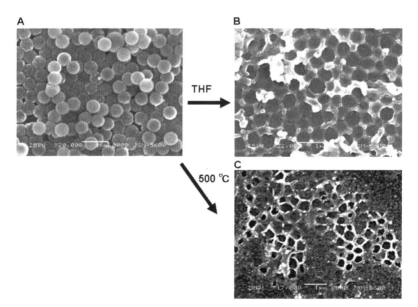

Fig. 1.12. SEM images for the nanocomposites by precipitation (**a**) before (bar: 5 µm) and (**b**) after the removal of PS templates by heating (bar: 10 µm) and solvent extraction by THF (bar: 1 µm)

single type of nanoparticle into self-assembled nanostructures can provide an opportunity to utilize unique properties of the different types of the nanoparticles. The range of properties can be greatly enhanced by organizing two kinds of metal nanoparticles to generate intermetallic nanocomposites due to synergistic effects [158]. Even a physical mixture of two types of metal nanoparticles shows higher catalytic activity than the corresponding monometallic nanoparticles [124, 159–161]. This also proved that combing nanoparticles of different materials allows the manufacture of novel nanocomposite materials. Stable nanostructures, originating from spontaneous self-organization of two kinds of nanoparticles to form nanocomposites, have recently attracted some interest [162–166]. By combining gold nanoparticles stabilized by a carboxylic acid and silica nanoparticles functionalized by a primary amine, acid–base chemistry followed by immediate charge pairing generates electrostatically bound mixed-colloid constructs [162]. Directed self-assembly of two kinds of nanoparticles was performed on a block copolymer micellar template by first introducing Au nanoparticles physically around hexagonally ordered micelles, and then synthesizing chemically Fe_2O_3 nanoparticles in the core area of the ordered micelles, resulting in Fe_2O_3 nanoparticles surrounded by the Au nanoparticles [163]. Self-organization of Au and CdS nanoparticles by electrostatic interaction led to complex-like structures [164]. Few investigations on self-assembly of two types of metal nanoparticles have been reported except that indirect self-assembly of Au and Ag nanoparticles into bimetallic 3D networks using recognition properties of surface-attached antibodies with bivalent antigens of appropriate double-headed functionalities [165], and evaporation of a mixed colloidal solution of thiol-functionalized Au and Ag nanoparticles on a flat substrate led to self-organization into ordered colloidal superlattices [166]. However, both of them were manipulated on rather limited spatial scales.

When the spherical aggregates of Pd nanoparticles with a mean diameter of 80 nm ($Pd–NH_3^+$), which were produced by stirring palladium(II) acetate with OAPOSS in methanol at room temperature via self-organized spherical templates of palladium ions and OAPOSS that are involved in electrostatic interaction with $Au–COO^-$ as a counterpart building block when spontaneous formation of microporous nanostructures may occur (Fig. 1.13). Microporous nanocomposites of palladium and gold nanoparticles were generated by utilizing electrostatic interaction between oppositely charged gold nanoparticles coated with carboxylate groups ($Au–COO^-$) and spherical aggregates of palladium nanoparticles($Pd–NH_3^+$) with a mean diameter of 80 ± 20s nm stabilized and crosslinked by OAPOSS. Amide bonds were formed between the reactive ion couples well defined in the Pd–Au colloidal nanocomposites during a subsequent chemical reaction to generate more stable nanocomposites with improved chemical and physical properties. Such nanostructures are expected to exhibit excellent catalytic properties for their great surface and synergistic effects of the palladium and gold nanoparticles. Furthermore, amide bonds were formed between the reactive ion couples well defined in the Pd–Au

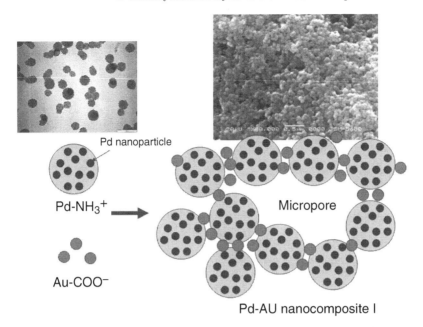

Fig. 1.13. Microporous nanocomposite of Pd and Au nanoparticles via electrostatic interaction between Pd–NH$_3^+$ and Au–COO$^-$. TEM image of Pd–NH$_3^+$ (scale bar: 100 nm) and electron micrographs of Pd–Au nanocomposite I by SEM (bar: 0.5 μm)

colloidal nanocomposites during a subsequent chemical reaction to generate more stable nanocomposites with improved chemical and physical properties as similar as Fig. 1.10. In this system, OAPOSS acts as a primary crosslinker for the palladium nanoparticles to construct Pd–NH$_3^+$, and Au–COO$^-$ was used as a secondary crosslinker for Pd–NH$_3^+$.

1.5 Organic–Metal 1D Nanostructures

One-dimensional (1D) nanostructures have attracted much attention due to their optical and electronic properties for potential uses as interconnects or active components in fabricating nanodevices [167]. Various chemical methods have been established to accomplish 1D growth of nanostructures [168], among which template-directed synthesis is the most widely used method for generating metallic or metal oxide nanowires [169]. Organic conducting 1D nanostructures have been studied with the expectation for potential application owing to their low density and flexibility for molecular design compared with the inorganic nanostructures [170–173]. Despite a variety of synthetic methods to the organic 1D nanostructures, creation of template-free and facile approaches are still required for future practical application [174, 175].

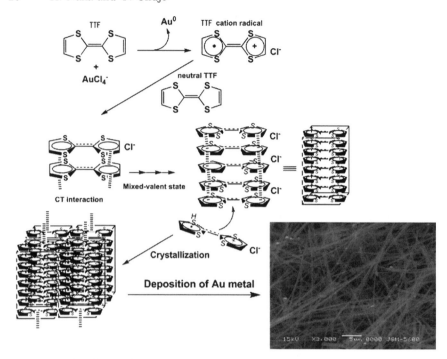

Fig. 1.14. Synthesis of organic–metal hybrid nanowires via electron-transfer reaction between TTF and a gold ion and a SEM image of the hybrid nanowires isolated as precipitates from the acetonitrile solution of TTF and HAuCl$_4$ after 24 h

Naka and Chujo [176] introduced the synthesis of organic–metal hybrid nanowires via electron-transfer reaction between tetrathiafulvalene (TTF) and a gold ion. The reduction of gold ions by TTF as a reducing reagent is a key process for the formation of the hybrid nanowires as illustrated in Fig. 1.14. An electron-transfer reaction from TTF to gold ions led to the formation of TTF radical cation and zero valent gold (Au^0) [177, 178]. Since no gold precipitates were observed, TTF might stabilize gold clusters consisting of Au^0 at its S sites in the solution. This might act as seeds for the subsequent crystallization of TTF [179]. The 1D crystallization of the intermediate would proceed through the interaction between the neutral and oxidized TTF along the stacking axis of the crystals with chloride anions to form nanocrystallites of TTF chloride. Adsorption and subsequent self-assembly of the resulting gold clusters on lateral dimension of the TTF nanocrystals kinetically controlled 1D crystal growth of TTF, during which the gold clusters acted as a capping agent to inhibit the lateral growth of the TTF nanocrystals by bonding gold surfaces at its S sites (Au–S bonds). The hybrid nanowires as shown in Fig. 1.14 were constructed by cooperative self-organization of the metal and TTF. To our knowledge, this is the first example for creating metal-containing TTF-based hybrid nanowires. The hybrid nanowires are expected to have unique chemical and physical properties, due to hybridization of metals and organic π-conjugated molecules.

1.6 Flocculation of Metal Nanoparticles by Stimuli

Of particular interest is the possibility of tailoring the metal nanoparticles surface with a molecular arrangement consisting of organic molecules that possess responsive properties to certain stimulation. Attempts at controlling the aggregation process have centered on carboxylate-functionalized colloids, which display flocculation behavior as a function of pH [180–182] or cationic species such as metal ions [4,183]. The ability to functionalize gold nanoparticles with stimuli-responsive property has opened new avenues to utilize these nanomaterials in optical and electronic applications.

Efficient methodologies for responsive self-organization of the metal nanoparticles involve selective control of noncovalent interactions. Several approaches such as hydrogen bonding [180–182], π–π interaction [184], host–guest interaction [7,185], van der Waals forces [186], electrostatic forces [187], antibody–antigen recognition [188], and charge-transfer interaction [189,190] have been described for the flocculation of the metal nanoparticles (Fig. 1.15). In the following section, several new concepts for the responsive flocculation of metal nanoparticles are described.

1.6.1 Photoresponsive Aggregation

Photoactive metal nanoparticles are important for designing light-energy-harvesting devices of nanometric dimension and photocatalysts [191–195]. For examples, Fox and coworkers [193] have demonstrated the ability of gold nanoparticles to preserve the photoreactivity of the *trans*-stilbene and *o*-nitrobenzyl ether moieties similar to the one observed in solution phase. Apart from these studies of discrete nanoparticles, it is also very important to design the photoresponsive aggregation system of metal nanoparticles. Irradiation of UV light to a solution of the gold nanoparticles modified with thymine units resulted in the formation of aggregates comprising chemical crosslinking gold nanoparticles through the photodimerization of the thymine units (Fig. 1.15) [196]. It is well known that thymine bases photodimerize upon the irradiation above 270 nm and revert back to be thymine again upon the irradiation below 270 nm [197–200].

The gold nanoparticles modified with the thymine units were prepared by the same protocol of Brust's method. A water–chloroform two-phase system was used instead of the water–toluene system. The mixture of ω-functionalized alkanethiol with the thymine unit and dodecanethiol was added to the chloroform phase to suppress intramolecular photoreaction of the thymine units on the nanoparticle surface. Photoirradiation on the solution of the

28   K. Naka and Y. Chujo

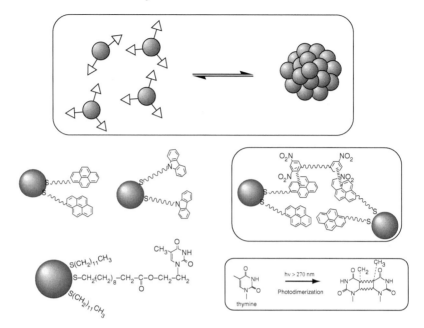

Fig. 1.15. Flocculation of metal nanoparticles by stimuli and examples. Thermally reversible self-assembly of metal nanoparticles modified with pyrenyl units or carbazolyl units with the bivalent *m*-dinitrophenyl linker by charge-transfer interaction (*middle*), and photochemical assembly of gold nanoparticles by photodimerization of the thymine units (*lower*)

thymine-functionalized gold nanoparticles was carried out to induce the photodimerization of the thymine. The solution color gradually changed from red to purple with an increase in the reaction time. The surface plasmon band changed from 496 nm to 525, 538, and 544 nm after the photoirradiation for 22, 32, and 46 h, respectively. The precipitation of black powders was observed after 72 h, indicating that the nanoparticle aggregates eventually became too large to remain in the solution. TEM images after the photoirradiation to the solution containing the gold nanoparticles showed the formation of aggregates. The formation of the aggregates with an average diameter of 0.15, 0.25, and 1 µm was observed after 6, 22, and 72 h, respectively. Thus, the degree of colloidal association would be controlled by adjusting photoirradiation time. Aggregation rate was also controlled by turning the thymine unit density on the nanoparticle surface.

1.6.2 Metal Nanoparticles Modified with an Ionic Liquid Moiety

The aggregation-induced color changes of the gold nanoparticles in aqueous solutions were demonstrated by using gold nanoparticles modified with an ionic liquid moiety based on imidazolium cation (Fig. 1.16) [201]. The key

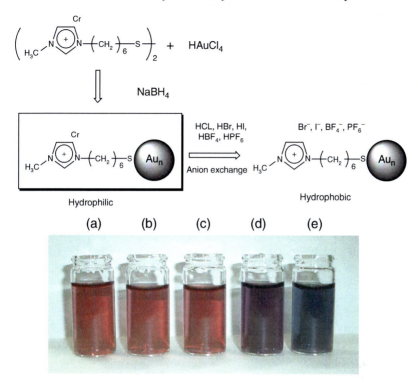

Fig. 1.16. Preparation of gold nanoparticle modified with ionic liquid based on the imidazolium cation and their anion exchange

concept of this system is an anion exchange of the imidazolium moiety to control the solubility in solutions. Hydrophilic and hydrophobic properties are tuned by anion exchange of the ionic liquid moiety. Use of the aggregation-induced color changes of the gold nanoparticles in aqueous solutions provides an optical sensor for anions via anion exchange of the ionic liquid moiety. It is well known that the addition of HX (X = BF_4^-, PF_6^-, and so on) to ionic liquids based on methylimidazolium chloride results in the anion exchange from Cl^- to BF_4^- or PF_6^-, respectively [202, 203]. The reaction of the imidazolium cation-modified gold nanoparticles with various anions was followed as a function of time through optical changes in the surface plasmon resonance in an UV-vis absorption spectrum. In the case of HI and HPF_6, the solution color changed dramatically from red to purple and blue, respectively. A TEM image of the obtained solution after addition of HI and HPF_6 clearly indicated particles aggregates. Elemental analysis of the modified gold nanoparticles after the addition of HPF_6 showed that 30% of the imidazolium cation immobilized on the nanoparticle surface changed from Cl^- to PF_6^- by the anion exchange. That is, the surface property of the gold nanoparticles changed from hydrophilic to hydrophobic by the anion exchange of the ionic liquid, which lead to the nanoparticle aggregation in water. In fact, the imidazolium cation containing Cl^- is soluble in water. On the other hand, the imidazolium cation containing PF_6^- is immiscible with water. The strength of hydrogen bonding between water molecules and anions in ionic liquids increases in the order of $PF_6^- < I^- < Cl^-$ [204, 205]. Based on the TEM and elemental analysis results, the spectral changes were dominantly induced by the anion exchange under the present conditions, demonstrating that this system can be applied for optical anion sensing.

A new type of gold nanoparticle with a zwitterionic liquid function was prepared using an imidazolium sulfonate-terminated thiol as a capping agent [206]. The zwitterionic liquid is composed of covalently tethered cations and anions, in which the ions do not migrate along a potential gradient. Contrary to the anion-responsive behavior of the gold nanoparticle with the imidazolium cation described above [201], the gold nanoparticle with the zwitterionic liquid were remarkably stable in high concentration of aqueous electrolyte. Lee and coworkers [207] reported that thiol-functionalized imidazolium ionic liquids acted as a highly effective medium for the preparation and stabilization of gold and platinum nanoparticles under a water-phase synthesis. The particle size and uniformity depended on the number of thiol groups and their positions in the thiol-functionalized imidazolium ionic liquids.

Room-temperature ionic liquids are attracting much interest in many fields of chemistry and industry, due to their potential as a "green" recyclable alternative to the traditional organic solvents [208, 209]. They are nonvolatile and provide an ultimate polar environment for chemical synthesis. Among various known ionic liquids, ionic liquids containing imidazolium cation and PF_6^- has a particularly useful set of properties, being virtually insoluble in water [204, 205]. Such biphasic ionic liquid systems have been used to enable simple extraction of products. The modified gold nanoparticles with the imidazolium cations in aqueous phase were transferred across a phase boundary (water to ionic liquid) via the anion exchange of the ionic liquid moiety by addition of HPF_6 [201]. When HPF_6 was added to the aqueous solution containing the modified gold nanoparticles with stirring, the ionic liquid phase (1-methyl-3-hexylimidazolium hexafluorophosphate) quickly became colored, drawing from the original deep red color of the aqueous nanoparticles solution. The complete phase transfer of the gold nanoparticles was achieved.

Phase transfer of metal nanoparticles from an aqueous phase to an ionic liquid phase was also reported by Wei and coworkers [210], which does not require the use of a thiol, for the phase transfer of the gold nanoparticles from an aqueous medium to an ionic liquid. An aqueous solution of the gold nanoparticles formed upon the reduction of a solution of $HAuCl_4$ with citrate was added to the organic phase. The water-immiscible ionic liquid, 1-butyl-3-methylimidazolium hexafluorophosphate is a solvent medium that allows complete transfer of the gold nanoparticles from an aqueous phase into an organic phase. Water-soluble cationic CdTe nanocrystals, which were prepared by an aqueous synthetic approach, were efficiently extracted from an aqueous phase into a water-immiscible ionic liquid [211]. The ionic liquid containing the transferred metal nanoparticles has potential for the recyclable biphasic catalysis process.

pH-Responsive system in aqueous solution has extremely promising prospect. Many biological phenomena, e.g., cellular recognition and transportation in tissues and organs, are largely concerned to pH at which they occur. There have been several reports for pH-responsive metal nanoparticles [182,212–215]. Toshima and coworkers synthesized 3-mercaptopropionic acid-modified gold nanoparticles and reversibly controlled the colloidal dispersions by using interparticle hydrogen bonds or electrostatic repulsion depending on pH. Lee and coworkers [214] synthesized gold nanoparticles coated by a hydrogel consisting of poly(acrylic acid-co-N-isopropylacrylamide) and acted as an interrupter of surface plasmon absorbance depending on pH and temperature.

Naka and Chujo [216] reported pH-responsive control of the colloidal dispersions of the imidazolium cation-modified gold nanoparticles using a combination of poly(acrylic acid) (PAA) in aqueous solution. The strategy is based on the electrostatic interaction between the imidazolium cations on the surface of the gold nanoparticles and the carboxylate anions of PAA. The resulting precipitate after the imidazolium cation-modified gold nanoparticles was mixed with PAA (MW $= 25{,}000$) in aqueous solution, as described above, was dissolved into a clear red colored solution when the pH dropped below 1.9 by the addition of 1.0 M HCl. Addition of 1.0 M NaOH to the acidified solution caused a precipitate again at the pH of 2.2, and the precipitate was redissolved when the pH rose above 5.5 (Fig. 1.17). Addition of 1.0 M HCl to the final solution reproduced the precipitation–redispersion process until the pH dropped below 1.9. The absorption maximum of the solution at pH 6.0 was redshifted compared with that in the case of the solution at pH 1.7 and the pristine solution of the imidazolium cation-modified gold nanoparticles without PAA ($\lambda_{\mathrm{max}} = 514$ nm). The redshift can be attributed to the coupled plasmon absorbance of the gold nanoparticles in closer contact, which indicates a formation of particle aggregates in the aqueous solution at pH 6.0. The TEM image of the sample at pH 6.0 shows the aggregates of the gold nanoparticles and the size of the aggregates was ranging from 40 to 150 nm. A DLS measurement suggested the formation of the aggregates in aqueous solution with a diameter of 115 ± 21 nm.

There are several reports about the phenomena of nanoparticles-based flocculation by charged polymers [217–223]. Although extensive experimental and theoretical works are reported for understanding the flocculation processes and construct nanoparticles-based flocculated materials with controlled structures, studies for pH-responsive phenomena of nanoparticles-based flocculated materials were limited. Sehgal et al. [223] reported a precipitation–redispersion mechanism for complexation of short chain PAA with cerium oxide nanoparticles. They showed that addition of PAA to a cerium oxide solutions leads to macroscopic precipitation and the solution redispersed into a clear sols of single particles with an anionic PAA corona as the pH increased.

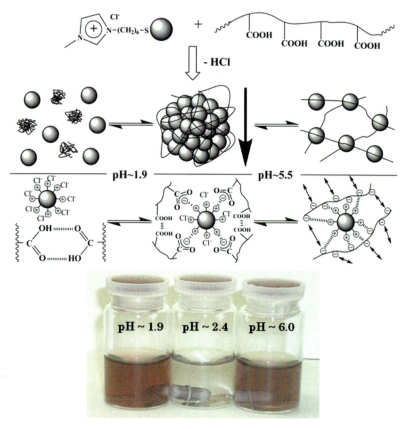

Fig. 1.17. Photograph of various pH solutions of the imidazolium cation-modified gold nanoparticles with PAA (MW = 25,000) and proposed mechanism of the flocculation and the pH-responsive aggregation of the imidazolium cation-modified gold nanoparticles with PAA

We found a different precipitation–redispersion behavior for the complexes of the imidazolium cations-modified gold nanoparticles with PAA [216].

1.7 Conclusion and Outlook

Metal nanoparticles can be functionalized with a wide variety of structural units using simple chemical process under moderate condition. Stabilization of metal nanoparticles by ligand units such as thiols or amines enable further manipulation of usual compounds. Since close-packed assembly of metal nanoparticles is expected to produce different electronic, magnetic, and optical functions based on quantum mechanical coupling of conduction electron localized in each metal nanoparticles, development of simple and easy methods for the organization of metal nanoparticles is indispensable for preparing new nanodevices.

To exploit metal nanoparticle properties for future device fabrication, self-organization of nanoparticles in controlled manner is required from the standpoint of the "bottom-up" approach. Dendritic molecules mediate assembly of the metal nanoparticles shows potential for controlling into one-, two-, and three-dimensional structures. It is well established that dendrimers have acted as excellent hosts for the formation and stabilization of metal nanoparticles. However, research direction of using dendrimers as linkers for metal nanoparticles is limited. Since dendrimers are monodisperse macromolecules prepared by completely designed chemical synthesis, combination of the metal nanoparticle-based building blocks and the dendrimers would provide a variety of nanocomposites in controlled self-organization. Only one disadvantage of using dendrimers is their multistep chemical synthesis. Polyhedral oligosilsesquioxanes, however, should be excellent candidates for this purpose due to their easy preparation.

Particular interest is focused on the functionalized metal nanoparticles that possess responsive properties to certain stimulation such as pH, anion, cation, temperature, light, small molecules, and biological molecules. Methodologies for responsive self-organization usually involve using noncovalent interaction, such as electrostatic interaction, hydrogen bonding, π–π interaction, charge-transfer interaction, and host–guest interaction. Using reversible chemical bond formation is also available. Changing solubility, such as hydrophobic and hydrophilic properties, of metal nanoparticles is an alternative way to self-assemble especially in aqueous solution. Use of the aggregation-induced color changes of gold nanoparticles provides these responsive colloidal solutions as an optical sensor for guest ions and molecules.

Key and basic requirements for future development in this field include further seeking and optimization of controlled synthesis of the nanoparticle-based building blocks. Although lowering size distribution of the nanoparticles with controlled diameter should be a central target, design and synthesis of functional materials with simple coating needs contribution of organic and polymer synthetic chemists. We expect that the continuous cooperation of organic and polymer chemists with inorganic and physical chemists, which is desirable to fabricate the metal nanoparticles-based hybrid materials, will lead to the next industrial revolution.

References

1. A.C. Templeton, W.P. Wuelfing, R.W. Murray, Acc. Chem. Res. **33**, 27 (2000)
2. M.-C. Daniel, D. Astruc, Chem. Rev. **104**, 293 (2004)
3. S.O. Obare, R.E. Holowell, C.J. Murphy, Langmuir **18**, 10407 (2002)
4. Y. Kim, R.C. Johnson, J.T. Hupp, Nano Lett. **1**, 165 (2001)
5. S. Watanabe, M. Sonobe, M. Arai, Y. Tazume, T. Matsuo, T. Nakamura, K. Yoshida, Chem. Commun. 2866 (2002)

34   K. Naka and Y. Chujo

6. C.A. Mirkin, R.L. Letsinger, R.C. Mucic, J.J. Storhoff, Nature **382**, 607 (1996)
7. S.Y. Lin, S.W. Liu, C.M. Lin, C.H. Chen, Anal. Chem. **74**, 330 (2002)
8. G. Chumanov, K. Sokolov, B.W. Gregory, T.M. Cotton, J. Phys. Chem. **99**, 9466 (1995)
9. S. Nie, S.R. Emory, Science **275**, 1102 (1997)
10. J.A. Dahl, B.L.S. Maddux, J.E. Hutchison, Chem. Rev. **107**, 2228 (2007)
11. H. Hirai, Y. Nakao, N. Toshima, K. Adachi, Chem. Lett. 905 (1976)
12. H. Hirai, Y. Nakao, N. Toshima, J. Macromol. Sci. Chem. A **12**, 1117 (1978)
13. R.P. Andres, J.D. Bielefeld, J.I. Henderson, D.B. Janes, V.R. Kolagunta, C.P. Kubiak, W.J. Mahoney, R.G. Osifchin, Science **273**, 1690 (1996)
14. C.P. Collier, R.J. Saykally, J.J. Shiang, S.E. Henrichs, J.R. Heath, Science **277**, 1978 (1997)
15. M. Brust, M. Walker, D. Bethell, D.J. Schiffrin, R. Whyman, J. Chem. Soc., Chem. Commun. 801 (1994)
16. M.J. Hostetler, C.-J. Zhong, B.K.H. Yen, J. Anderegg, S.M. Gross, N.D. Evans, M. Porter, R.W. Murray, J. Am. Chem. Soc. **120**, 922 (1998)
17. A.C. Templeton, M.J. Hostetler, C.T. Kraft, R.W. Murray, J. Am. Chem. Soc. **120**, 1906 (1998)
18. M.J. Hostetler, A.C. Templeton, R.W. Murray, Langmuir **15**, 3782 (1999)
19. A.C. Templeton, S. Chen, S.M. Gross, R.W. Murray, Langmuir **15**, 66 (1999)
20. T. Yonezawa, T. Kunitake, Colloids Surf. A, Physicochem. Eng. Asp. **149**, 193 (1999)
21. D.V. Leff, L. Brandt, J.R. Heath, Langmuir **12**, 4723 (1996)
22. N.R. Jana, X. Peng, J. Am. Chem. Soc. **125**, 14280 (2003)
23. H. Hiramatsu, F.E. Osterloh, Chem. Mater. **16**, 2509 (2004)
24. M. Aslam, L. Fu, M. Su, K. Vijayamohanan, V.P. David, J. Mater. Chem. **14**, 1795 (2004)
25. G. Schmid, V. Maihack, F. Lantermann, S. Peschel, J. Chem. Soc., Dalton Trans. 589 (1996)
26. M.N. Vargaftik, V.P. Zagorodnikov, I.P. Stolyarov, I.I. Moiseev, V.A. Likholobov, D.I. Kochubey, A.L. Chuvilin, V.I. Zaikovsky, K.I. Zamaraev, G.I. Timofeeva, J. Chem. Soc., Chem. Commun. 937 (1985)
27. G. Schmid, M. Harms, J. Malm, J. Bovin, J. Ruitenbeck, H.W. Zandbergen, W.T. Fu, J. Am. Chem. Soc. **115**, 2046 (1993)
28. G. Schmid, B. Morum, J. Malm, Angew. Chem. Int. Ed. Engl. **28**, 778 (1989)
29. A. Duteil, G. Schmid, W. Meyer-Zaika, J. Chem. Soc., Chem. Commun. 31 (1995)
30. G. Schmid, R. Pfeil, R. Boese, F. Bandermann, S. Meyer, G.H.M. Calis, J.W.A. Van der Velden, Chem. Ber. **114**, 3634 (1981)
31. W.W. Weare, S.M. Reed, M.G. Warner, J.E. Hutchison, J. Am. Chem. Soc. **122**, 12890 (2000)
32. J. Petroski, M.H. Chou, C. Creutz, Inorg. Chem. **43**, 1597 (2004)
33. G.H. Woehrle, L.O. Brown, J.E. Hutchison, J. Am. Chem. Soc. **127**, 2172 (2005)
34. R. Balasubramanian, R. Guo, A.J. Mills, R.W. Murray, J. Am. Chem. Soc. **127**, 8126 (2005)
35. Y. Shichibu, Y. Negishi, T. Tsukuda, T. Teranishi, J. Am. Chem. Soc. **127**, 13464 (2005)
36. J.G. Worden, A.W. Shaffer, Q. Huo, Chem. Commun. 518 (2004)

37. K.-M. Sung, D.W. Mosley, B.R. Peelle, S. Zhang, J.M. Jacobson, J. Am. Chem. Soc. **126**, 5064 (2004)
38. R. Sardar, T.B. Heap, J.S. Shumaker-Parry, J. Am. Chem. Soc. **129**, 5356 (2007)
39. Q. Dai, J.G. Worden, J. Trullinger, Q. Huo, J. Am. Chem. Soc. **127**, 8008 (2005)
40. G. Ramakrishna, Q. Dai, J. Zou, Q. Huo, T. Goodson III, J. Am. Chem. Soc. **129**, 1848 (2007)
41. A.M. Jackson, J.W. Myerson, F. Stellacci, Nat. Mater. **18**, 330 (2004)
42. A.M. Jackson, Y. Hu, P.J. Silva, F. Stellacci, J. Am. Chem. Soc. **128**, 11135 (2006)
43. G.A. DeVries, M. Brunnbauer, Y. Hu, A.M. Jackson, B. Long, B.T. Neltner, O. Uzun, B.H. Wunsch, F. Stellacci, Science **315**, 358 (2007)
44. K. Naka, M. Yaguchi, Y. Chujo, Chem. Mater. **11**, 849 (1999)
45. A.B.R. Mayer, J.E. Mark, PMSE Preprints **73**, 220 (1995)
46. W.P. Wuelfing, S.M. Gross, D.T. Miles, R.W. Murray, J. Am. Chem. Soc. **120**, 12696 (1998)
47. J. Shan, M. Nuopponen, H. Jiang, E. Kauppinen, H. Tenhu, Macromolecules **36**, 4526 (2003)
48. C. Mangeney, F. Ferrage, I. Aujard, V. Marchi-Artzner, L. Jullien, O. Ouari, E.D. Rekai, A. Laschewsky, I. Vikholm, J.W. Sadowski, J. Am. Chem. Soc. **124**, 5811 (2002)
49. M.K. Corbierre, N.S. Cameron, M. Sutton, S.G.J. Mochrie, L.B. Lurio, A. Rhm, R.B. Lennox, J. Am. Chem. Soc. **123**, 10411 (2001)
50. J.J. Chiu, B.J. Kim, E.J. Kramer, D.J. Pine, J. Am. Chem. Soc. **127**, 5036 (2005)
51. E.R. Zubarev, J. Xu, A. Sayyad, J.D. Gibson, J. Am. Chem. Soc. **128**, 4958 (2006)
52. K. Ohno, K. Koh, Y Tsujii, T. Fukuda, Macromolecules **35**, 8989 (2002)
53. T.K. Mandal, M.S. Fleming, D.R. Walt, Nano Lett. **2**, 3 (2002)
54. J. Raula, J. Shan, M. Nuopponen, A. Niskanen, H. Jiang, E.I. Kauppinen, H. Tenhu, Langmuir **19**, 3499 (2003)
55. D. Rutot-Houze, W. Fris, P. Degee, P. Dubois, J. Macromol. Sci. Pure Appl. Chem. A **41**, 697 (2004)
56. H. Itoh, K. Naka, Y. Chujo, Polym. Bull. **46**, 357 (2001)
57. R. Gangopadhyay, A. De, Chem. Mater. **12**, 608 (2000)
58. K.G. Neoh, T.T. Young, N.T. Looi, E.T. Kang, K.L. Tan, Chem. Mater. **9**, 2906 (1997)
59. K.V. Sarathy, K.S. Narayan, J. Kim, J.O. White, Chem. Phys. Lett. **318**, 543 (2000)
60. T. Hertel, E. Knoesel, M. Wolf, G. Ertl, Phys. Rev. Lett. **76**, 535 (1996)
61. F. Ficicoglu, F. Kadirgan, J. Electroanal. Chem. **451**, 95 (1998)
62. C.S.C. Bose, K. Rajeshwar, J. Electroanal. Chem. **333**, 235 (1992)
63. S.W. Huang, K.G. Neoh, C.W. Shih, D.S. Lim, E.T. Kang, H.S. Han, K.L. Tan, Synth. Met. **96**, 117 (1998)
64. K.G. Neoh, K.K. Tan, P.L. Goh, S.W. Huang, E.T. Kang, K.L. Tan, Polymer **40**, 887 (1999)
65. S.W. Huang, K.G. Neoh, E.T. Kang, H.S. Han, K.L. Tan, J. Mater. Chem. **8**, 1743 (1998)

36   K. Naka and Y. Chujo

66. Y. Zhou, H. Itoh, T. Uemura, K. Naka, Y. Chujo, Chem. Commun. 613 (2001)
67. Y. Zhou, H. Itoh, T. Uemura, K. Naka, Y. Chujo, Langmuir **18**, 277 (2002)
68. K. Naka, T. Uemura, Y. Chujo, Macromolecules **32**, 4641 (1999)
69. M.A. Hempenius, B.M.W. Langeveld-Voss, J.A.E.H. van Haare, R.A.J. Janssen, S.S. Sheiko, J.P. Spatz, M. Möller, E.W. Meijer, J. Am. Chem. Soc. **120**, 2798 (1998)
70. J.H. Youk, J. Locklin, C. Xia, M.-K. Park, R. Advincula, Langmuir **17**, 4681 (2001)
71. T. Sanji, Y. Ogawa, Y. Nakatsuka, M. Tanaka, H. Sakurai, Chem. Lett. **32**, 980 (2003)
72. G. Schmid, M. Bämle, M. Geerkens, I. Heim, C. Osemann, T. Sawitowski, Chem. Soc. Rev. **28**, 179 (1999)
73. C.N.R. Rao, G.U. Kulkarni, P. John Thomas, P.P. Edwards, Chem. Soc. Rev. **29**, 27 (2000)
74. G. Schmid, L.F. Chi, Adv. Mater. **10**, 515 (1998)
75. Y. Xia, J.A. Rogers, K.E. Paul, G.M. Whitesides, Chem. Rev. **99**, 1823 (1999)
76. R. Elghanian, J.J. Storhoff, R.C. Mucic, R.L. Letsinger, C.A. Mirkin, Science **277**, 1078 (1997)
77. C. Stowell, B.A. Korgel, Nano Lett. **1**, 595 (2001)
78. P.C. Ohara, W.M. Gelbart, Langmuir **14**, 3418 (1998)
79. M. Sastry, M. Rao, K.N. Ganesh, Acc. Chem. Res. **35**, 847 (2002)
80. A.K. Boal, F. Llhan, J.E. Derouchey, T. Thurn-Albrecht, T.P. Russell, V.M. Rotello, Nature **404**, 746 (2000)
81. A.K. Boal, V.M. Rotello, J. Am. Chem. Soc. **122**, 734 (2000)
82. J.J. Storhoff, A.A. Lazarides, R.C. Mucic, C.A. Mirkin, R.L. Letsinger, G.C. Schatz, J. Am. Chem. Soc. **122**, 4640 (2000)
83. M. Brust, D. Bethell, D.J. Schiffrin, C.J. Kiely, Adv. Mater. **7**, 795 (1995)
84. R. Shenhar, T.B. Norsten, V.M. Rotello, Adv. Mater. **17**, 657 (2005)
85. M. Antonietti, C. Göltner, Angew. Chem. Int. Ed. Engl. **36**, 910 (1997)
86. M.D. Musick, C.D. Keating, M.H. Keefen, M.J. Natan, Chem. Mater. **9**, 1499 (1997)
87. D. Bethell, M. Brust, D.J. Schiffrin, C. Kiely, Electroanal. Chem. **409**, 137 (1996)
88. A. Doron, E. Katz, I. Willner, Langmuir **11**, 1313 (1995)
89. R.G. Freeman, K.C. Grabar, K.J. Allison, R.M. Bright, J.A. Davis, A.P. Guthrie, M.B. Hommer, M.A. Jackson, P.C. Smith, D.G. Walter, M.J. Natan, Science **267**, 1629 (1995)
90. K.C. Grabar, R.G. Freeman, M.B. Hommer, M.J. Natan, Anal. Chem. **67**, 735 (1995)
91. J.R. Heath, C.M. Knobler, D.V. Leff, J. Phys. Chem. B **101**, 189 (1997)
92. B.O. Dabbousi, C.B. Murray, M.F. Rubner, M.G. Bawendi, Chem. Mater. **6**, 216 (1994)
93. M. Giersig, P. Mulvaney, Langmuir **9**, 3408 (1993)
94. T. Teranishi, M. Hosoe, M. Miyake, Adv. Mater. **9**, 65 (1997)
95. J. Schmitt, G. Decher, W.J. Dressick, S.L. Brandow, R.E. Geer, R. Shashidhar, J.M. Calvert, Adv. Mater. **9**, 61 (1997)
96. N.A. Kotov, I. Dékány, J.H. Fendler, J. Phys. Chem. **99**, 13065 (1995)
97. T. Yonezawa, S. Onoue, T. Kunitake, Chem. Lett. 1061 (1999)
98. G. Schmid, M. Bäumle, N. Beyer, Angew. Chem. Int. Ed. **39**, 181 (2000)

1 Nanohybridized Synthesis of Metal Nanoparticles   37

99. G.L. Hornyak, M. Krll, R. Pugin, T. Sawaitowski, G. Schmid, J.-O. Bovin, G. Karsson, H. Hofmeister, S. Hopfe, Chem. Eur. J. **3**, 1951 (1997)
100. G. Schmid, R. Pugin, J.-O. Malm, J.-O. Bovin, Eur. J. Inorg. Chem. **6**, 813 (1998)
101. T. Teranishi, A. Sugawara, T. Shimizu, M. Miyake, J. Am. Chem. Soc. **124**, 4210 (2002)
102. T. Oku, K. Suganuma, Chem. Commun. 2355 (1999)
103. E. Fort, C. Ricolleau, J. Sau-Pueyo, Nano Lett. **3**, 65 (2003)
104. E. Braun, Y. Eichen, U. Sivan, G. Ben-Yoseph, Nature **391**, 775 (1998)
105. J. Richter, R. Seidel, R. Krisch, M. Mertig, W. Pompe, J. Plaschke, H.K. Schackert, Adv. Mater. **12**, 507 (2000)
106. W.E. Ford, O. Harnack, A. Yasuda, J.M. Wessels, Adv. Mater. **13**, 1793 (2001)
107. M. Mertig, L. Colombi Ciacchi, R. Seidel, W. Pompe, A. De Vita, Nano Lett. **2**, 841 (2002)
108. O. Harnack, W.E. Ford, A. Yasuda, J.M. Wessels, Nano Lett. **2**, 919 (2002)
109. A. Kumar, M. Pattarkine, M. Bhadbhade, A.B. Mandale, K.N. Ganesh, S.S. Datar, C.V. Dharmadhikari, M. Sastry, Adv. Mater. **13**, 341 (2001)
110. M.G. Warner, J.E. Hutchison, Nat. Mater. **2**, 272 (2003)
111. C.M. Niemeyer, W. Bürger, J. Peplies, Angew. Chem. Int. Ed. **37**, 2265 (1998)
112. R. Djalali, Y.-F. Chen, H. Matsui, J. Am. Chem. Soc. **124**, 13660 (2002)
113. C.A. Berven, L. Clarke, J.L. Mooster, M.N. Wybourne, J.E. Hutchison, Adv. Mater. **13**, 109 (2001)
114. E. Dujardin, C. Peet, G. Stubbs, J.N. Culver, S. Mann, Nano Lett. **3**, 413 (2003)
115. S.L. Burkett, S. Mann, Chem. Commun. 321 (1996)
116. S. Fullam, D. Cottel, H. Rensmo, D. Fitzmaurice, Adv. Mater. **12**, 1430 (2000)
117. K. Jiang, A. Eitan, L.S. Schadler, P.M. Ajayan, R.W. Siegel, N. Grobert, M. Mayne, M. Reyes-Reyes, H. Terrones, M. Terrones, Nano Lett. **3**, 275 (2003)
118. A. Kirity, S. Minko, G. Gorodyska, M. Stamm, W. Jaeger, Nano Lett. **2**, 881 (2002)
119. B.W. Maynor, Y. Li, J. Liu, Langmuir **17**, 2575 (2001)
120. L.M. Demers, C.A. Mirkin, Angew. Chem. Int. Ed. **40**, 3069 (2001)
121. H. Zhang, Z. Li, C.A. Mirkin, Adv. Mater. **14**, 1472 (2002)
122. M.F. Ottaviani, S. Bossmann, N.J. Turro, D.A. Tomalia, J. Am. Chem. Soc. **116**, 661 (1994)
123. M. Zhao, L. Sun, R.M. Crooks, J. Am. Chem. Soc. **120**, 4877 (1998)
124. K. Esumi, A. Suzuki, N. Aihara, K. Usui, K. Torigoe, Langmuir **14**, 3157 (1998)
125. K. Esumi, A. Suzuki, A. Yamashita, K. Torigoe, Langmuir **16**, 2604 (2000)
126. M.C. Garcia, L.A. Baker, R.M. Crooks, Anal. Chem. **71**, 256 (1999)
127. K. Hayakawa, T. Yoshimura, K. Esumi, Langmuir **19**, 5517 (2003)
128. R.M. Laine, C. Zhang, A. Sellinger, L. Viculis, Appl. Organomet. Chem. **12**, 715 (1998)
129. J. Choi, J. Harcup, A.F. Yee, Q. Zhu, R.M. Laine, J. Am. Chem. Soc. **123**, 11420 (2001)
130. K. Naka, M. Fujita, K. Tanaka, Y. Chujo, Langmuir **23**, 9057 (2007)
131. F.J. Feher, K.D. Wyndham, Chem. Commun. 323 (1998)
132. M.C. Gravel, C. Zhang, M. Dinderman, R.M. Laine, Appl. Organomet. Chem. **13**, 329 (1999)
133. H. Itoh, K. Naka, Y. Chujo, Bull. Chem. Soc. Jpn. **77**, 1767 (2004)

134. B.L. Frankamp, A.K. Boal, V.M. Rotello, J. Am. Chem. Soc. **124**, 15146 (2002)
135. K. Naka, H. Itoh, Y. Chujo, Nano Lett. **2**, 1183 (2002)
136. H. Bönnemann, G. Braun, W. Brijoux, R. Brinkmann, A.S. Tilling, K. Seevogel, K. Siepen, J. Organomet. Chem. **520**, 142 (1996)
137. A.B.R. Mayer, J.E. Mark, R.E. Morris, Polym. J. **30**, 197 (1998)
138. H. Tanaka, S. Koizumi, T. Hashimoto, H. Itoh, M. Satoh, K. Naka, Y. Chujo, Macromolecules **40**, 4327 (2007)
139. K. Naka, H. Itoh, Y. Chujo, Chem. Lett. **33**, 1236 (2004)
140. J.B. Carroll, B.L. Frankamp, V.M. Rotello, Chem. Commun. 1892 (2002)
141. J.B. Carroll, B.L. Frankamp, S. Srivastava, V.M. Rotello, J. Mater. Chem. **14**, 690 (2004)
142. X. Wang, K. Naka, H. Itoh, Y. Chujo, Chem. Lett. **33**, 216 (2004)
143. T. Cassagneau, F. Caruso, J. Am. Chem. Soc. **124**, 8172 (2002)
144. D.G. Shchukin, R.A. Caruso, Chem. Commun. 1478 (2003)
145. M.D. Musick, C.D. Keating, M.H. Keefe, M.J. Natan, Chem. Mater. **9**, 1499 (1997)
146. G. Decher, J.D. Hong, J. Schmitt, Thin Solid Films **210/211**, 831 (1992)
147. Y. Lvov, G. Decher, H. Moehwald, Langmuir **9**, 481 (1993)
148. D.I. Svergun, M.B. Kozin, P.V. Konarev, E.V. Shtykova, V.V. Volkov, D.M. Chernyshov, P.M. Valetsky, L.M. Bronstein, Chem. Mater. **12**, 3552 (2000)
149. J. Zhang, L. Bai, K. Zhang, Z. Cui, G. Zhang, B. Yang, J. Mater. Chem. **13**, 514 (2003)
150. M.D. Musick, C.D. Keating, L.A. Lyon, S.L. Botsko, D.J. Peña, W.D. Holliway, T.M. McEvoy, J.N. Richardson, M.J. Natan, Chem. Mater. **12**, 2869 (2000)
151. F.P. Zamborini, J.F. Hicks, R.W. Murray, J. Am. Chem. Soc. **122**, 4514 (2000)
152. N. Krasteva, I. Besnard, B. Guse, R.E. Bauer, K. Mülen, A. Yasuda, T. Vossmeyer, Nano Lett. **2**, 551 (2002)
153. X. Wang, K. Naka, M. Zhu, H. Kuroda, H. Itoh, Y. Chujo, J. Inorg. Organomet. Polym. Mater. **17**, 447 (2007)
154. D.M. Shchukin, R.A. Caruso, Chem. Commun. 1478 (2003)
155. B. Rodríguez-González, V. Salgueirio-Maceira, F. García-Santamaría, L.M. Liz-Marzán, Nano Lett. **2**, 471 (2002)
156. Y. Zhou, S.-H. Yu, A. Thomas, B.-H. Han, Chem. Commun. 262 (2003)
157. M.S. Wong, J.N. Cha, K.-S. Choi, T.J. Deming, G.D. Stucky, Nano Lett. **2**, 583 (2002)
158. S. Darby, T.V. Mortimer-Jones, R.L. Johnston, C. Roberts, J. Chem. Phys. **116**, 1536 (2002)
159. N. Toshima, K. Hirakawa, Appl. Surf. Sci. **121/122**, 534 (1997)
160. K. Hirakawa, N. Toshima, Chem. Lett. **32**, 78 (2003)
161. T. Sehayek, M. Lahav, R. Popovltz-Biro, A. Vaskevich, I. Rubinstein, Chem. Mater. **17**, 3743 (2005)
162. T.H. Galow, A.K. Boal, V.M. Rotello, Adv. Mater. **12**, 576 (2000)
163. B.-H. Sohn, J.-M. Choi, S.I. Yoo, S.-H. Yun, W.-C. Zin, J.-C. Jung, M. Kanehara, T. Hirata, T. Teranishi, J. Am. Chem. Soc. **125**, 6368 (2003)
164. J. Kolny, A. Kornowski, H. Weller, Nano Lett. **2**, 361 (2002)
165. W. Shenton, S.A. Davis, S. Mann, Adv. Mater. **11**, 449 (1999)
166. C.J. Kiely, J. Fink, J.G. Zheng, M. Brust, D. Bethell, D.J. Schiffrin, Adv. Mater. **12**, 640 (2000)
167. M.Y. Han, C.H. Quek, Langmuir **16**, 362 (2000)

1 Nanohybridized Synthesis of Metal Nanoparticles   39

168. T.S. Ahmadi, Z.L. Wang, A. Henglein, M.A. El-Sayed, Science **272**, 1924 (1996)
169. Y. Xia, P. Yang, Y. Sun, B. Mayers, B. Gates, Y. Yin, F. Kim, H. Yan, Adv. Mater. **15**, 353 (2003)
170. J. Haung, S. Virji, B.H. Weiller, R.B. Kaner, J. Am. Chem. Soc. **125**, 314 (2003)
171. J.-J. Chiu, C.-C. Kei, T.-P. Perng, W.-S. Wang, Adv. Mater. **15**, 1361 (2003)
172. H. Liu, Q. Zhao, Y. Li, Y. Liu, F. Lu, J. Zhuang, S. Wang, L. Jiang, D. Zhu, D. Yu, L. Chi, J. Am. Chem. Soc. **127**, 1120 (2005)
173. H. Gan, H. Liu, Y. Li, Q. Zhao, Y. Li, S. Wang, T. Jiu, N. Wang, X. He, D. Yu, D. Zhu, J. Am. Chem. Soc. **127**, 12452 (2005)
174. T. Akutagawa, T. Ohta, T. Hasegawa, T. Nakamura, C.A. Christensen, Proc. Natl. Acad. Sci. USA **99**, 5028 (2002)
175. K. Yasui, N. Kimizuka, Chem. Lett. **34**, 248 (2005)
176. K. Naka, D. Ando, X. Wang, Y. Chujo, Langmuir **23**, 3450 (2007)
177. X. Wang, H. Itoh, K. Naka, Y. Chujo, Chem. Commun. 1300 (2002)
178. X. Wang, H. Itoh, K. Naka, Y. Chujo, Langmuir **19**, 6242 (2003)
179. F. Favier, H. Liu, R.M. Penner, Adv. Mater. **13**, 1567 (2001)
180. J. Simard, C. Briggs, A.K. Boal, V.M. Rotello, Chem. Commun. 1943 (2000)
181. M. Sastry, K.S. Mayya, K. Bandyopadhyay, Colloids Surf. A **127**, 221 (1997)
182. Y. Shiraishi, D. Arakawa, N. Toshima, Eur. Phys. J. E **8**, 377 (2002)
183. A.N. Shipway, M. Lahav, R. Gabai, I. Willner, Langmuir **16**, 8789 (2000)
184. J. Jin, T. Iyoda, C. Cao, Y. Song, L. Jiang, T.J. Li, D.B. Zhu, Angew. Chem. Int. Ed. **40**, 2135 (2001)
185. J. Liu, S. Mendoza, E. Roman, M.J. Lynn, R. Xu, A.E. Kaifer, J. Am. Chem. Soc. **121**, 4304 (1999)
186. V. Patil, K.S. Mayya, S.D. Pradhan, M. Sastry, J. Am. Chem. Soc. **119**, 9281 (1997)
187. F. Caruso, R.A. Caruso, H. Mohwald, Science **282**, 1111 (1998)
188. W. Shenton, W.A. Davis, S. Mann, Adv. Mater. **11**, 449 (1999)
189. K. Naka, H. Itoh, Y. Chujo, Langmuir **19**, 5496 (2003)
190. K. Naka, H. Itoh, Y. Chujo, Bull. Chem. Soc. Jpn. **78**, 501 (2005)
191. A. Manna, P.-L. Chen, H. Akiyama, T.-X. Wei, K. Tamada, W. Knoll, Chem. Mater. **15**, 20 (2003)
192. B.I. Ipe, S. Mahima, K.G. Thomas, J. Am. Chem. Soc. **125**, 7174 (2003)
193. J. Hu, J. Zhang, F. Liu, K. Kittredge, J.K. Whitesell, M.A. Fox, J. Am. Chem. Soc. **123**, 1464 (2001)
194. J. Zhang, J.K. Whitesell, M.A. Fox, Chem. Mater. **13**, 2323 (2001)
195. P.V. Kamat, J. Phys. Chem. B **106**, 7729 (2002)
196. H. Itoh, A. Tahara, K. Naka, Y. Chujo, Langmuir **20**, 1972 (2004)
197. C.G. Overberger, Y. Inaki, J. Polym. Sci., Polym. Chem. Ed. **17**, 1739 (1979)
198. M.J. Moghaddam, S. Hozumi, Y. Inaki, K. Takemoto, Polym. J. **21**, 1739 (1989)
199. M.J. Moghaddam, K. Kanbara, S. Hozumi, Y. Inaki, K. Takemoto, Polym. J. **22**, 369 (1990)
200. M.J. Moghaddam, S. Hozumi, Y. Inaki, K. Takemoto, Polym. J. **21**, 203 (1989)
201. H. Itoh, K. Naka, Y. Chujo, J. Am. Chem. Soc. **126**, 3026 (2004)
202. J.G. Huddleston, H.D. Willauer, R.P. Swatloski, A.E. Visser, R.D. Rogers, Chem. Commun. 1765 (1998)
203. C.J. Mathews, P.J. Smith, T. Welton, A.J.P. White, D.J. Williams, Organometallics **20**, 3848 (2001)

204. L. Cammarata, S.G. Kazarian, P.A. Salter, T. Welton, Phys. Chem. Chem. Phys. **3**, 5192 (2001)
205. S. Carda-Broch, A. Berthod, D.W. Armstrong, Anal. Bioanal. Chem. **375**, 191 (2003)
206. R. Tatumi, H. Fujihara, Chem. Commun. 83 (2005)
207. K.-S. Kim, D. Demberelnyamba, H. Lee, Langmuir **20**, 556 (2004)
208. T. Welton, Chem. Rev. **99**, 2071 (1999)
209. J. Dupont, R.F. de Souza, P.A.Z. Suarez, Chem. Rev. **102**, 3667 (2002)
210. G.T. Wei, Z.S. Yang, C.Y. Lee, H.Y. Yang, C.R.C. Wang, J. Am. Chem. Soc. **126**, 5036 (2004)
211. T. Nakashima, T. Kawai, Chem. Commun. 1643 (2005)
212. J. Simard, C. Briggs, A.K. Boal, V.M. Rotello, Chem. Commun. 1943 (2000)
213. C.-H. Su, P.-L. Wu, C.-S. Yeh, Bull. Chem. Soc. Jpn. **77**, 189 (2004)
214. J.-H. Kim, T.R. Lee, Chem. Mater. **16**, 3647 (2004)
215. P.R. Selvakannan, S. Mandel, S. Phadtare, R. Pasricha, M. Sastry, Langmuir **19**, 3545 (2003)
216. K. Naka, H. Tanaka, Y. Chujo, Polym. J. **39**, 1122 (2007)
217. T. Yonezawa, S. Onoue, N. Kimizuka, Chem. Lett. 1172 (2002)
218. A. Larsson, C. Walldal, S. Wall, Colloids Surf. A, Physicochem. Eng. Asp. **191**, 65 (1999)
219. Z. Zhong, S. Patskovskyy, P. Bouvrette, J.H.T. Luong, A. Gendanken, J. Phys. Chem. B **108**, 4046 (2004)
220. J.N. Cha, H. Birkedal, L.E. Euliss, M.H. Bartl, M.S. Wong, T.J. Deming, G.D. Stucky, J. Am. Chem. Soc. **125**, 8285 (2003)
221. V.S. Murthy, J.N. Cha, G.D. Stucky, M.S. Wong, J. Am. Chem. Soc. **126**, 5292 (2004)
222. S. Ulrich, M. Seijo, A. Laguecir, S. Stoll, J. Phys. Chem. B **110**, 20954 (2006)
223. A. Sehgal, Y. Lalatonne, J.-F. Berret, M. Morvan, Langmuir **21**, 9359 (2005)

2

Organic–Inorganic Hybrid Liquid Crystals: Innovation Toward "Suprahybrid Material"

Kiyoshi Kanie and Atsushi Muramatsu

2.1 Materials Innovation Toward "Suprahybrid Material" by the Utilization of Organic–Inorganic Hybridization

Organic–inorganic hybrid material, fabricated by the nano- or molecular-level interactions between organic and inorganic matters, has attracted great deal of attention for the introduction of novel functions [1]. One of the most important research topics in this area is the generation of synergistic functions between organic and inorganic matters. Recently, extensive efforts have been carried out for the synergistic adaption of the cross functions such as flexibility of organic materials and high hardness of inorganic materials into the resulting hybrids. Furthermore, on-demand appearance of the desired cross functions between organic and inorganic matters as required is an important research topic in materials development. The difference between organic–inorganic hybrid materials and organic–inorganic complex materials is the presence or absence of molecular-level interactions between organic and inorganic interfaces. For example, glass fiber-reinforced plastics (FRPs), which were obtained by the mixing of organic polymer matrices with inorganic fibrous matters, are one of the most representative advanced materials. The resulting FRPs show cross functions such as flexibility and durability; however, the original characters of organic and inorganic matters decrease in proportion to the mixing ratio. Furthermore, the FRPs have micrometer-level phase segregated structures between organic and inorganic domains, and there are no molecular-level interactions between organic and inorganic interfaces. Therefore, such FRPs can be classified as organic–inorganic complex materials. Recently, utilization of nanoparticles have attracted great deals of attention because nanoparticles show interesting properties such as magnetic, electric, fluorescence, and catalytic properties and surface Plasmon resonance behavior, and it is applicable for the development of novel organic–inorganic hybrid materials. The preparation method of the organic–inorganic hybrid materials using nanoparticles is divided roughly into the following two categories. The one is the direct deposition of surface-modified inorganic nanoparticles by organic matter in reaction solvents or organic matrices. The representative procedures for the direct deposition are sol-gel method [1], supercritical solvent method [2], and thermal decomposition method of organometallics [3]. The other is the surface modification of nanoparticles by organic matters. In this case, the desired nanoparticles are separately prepared, so that their resulting surfaces are modified by silica or polymer coating and adsorption of ions or organic molecules for the formation of organic–inorganic interfaces. Induction of liquid crystallinity into inorganic nanoparticles [4,5] and introduction of photoresponsive molecules [6] are the representative examples of the hybrid materials obtained by the surface modification of nanoparticles.

As mentioned above, the organic–inorganic hybrid materials are the future materials focused on the control of organic–inorganic interfaces for the induction of on-demand cross functions and generation of synergistic functions between organic and inorganic matters. In the near future, if we can succeed in the development of the organic–inorganic hybrid material showing on-demand cross functions, synergistically, it can be named as "Suprahybrid material." Further research dealing with the hybrid materials based on precious control of organic–inorganic nanointerfaces would lead to novel findings in materials sciences.

2.2 Organic Liquid Crystals and Lyotropic Liquid–Crystalline Inorganic Fine Particles

Liquid crystal (LC) is one of the most representative functional materials, it shows both fluidity and crystallinity. Material showing liquid–crystalline property induced by the changes of the temperatures is called as thermotropic LC. On the other hand, matter exhibiting such property by the changes of the concentration of the solutions is named lyotropic LCs. The discovery of the LC goes back to the report about a cloudy liquid state of cholesteryl benzoate on heating, observed by Reinitzer in 1888. This is the first example of the thermotropic LC and a pioneering work about LCs. Further detailed descriptions about the histories and the general characters of LCs are well summarized in the literatures [7–10].

In general, LCs used as devices of LC displays are organic thermotropic LCs with low molecular weight that consist of calamitic shapes, and exhibit electric responsivity and optical anisotropy. In this regard, extensive efforts have been taken for the development of novel types of organic LCs applicable to high-performance electro-optic devices, and widely varied types of organic LCs have been reported so far [11]. Recent remarkable progress in organic synthetic chemistry might also contribute to such rapid development in the fields of organic LCs.

2 Organic–Inorganic Hybrid Liquid Crystals   43

As other examples of LCs, except organic LC systems, lyotropic liquid–crystalline dispersions of inorganic fine particles are included. The lyotropic LC dispersion shows optical anisotropy under fluidity and electric and/or magnetic fields. Such colloidal dispersions have been called before as *tactosols*. The pioneering study, concerning about iron oxyhydroxide solutions with optical anisotropy under magnetic field, has been carried out by Majorana [12] in 1902. Researches on V_2O_5 sols showing optical anisotropy under fluidity and electric field, reported by Freundlich et al. [13, 14] in 1915 and 1916, also provide important knowledge at the dawn of these fields. As mentioned above, studies both organic LCs and LC dispersions of inorganic particles have been began almost at the same time. In spite of the remarkable progress of the studies on organic LCs, only a few examples of the LC dispersions of inorganic particles were reported until the 1990s. When compared with the synthetic organic chemistry, the difference might be due to following reasons (1) the knowledge concerning precise syntheses of inorganic fine particles had lagged behind and (2) there had been no discovery of industrial applications of LC dispersions of inorganic particles compared with organic LCs. However, researches on LC dispersions of inorganic particles that showed lyotropic LC behavior have been reported rapidly for these several years. The introduction to details concerning these lyotropic LC dispersions of inorganic particles is well summarized in the literatures [15–19]. In addition, interesting researches on lyotropic liquid crystallinity of organometallic clusters [20, 21] and self-assembling behavior of gold nanorods [22] have been reported in recent years. Lyotropic LC dispersions, except inorganic systems as mentioned above, have also been observed. For example, dispersions of self-assembled rod-like viruses [23], monodispersed polypeptides [24], carbon nanotubes [25, 26], fullerene derivatives [27], and cellulose fine particles [28] show lyotropic LC states, spontaneously. As a new concept to form lyotropic lamellar structure, hybridization of virus and zinc oxides nanoparticle has recently been demonstrated [29].

Here, let us consider the different points of organic LCs with LC dispersions of inorganic particles. First of all, the digit of lattice constants of LC structures such as lamellar and columnar is different. That is, even if it is large, the lattice constants of organic LCs are about ∼5 nm. This is due to the size of the organic compounds or the molecular self-assembled structures form the LC phases. The virtual limits of the molecular sizes of the organic compounds, obtained by organic synthesis except the polymer and the oligomer, are several nanometers. Furthermore, it is difficult to form the organic LC phases that consist of the lattice parameters of more than 10 nm. On the other hand, even if it is small, the lattice parameters of lyotropic LC dispersions obtained so far are 10 nm or more. It is due to the sizes of the inorganic particles that form the LC dispersions. In order to obtain LC order in the LC dispersions, utilization of inorganic particles with shape anisotropy such as spindle-, rod-, sheet-, and disk-like structures with high aspect ratios is essential. In general, such shape control has been achieved by kinetically controlled particle growth. The

44   K. Kanie and A. Muramatsu

particle growth mechanism means that the size of the resulting particle toward the long axis reaches at least more than 10 nm. Recent remarkable progress in synthetic inorganic fine and nanoparticles is becoming useful to obtain various types of monodispersed nanoparticles with shape anisotropies. In near future, we can expect that we will have opportunities to meet new horizon of novel functional materials using size- and shape-controlled inorganic nanoparticles as key components.

On the other hand, organic–inorganic hybrid materials have attracted a great deal of interest especially in the fields of materials science and nanotechnology because it is not rare to find unexpected novel properties with such materials. Here, we describe about the "organic–inorganic hybrid LCs," which we have first succeeded in the introduction of thermotropic liquid crystallinity into monodispersed inorganic fine particles with shape anisotropy [4, 5]. In our studies, we applied the "Gel-Sol method" first reported by Sugimoto and Muramatsu in 1991 for the synthesis of monodispersed fine particles in large quantities at low cost [30]. The invention was based on the idea that highly viscous gels such as metal hydroxide gel network works as a matrix for holding the nuclei and growing particles of the final product to protect them from aggregation even in a high concentration of electrolyte, and sometimes also works as a reservoir of precursory metal complex. The particles with high monodispersity in size and shape, having a variety of resulting shapes, obtained by this method, are quite suitable for the development of "organic–inorganic hybrid LCs."

2.3 Development of Organic–Inorganic Hybrid Liquid Crystals

2.3.1 Hybridization of Calamitic Liquid–Crystalline Amines with Monodispersed TiO$_2$ Nanoparticles [4]

By the utilization of the "Gel-Sol method," precise control of anatase TiO$_2$ nanoparticles in size and shape are readily achieved, so that highly monodispersed TiO$_2$ with ellipsoidal, acicular, spherical, or cubic shapes are prepared in large quantities [31, 32]. We have been focusing our attention on the anisotropic shapes of the TiO$_2$ particles for the induction of thermotropic liquid crystallinity into the particles by the organic–inorganic hybridization. For the nanometer-level hybridization between organic molecules and inorganic nanoparticles, introduction of interaction moiety between them might be one of the most useful ways. Here, if we use electric and magnetic fields and light-responsive functional molecules as organic molecules for the hybridization, addition of such functions into TiO$_2$ nanoparticles might also be expected. Based on this idea, we focus on organic LCs as electro-optical materials, and designed organic LCs having adsorption moieties on the surfaces of TiO$_2$ nanoparticles. The key for the anisotropic shapes of TiO$_2$ nanoparticles obtained by the "Gel-Sol method" is specific adsorption of organic small

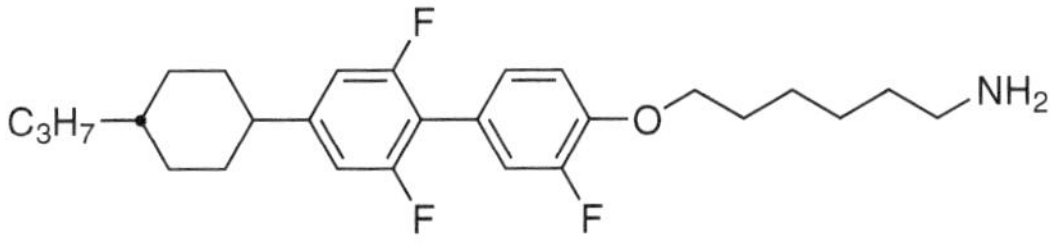

L1: Cr 35 S_X 73 N 101 Iso

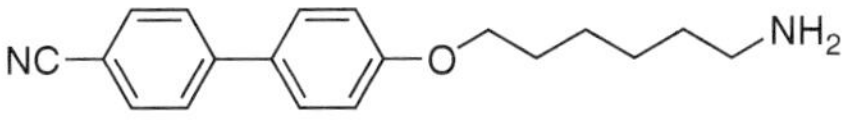

L2: Cr 72 N 92 Iso

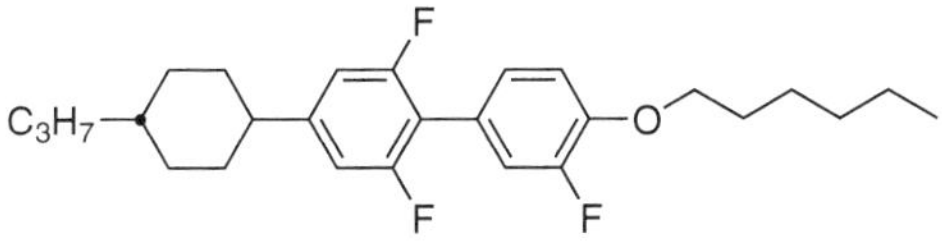

L3: Cr 58 S_X 64 N 100 Iso

Fig. 2.1. Structures of thermotropic LCs **L1**–**L3** for hybridization with TiO_2 particles. *Cr* crystal, S_X higher order smectic, *N* nematic, *Iso* isotropic phases. The figures in-between are transition temperatures in °C

molecules and/or anions used as a shape controller. For example, the acicular shape of TiO_2 nanoparticles is induced by the specific adsorption of ammonia or primary amines onto surfaces parallel to the *c*-axis of tetragonal anatase TiO_2 particles during their growth. The strong adsorption of amines gives us a hint that we would obtain novel organic–inorganic hybrid LCs by the adsorption of amino-substituted mesogenic organic LC molecules to the surfaces of the anisotropic TiO_2 nanoparticles.

We designed and synthesized amino-substituted **L1** and **L2**, as well as **L3**, the same one as **L1** but without an amino group, as shown in Fig. 2.1. The phase transition temperatures and LC phases of **L1**–**L3** were determined by differential scanning calorimetry (DSC) and polarized optical microscopy (POM) with a hot stage. Both fluorophenyl-substituted **L1** and cyanobiphenyl-substituted **L2** show thermotropic nematic (*N*) phases. On the other hand, TiO_2 particles with various shapes, in the transmission electron micrographs of Fig. 2.2, were prepared by the "Gel-Sol method" and thoroughly washed with 2 M HNO_3, 0.01 M NaOH, and distilled water by centrifugation to completely remove the shape controllers. The rather polydispersed and irregular-shaped particles **T4** were prepared by mixing uniform TiO_2 particles of different shapes. All TiO_2 powders were confirmed as an anatase TiO_2 by X-ray diffraction analysis, and their surfaces were all of pure anatase titania as revealed by X-ray photoelectron spectroscopy.

Hybridization of **L1**/**T1**,**2**,**3**,**4** and **L2**,**3**/**T1** was carried out as follows. Twenty milligram of **L** and the same weight of **T** were mixed together and dispersed in 2 mL of methanol by ultrasonication for 30 min. Then, the solvent was removed by evaporation at 50°C and subsequent suction under a reduced pressure. Figure 2.3 exhibits POM images of a hybrid **L1** and **T1** (**L1**/**T1**)

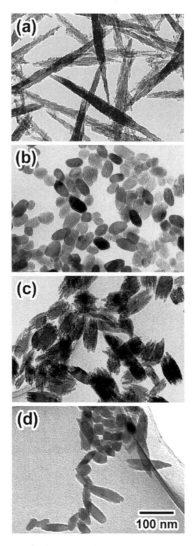

Fig. 2.2. Growth striation of silicon single crystal grown under microgravity, where buoyancy flow is suppressed but the Marangoni flow is dominant [11]

at different temperatures. The hybrid melted at 48°C, and a mesophase with strong birefringence was formed as shown in Fig. 2.3a. Phase transition of the mesophase occurred at 73°C as revealed from a differential scanning calorimeter (DSC), and a schlieren texture as a characteristic of an N phase became observed up to 113°C (Fig. 2.3b). On reaching 113°C, the schlieren texture disappeared and an optically isotropic mesophase was seen up to ca. 130°C (Fig. 2.3c), at which a schlieren texture in a fluidized state was formed again (Fig. 2.3d). The mesophase was readily transformed to a uniaxially aligned

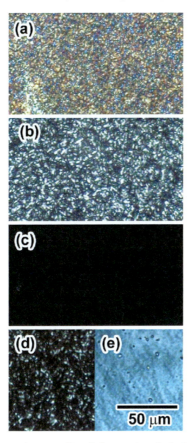

Fig. 2.3. Polarized photomicrographs of thermotropic hybrid LC phases of hybrid **L1/T1** on the heating: (**a**) 70°C; (**b**) 100°C; (**c**) 120°C; (**d**) 140°C, polydomain; and (**e**) 140°C, sheared monodomain

monodomain by shearing as shown in Fig. 2.3e. The birefringence was observed up to 250°C. Interestingly, such liquid crystallinity owing to the hybridization was not observed at all with hybrids **L1/T2,3,4**. Thus, the aspect ratio of the TiO$_2$ particles and their uniformity in morphology are decisive factors for the formation of the hybrids of the LCs. None of the mesomorphic phenomena was seen either with hybrid **L2/T1** until **L2** of the **L2/T1** hybrid was decomposed (ca. 240°C). Hence, in this case, the original liquid crystallinity of **L2** was lost by the hybridization with **T1**. Surface situations of fluorophenyl-covered **L1/T1** and cyanobiphenyl-covered **L2/T1** become lipophilic and hydrophilic states, respectively. The fluidity, essential for the induction of the liquid crystallinity, might be brought by the lipophilization of the surfaces of the TiO$_2$ nanoparticles. Furthermore, since the phase separation was observed only with the hybrid **L3/T1** through POM, amino group of **L1** was likely to play an essential role in the formation of the hybrid LC of **L1** with the anisotropic TiO$_2$ particles. The definite interaction between the amino group of **L1** and the surfaces of **T1** was confirmed by temperature variable infrared spectroscopy. The mixing ratio of **L1** and **T1** is also influenced by the formation of hybrid LC phases. The **L1/T1** $= 1/2$ hybrid exhibited no liquid crystallinity on heating, and only showed a solid state by the coagulation of **T1**. In contrast, the **L1/T1** $= 2/1$ hybrid formed thermotropic LC state, and the phase transition behavior on the heating is as follows: a mesomorphic phase (50–72°C); an N phase (72–121°C); an isotropic phase (121–150°C); and a hybrid N phase (150–250°C). Further increase of the **L1** component in the **L1/T1** hybrids results in the partial phase separation. The structural characterization of the hybrids are described in the literature [4].

2.3.2 Hybridization of Calamitic Liquid–Crystalline Phosphates with Monodispersed α-Fe$_2$O$_3$ Fine Particles [5]

By using the "Gel-Sol method," we have prepared monodispersed α-Fe$_2$O$_3$ particles with various shapes and sizes by the addition of shape controllers and seed particles [30,33–35]. For example, in the presence of PO$_4^{3-}$ ions, monodispersed spindle-type α-Fe$_2$O$_3$ particles have been readily obtained [33–35]. The morphological control was derived from the specific adsorption of the PO$_4^{3-}$ ions on the crystal planes parallel to the c-axis, which retarded the particle growth perpendicular to the c-axis. Such strong adsorption of PO$_4^{3-}$ ions gives us a possibility that we would obtain novel types of organic–inorganic hybrid LCs [5] by the adsorption of organic LC molecules with a phosphate moiety to the surfaces of α-Fe$_2$O$_3$ particles with different shapes. Based on this idea, α-Fe$_2$O$_3$ particles **H1–H6** with different sizes and morphologies were prepared (Fig. 2.4). Seeding technique [33, 34] enabled us the control of particle mean size of spindle-type **H1–H3** without changing the aspect ratio. Monodispersed cuboidal **H4**, polydispersed **H5**, and hexagonal platelet **H6** with {0 0 1} as a basal plane, prepared by using the specific adsorption of OH$^-$ ions on the surfaces to the c-axis [33, 34], were used in our study. All particles were thoroughly washed with 1 M NH$_3$ and water by centrifugation to completely remove impurities and freeze dried. The resulting powders **H1–H6** were confirmed as α-Fe$_2$O$_3$ by X-ray diffraction analysis, and no difference in their surface conditions was observed by X-ray photoelectron spectroscopy.

Also, we designed and synthesized **PL1** and **PL3** with a phosphate moiety as shown in Fig. 2.5. Phosphate group-free **PL2** was also prepared. The phase transition temperatures and **LC** phases of **PL1–PL3** were determined by POM with a hot stage, DSC, and X-ray diffraction. The monodispersed spindle-type **H1–H3** are expected to adsorb the phosphate moiety of **PL1** because they were obtained with PO$_4^{3-}$ ions as a shape controller [36,37].

Hybridization of **PL1** and **H1–H6** or **PL2–PL3** and **H2** was carried out as follows. Twenty milligram of **PL** and the same weight or 40 mg of **H** (**PL/H** $= 1/1$ or $1/2$) were mixed together and dispersed in 2 mL of methanol

2 Organic–Inorganic Hybrid Liquid Crystals   49

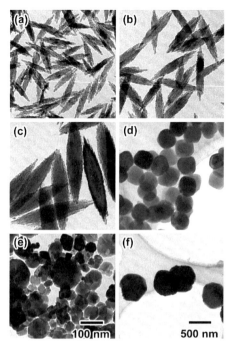

Fig. 2.4. TEM images of α-Fe$_2$O$_3$ particles with various shapes **H1–H6**. The *scale bar* in **H5** is common for **H1–H4**

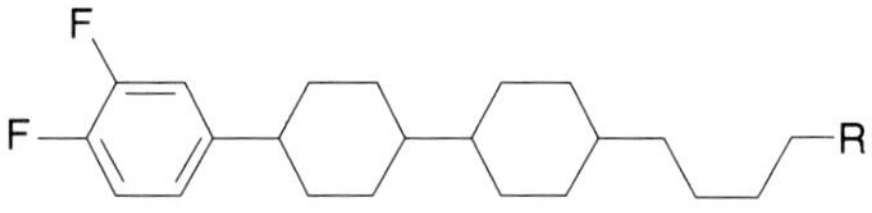

PL1: R = OPO$_3$H$_2$;
  G–50 M$_1$ 77 M$_2$ 129 S$_A$ 193 N 194 Iso
PL2: R = H;
  Cr 41 S$_B$ 45 N 121 Iso

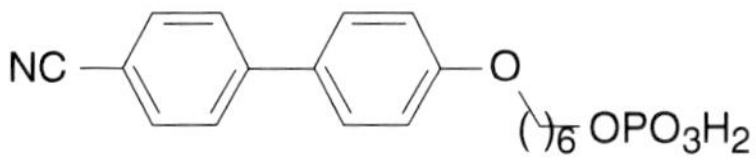

PL3: G–4 M 57 S$_A$ 146 Iso

Fig. 2.5. Structures of thermotropic LCs **PL1–PL3** for hybridization with α-Fe$_2$O$_3$ particles. *G* glass, *Cr* crystal, *M* mesomorphic, *S$_A$* smectic A, *N* nematic, *Iso* isotropic phases. The figures in-between are transition temperatures in °C

and 2 mL of CHCl$_4^{3-}$ by ultrasonication for 30 min followed by the removal of the solvents at 60°C under Ar flowing. Figure 2.6 exhibits optical microscopic images of 1/2 hybrids of **PL1/H2**, **PL1/H5**, and **PL1/H6** at 90°C in the presence of a cross polarizer (a) and its absence (b). For the **PL1/H2** hybrid,

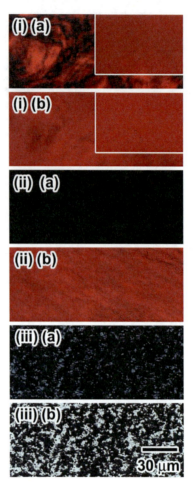

Fig. 2.6. Optical photomicrographs of thermotropic hybrid LC phases of 1/2 hybrids (i) **PL1/H2**, (ii) **PL1/H5**, and (iii) **PL1/H6** at 90°C (a) with a polarizer; (i) – (a) and (b) *insets*: sheared monodomain

a marbled texture as a characteristic of an N phase with strong birefringence and fluidity was observed. The mesomorphic phase was readily transformed to a uniaxially aligned monodomain by shearing (Fig. 2.6i(a), inset).

DSC measurement on the second heating scan revealed that the hybrid showed the glass–mesophase transition at −35°C. After an endothermic peak at 148°C, the hybrid showed a highly fluidized state with strong birefringence observed by POM. The birefringence was kept up to 250°C. Similar mesomorphic behavior was also seen for 1/2 hybrids of **PL1/H1,H3** and 1/1 hybrids of **PL1/H1–H3**. However, the further increase of the **PL1** component resulted in the partial phase separation of **PL1** with **H1–H3**. On the other hand, 1/2 as well as 1/1 hybrids of **PL1/H5** also formed a uniform fluidized material without phase segregation of **PL1** with **H5** (Fig. 2.6ii(b)), however no birefringence was found and formed an optically isotropic state. The **PL1/H4** $= 1/2$ and $1/1$ hybrids, which consists of monodispersed cuboidal α-Fe_2O_3 particle, also showed an optically isotropic state with fluidity. Thus, the aspect ratio of the α-Fe_2O_3 particles with different shapes was the decisive factor for the formation of the nematic LCs of the hybrids. Interestingly, birefringence owing to **PL1** single component and coagulated **H6** is seen for a $1/2$ mixture of **PL1/H6** as shown in Fig. 2.6iii.

It means phase segregation between **PL1** and **H6** without hybridization. The platelet **H6** particles have c-plane as a principal one, formed by the specific adsorption of OH^- on the basal plane [33, 34]. It shows that the factor for the hybridization might be brought by the specific adsorption of the phosphate group of **PL1** on a plane parallel to c-axis. None of the mesomorphic phenomena was seen with **PL3/H2** hybrids and it was decomposed at ca. $260°C$. Hence, in this case, the original liquid crystallinity of **PL3** was lost by the hybridization with **H2**. Furthermore, since the phase separation was observed only with **PL2/H2** hybrids through POM, the phosphate group of **PL1** played an essential role in the formation of the hybrid LC of **PL1**. The definite interaction between the phosphate group of **PL1** and the surfaces of **H1**–**H5** was confirmed by temperature variable infrared spectroscopy, and no free PO_3H_2 vibration was observed by the hybridization. The mesomorphic phase structures of the **PL1/H1**–**H5** hybrids were examined by small-angle X-ray scattering measurements. The profiles of the **PL1/H2** $= 1/2$ hybrid at 90 and $170°C$ suggested the existence of periodic particle interactions at intervals of 49.7 and 46.5 nm, respectively, corresponding to the total width of an **H2** covered with the **PL1** molecules toward the short axis of the spindle-type particles. Therefore, the mesomorphic phase seems to have a nematic-like 1D order in the direction of the long axis of the spindle-type particles. Such a periodic scattering was not observed for the single components of **H1**–**H6** and the **PL1/H5** hybrids. On the other hand, peaks of 45.6, 30.4, and 22.8 nm at $170°C$ were observed for the hybrid of **PL1/H4** $= 1/1$, which were corresponding to d_{200}, d_{300}, and d_{400} planes of a superlattice structure with a simple cubic LC structure. The lattice parameter could be assigned as $a = 91.2$ nm, which was consistent with the particle mean sizes of **H4** covered with the **PL1** molecules by the adsorption on their surfaces. The schematic illustration for the formation of hybrid LC phases is shown in Fig. 2.7.

2.4 Summary and Prospect

Recent remarkable progress in synthetic chemistry concerning monodispersed inorganic nano- and fine particles with various shapes such as the "Gel-Sol method" enabled us to obtain size- and shape-controlled TiO_2 and α-Fe_2O_3 particles in large quantities. By using these particles, we have succeeded in the development of the "organic–inorganic hybrid LCs" as a novel functional nanohybrid material. To date, LC behaviors are known for some inorganic

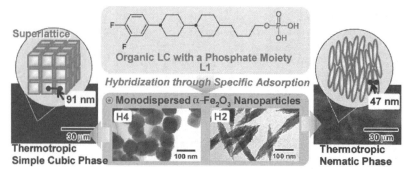

Fig. 2.7. Formation of organic–inorganic hybrid LC phases

particles with anisotropic shapes in highly dilute dispersions, as mentioned above. In such systems, the liquid–crystalline performance is governed by the balance of the repulsive and attractive forces among the anisotropic particles themselves. On the other hand, surface modification of rare-metal nanospheres by thiol derivatives of LC compounds has been examined [38–40]. However, optical anisotropy due to the particle alignment was not observed. Hence, to the best of our knowledge, our observations are the first results on new optically anisotropic LC phases with a novel organic–inorganic hybrid of a liquid–crystalline compound and monodispersed inorganic particles. In view of the almost infinite possibilities in the combination of organic and inorganic matters, such hybrid materials may be of unfathomable potentials as one of the most advanced functional materials in the future. For example, monodispersed α-Fe_2O_3 particles used in our study are readily converted into Fe_3O_4 and γ-Fe_2O_3 as magnetic materials, precisely keeping their morphology without sintering [34]. The nanolevel hybridization of inorganic nanoparticles with fair magnetic properties and LCs with good electric response will make them one of the most advanced functional materials with excellent multiresponsivity such as active devices.

References

1. G. Kickelbick (ed.), *Hybrid Materials: Synthesis, Characterization, and Applications* (Wiley-VCH, Weinheim, 2006)
2. T. Adschiri, Chem. Lett. **36**, 1188 (2007)
3. Y. Yin, A.P. Alivisatos, Nature **437**, 664 (2005)
4. K. Kanie, T. Sugimoto, J. Am. Chem. Soc. **125**, 10518 (2003)
5. K. Kanie, A. Muramatsu, J. Am. Chem. Soc. **127**, 11578 (2005)
6. R. Mikami, M. Taguchi, K. Yamada, K. Suzuki, O. Sato, Y. Einaga, Angew. Chem. Int. Ed. **43**, 6135 (2004)
7. Ekisho Binran Henshuu Iinkai (eds.), *Ekisho Binran* (Maruzen, Tokyo, 2000)
8. Japan Chemical Society (eds.), *Ekisho no Kagaku* (Japan Scientific Societies Press, Tokyo, 1994)

2 Organic–Inorganic Hybrid Liquid Crystals   53

9. N. Kusabayashi (ed.), *Ekisho Zairyo* (Kodan-shya Scientific, Tokyo, 1991)
10. P.J. Collings, M. Hird, *Introduction to Liquid Crystals* (Taylor & Francis, Bristol, PA, 1997)
11. http//liqcryst.chemie.uni-hamburg.de/en/lolas.php
12. Q. Majorana, Phys. Z. **4**, 145 (1902)
13. H. Diesselhorst, H. Freundlich, Phys. Z. **16**, 419 (1915)
14. H. Freundlich, Z. Elektrochem. **22**, 27 (1916)
15. K. Kanie, T. Sugimoto, *A New Phase on Organic/Inorganic Nanocomposite Materials* (NTS, Tokyo, 2004), pp. 175–188
16. P. Davidson, P. Batail, J.C.P. Gabriel, J. Livage, C. Sanchez, C. Bourgaux, Prog. Polym. Sci. **22**, 913 (1997)
17. A.S. Sonin, Colloid J. **60**, 129 (1998)
18. A.S. Sonin, J. Mater. Chem. **8**, 2557 (1998)
19. J.-C.P. Gabriel, P. Davidson, Adv. Mater. **12**, 9 (2000)
20. J. Sayettat, L.M. Bull, J.-C.P. Gabriel, S. Jobic, F. Camerel, A.-M. Marie, M. Fourmigué, P. Batail, R. Brec, R.-L. Inglebert, Angew. Chem. Int. Ed. **37**, 1711 (1998)
21. F. Camerel, M. Antonietti, C.F.J. Faul, Chem. Eur. J. **9**, 2160 (2003)
22. N.R. Jana, Angew. Chem. Int. Ed. **43**, 1536 (2004)
23. Z. Dogic, S. Fraden, Langmuir **16**, 7820 (2000)
24. S.M. Yu, V.P. Conticello, G. Zhang, C. Kayser, M.J. Fournier, T.L. Mason, D.A. Tirrell, Nature **389**, 167 (1997)
25. V.A. Davis, L.M. Ericson, A. Nicholas, G. Parra-Vasquez, H. Fan, Y. Wang, V. Prieto, J.A. Longoria, S. Ramesh, R.K. Saini, C. Kittrell, W.E. Billups, W.W. Adams, R.H. Hauge, R.E. Smalley, M. Pasquali, Macromolecules **37**, 154 (2004)
26. W. Song, I.A. Kinloch, A.H. Windle, Science **302**, 1363 (2003)
27. M. Sawamura, K. Kawai, Y. Matsuo, K. Kanie, T. Kato, E. Nakamura, Nature **419**, 702 (2002)
28. J. Araki, M. Wada, S. Kuga, T. Okano, Langmuir **16**, 2413 (2000)
29. S.-W. Lee, C. Mao, C.E. Flynn, A.M. Belcher, Science **296**, 892 (2002)
30. T. Sugimoto, *Monodispersed Particles* (Elsevier, Amsterdam, 2001)
31. T. Sugimoto, X. Zhou, A. Muramatsu, J. Colloid Interface Sci. **259**, 53 (2003)
32. K. Kanie, T. Sugimoto, Chem. Commun. 1584 (2004)
33. T. Sugimoto, M.M. Khan, A. Muramatsu, H. Itoh, J. Colloids Surf. A **79**, 233 (1993)
34. T. Sugimoto, Y. Wang, J. Colloid Interface Sci. **207**, 137 (1998)
35. H. Itoh, T. Sugimoto, J. Colloid Interface Sci. **265**, 283 (2003)
36. T. Sugimoto, A. Muramatsu, J. Colloid Interface Sci. **184**, 626 (1996)
37. M. Ozaki, S. Kratohvil, E. Matijevic, J. Colloid Interface Sci. **102**, 146 (1984)
38. N. Kanayama, O. Tsutsumi, A. Kanazawa, T. Ikeda, Chem. Commun. 2640 (2001)
39. H. Yoshikawa, K. Maeda, Y. Shiraishi, J. Xu, H. Shiraki, N. Toshima, S. Kobayashi, Jpn. J. Appl. Phys. **41**, L1315 (2002)
40. L. Cseh, G.H. Mehl, J. Am. Chem. Soc. **128**, 13376 (2006)

3

Polymer-Assisted Composites of Trimetallic Nanoparticles with a Three-Layered Core–Shell Structure for Catalyses

Naoki Toshima

3.1 Introduction

Metal nanoparticles, nanoscopic metal particles, or fine particles of metal in a nanometer size have received much attention recently because they have specific chemical and physical properties much different from those of bulk metal [1, 2]. They are expected to be useful in various ways such as catalysts [3–7], magnetic materials [8–11], semiconductors [12, 13], electro-optic materials [14, 15], drug delivery materials [16], and so on, and thus attractive to many researchers. These metal nanoparticles are often surrounded by organic corona, which stabilizes the dispersion of nanoparticles and provide additional novel functions to the core nanoparticles. Polymers are useful materials as the organic corona because of their strong stabilizing function due to multidentate interaction. Thus, these metal nanoparticles are prepared and used as composites with organic materials like polymers, when they are applied to various fields.

Among these applications, their use as catalysts has been most commonly investigated by many researchers because the metal nanoparticle catalysts are much different from practical heterogeneous and homogeneous ones. The peculiar catalytic properties of metal nanoparticles are considered to be attributed to their high specific surface area and special surface structure. The polymer-protected metal nanoparticles can be used as catalysts in the form of colloidal dispersions in solution, and solids supported on inorganic supports in gas phase. From the viewpoint of the catalyst design [3–7] it is of great importance to develop metal catalysts arranged on the nanometer scale in order to provide superior functions to metal nanoparticles with uniform size and narrow size distribution.

In practical metal catalysts, addition of other elements can sometimes improve the catalytic activity and selectivity. From this point of view bimetallic and trimetallic systems are often investigated. The effect of the second element can be explained by a ligand (electronic) effect and an ensemble (geographic) effect [17]. In the electronic effect the added metal can electronically affect the electron density of metal at catalytic sites, which improves or deactivates the catalytic activity and selectivity. In this case the atom of the added metal should be adjacent to that of the catalytic metal. The reaction substrates interact only with catalytic metal atoms without direct interaction with added metal atoms. In the case of a geographic effect, on the other hand, the substrate should interact not only with the catalytic metal atoms but also with the added metal atoms. Thus, both kinds of metal atoms can directly interact with a substrate and affect the activity and selectivity of the reaction.

Under these circumstances preparation of bimetallic and trimetallic nanoparticles has been investigated with great interest. If the two kinds of metal ions are simultaneously reduced in the presence of protective polymer, there are two possibilities. One is separate formation of two kinds of metal nanoparticles and the other, formation of bimetallic nanoparticles. In our first attempt to reduce $PdCl_2$ and H_2PtCl_6 by alcohol in the presence of poly(N-vinyl-2-pyrrolidone) (PVP), formation of bimetallic nanoparticles having a Pt-core/Pd-shell structure was fortunately discovered [18].

After the discoveries we have recently developed a novel method to prepare the core/shell structure under very mild conditions [19]. This is a so-called self-organization method, in which only mixture of two kinds of dispersions of metal nanoparticles can produce bimetallic nanoparticles having a core/shell structure. This mild preparation method can give us a chance to fabricate trimetallic nanoparticles having a three-layered core/shell structure. In this chapter, details of the preparation procedure and the characterization of such trimetallic nanoparticles will be described, in addition to a new synthetic method (a self-organization method) of bimetallic nanoparticles with a core/shell structure.

3.2 Fabrication of a Core/Shell Structure

Bimetallic nanoparticles, which are composed of two metal elements in a particle, often have higher catalytic activity and/or selectivity than the monometallic ones. In this case the catalytic activity depends not only on the composition but also on the structure of bimetallic nanoparticles. Among various structures of bimetallic nanoparticles the most interesting one could be a core/shell structure, in which atoms of one element form a core and those of the other element cover the core to form a shell.

We have already reported the formation of core/shell-structured bimetallic nanoparticles by a simultaneous reduction (coreduction) [20] and a sacrificial hydrogen reduction method [21], before the present self-organization (physical mixture) method (Fig. 3.1). In the core/shell-structured bimetallic nanoparticles thus prepared, the catalytic activity of shell atoms can be electronically affected by the core atoms. Among three fabrication methods for a core/shell structure the so-called self-organization method, in which two kinds of dispersions of metal nanoparticles are mixed at room temperature in the solution,

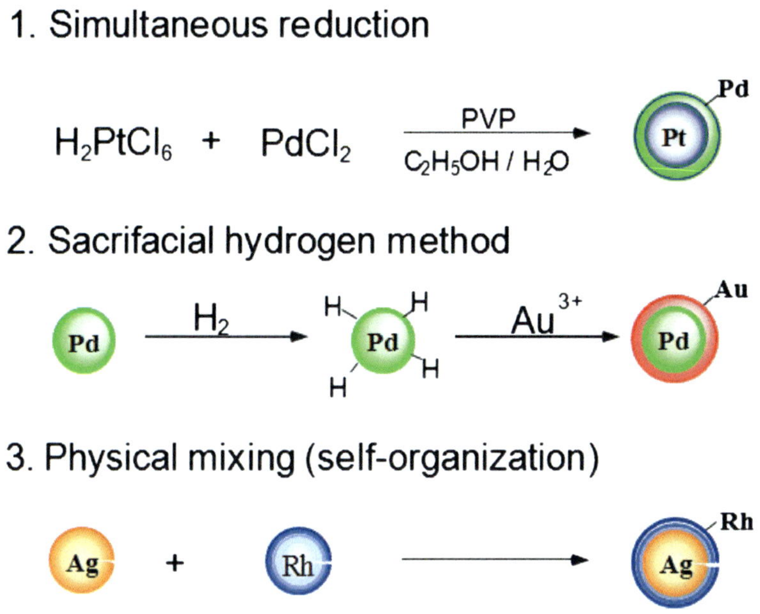

1. Simultaneous reduction

$$H_2PtCl_6 + PdCl_2 \xrightarrow[C_2H_5OH / H_2O]{PVP}$$ (Pt/Pd)

2. Sacrifacial hydrogen method

Pd $\xrightarrow{H_2}$ Pd(H) $\xrightarrow{Au^{3+}}$ (Pd/Au)

3. Physical mixing (self-organization)

Ag + Rh $\longrightarrow$ (Ag/Rh)

Fig. 3.1. Three methods to fabricate core/shell-structured bimetallic nanoparticles

will be carried out under the mildest conditions. Thus, details of this method will be described here.

This method was first discovered when dispersions of PVP-protected Ag nanoparticles and Rh nanoparticles were mixed at room temperature. [19] As well-known dispersions of Ag nanoparticles have a plasmon band and a yellow color, while those of Rh nanoparticles have a brownish black color and no absorption maximum in the visible region. Thus, when a dispersion of PVP-protected Ag nanoparticles was mixed with that of PVP-protected Rh nanoparticles at room temperature, plasmon absorption at ca. 480 nm due to Ag nanoparticles was observed at first. The strength of this plasmon absorption, however, decreased rapidly and disappeared completely in 30–60 min. The same reaction was observed between the dispersions of Ag nanoparticles and Pd nanoparticles.

Figure 3.2 shows transmission electron micrograph (TEM) images of meal nanoparticles of (a) Ag nanoparticles before the mixture, (b) Pd nanoparticles before the mixture, (c) the mixtures of Ag and Pd nanoparticles in 30 min after the mixture, and (d) the mixtures of Ag and Pd nanoparticles 60 min after the mixture. These TEM images indicate that Ag nanoparticles are large in size, while Pd nanoparticles, small in size, and aggregates are observed in the mixtures of Ag and Pd nanoparticles, in which small particles look to surround the large [19]. After 24 h, the mixture large particles and aggregates disappear to form rather small particles in moderately uniform size, suggesting that the spontaneous reaction may occur under mild reaction conditions at

58   N. Toshima

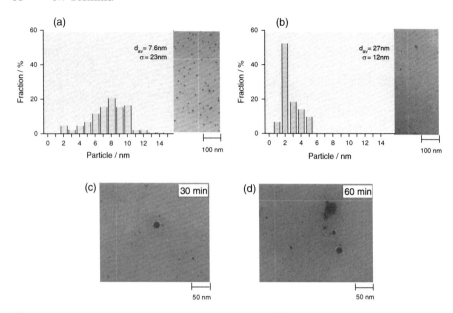

Fig. 3.2. Transmission electron micrograph of (**a**) Ag nanoparticles before the mixture, (**b**) Pd nanoparticles before the mixture, (**c**) mixtures of Ag and Pd nanoparticles in 60 min after the mixtures, and (**d**) the mixtures of Ag and Pd nanoparticles in 24 h after the mixture

room temperature in solution. This is surprising and very interesting, but in fact, microcalorimetric measurements have revealed that the mixture provides a exothermic reaction, for example, between Ag nanoparticles and Rh or Pd nanoparticles [22].

As for the bimetallic structure of nanoparticles produced by physical mixture of dispersions of Ag and Rh or Pd nanoparticles at room temperature in 24 h, mappings of Ag and Rh or Pd elements by using the energy-filtered TEM (EF-TEM) has showed that Ag is located near the center of particles, while Rh or Pd, near the edge of the particles. In other words, an Ag-core/Rh-shell or Ag-core/Pd-shell structure would be spontaneously produced by the mixture of the two metal nanoparticles. Thus, metal nanoparticles react with the other kinds of metal nanoparticles, where the structure can be rearranged to form the core/shell structure. This is a self-organizing process. We call this method to produce a core/shell structure "a self-organization method." This reaction looks very curious, but this process produces thermodynamically stable products like Ostwald ripening in colloid chemistry.

Here, we would like to emphasize that this self-organization is a very mild reaction to produce a core/shell structure, which is thermodynamically stable. Based on these observations, we have reached a conclusion that this self-organization method (physical mixture) can be applied to produce trimetallic nanoparticles having a three-layered core/shell structure.

3.3 Fabrication of Pd/Ag/Rh Trimetallic Nanoparticles

Based on the concept of an electronic effect in bimetallic nanoparticles having a core/shell structure, a three-layered core/shell structure, in which one element forms a core, the second element covers the core forming an interlayer, and the third element covers the interlayer forming a shell, has been proposed. A successive electronic effect is expected in trimetallic nanoparticle having this three-layered core/shell structure. This successive electronic effect is considered to be more effective than the electronic effect in a bilayered system due to the concept of sequential potential field [23]. However, only few reports have been published on trimetallic nanoparticles. Henglein [24] has reported the formation of Pd-core/Au-interlayer/Ag-shell trimetallic nanoparticles by successive reduction of the corresponding metal ions with γ-irradiation, although they have not been applied to catalysis.

We have briefly reported on the formation of the Pd/Ag/Rh trimetallic nanoparticles [25] and described the syntheses and characterization of Pd/Ag bimetallic and Pd/Ag/Rh trimetallic nanoparticles, as well as their catalytic activities toward hydrogenation of olefin [26]. Interestingly, the catalytic activity strongly depended on the composition. Only the trimetallic nanoparticles with strictly designed composition had very high activity, which suggested that the three-layered core/shell structure was spontaneously produced at the special composition.

For the fabrication of trimetallic nanoparticles having a three-layered core/shell structure, we attempted at first a simultaneous reduction or coreduction method. In the case of Pd/Ag/Rh trimetallic nanoparticles, where the ionization potentials of Pd, Ag, and Rh are 8.34, 7.576, and 7.46 eV, respectively, the simultaneous reduction of Pd^{2+}, Ag^+, and Rh^{3+} ions in refluxing alcohol resulted in random alloy nanoparticles. The trimetallic nanoparticles thus prepared had much less catalytic activity for hydrogenation of olefin than Rh nanoparticles. This means that all the Rh atoms do not serve as catalytic sites on the surface of alloy nanoparticles. Thus, we examined the physical mixture (self-organization) method, in which the reaction was carried out under mild conditions like at room temperature. The fabrication reaction was designed by the mixture of PVP-protected Pd-core/Ag-shell bimetallic nanoparticles with PVP-protected Rh nanoparticles as shown in Fig. 3.3.

Colloidal dispersions of PVP-protected Pd/Ag bimetallic nanoparticles with a Pd-core/Ag-shell structure were prepared by a so-called "sacrificial hydrogen reduction" method [21]. Polymer-protected Pd nanoparticles, prepared separately from palladium(II) acetate by an alcohol reduction method in ethylene glycol (an average diameter $= 4.4 \pm 1.2$ nm), were treated with hydrogen in advance, and then a solution of silver perchlorate was slowly added to the colloidal dispersion of hydrogen-treated seed Pd nanoparticles. No aggregates or precipitates were observed in the colloidal dispersions of PVP-protected Pd/Ag bimetallic nanoparticles thus prepared. The bimetallic nanoparticles were separated by filtration with an ultrafilter membrane. Since

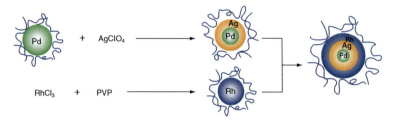

Fig. 3.3. Design to fabricate three-layered Pd-core/Ag-interlayer/Rh-shell trimetallic nanoparticles

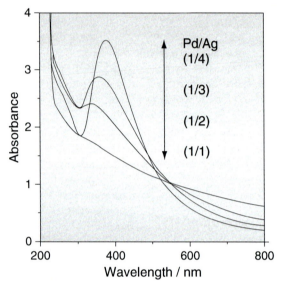

Fig. 3.4. UV-Vis spectra of the dispersions of PVP-protected Pd/Ag bimetallic nanoparticles at the atomic ratio Pd/Ag = 1/4, 1/3, 1/2, and 1/1, prepared by a sacrificial hydrogen reduction method [26]

the filtrates did not contain any metal ions according to qualitative analyses, we believe the yield of the bimetallic nanoparticles is almost quantitative.

Figure 3.4 shows the UV-Vis spectra of the colloidal dispersions of polymer-protected Pd/Ag bimetallic nanoparticles at various Pd/Ag atomic ratios. The absorption peak at 380 nm can be attributed to the surface plasmon absorption of Ag nanoparticles [27]. The surface plasmon absorption of Ag cannot be observed for the colloidal dispersions of Pd/Ag bimetallic nanoparticles at a Pd/Ag atomic ratio of 1/1. This means that the Ag ions added do not form large enough Ag nanoparticles to show plasmon absorption, but deposit on the surface of Pd nanoparticles to form only tiny spots, which are too small to show plasmon absorption. When the amount of Ag was larger than Pd, i.e., the atomic ratio of Pd/Ag was less than 1/2, the strong plasmon absorption was

3 Polymer-Assisted Composites of Trimetallic Nanoparticles 61

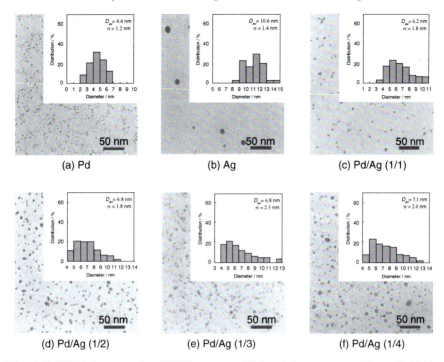

Fig. 3.5. TEM photographs of PVP-protected Pd and Ag monometallic, and Pd/Ag bimetallic nanoparticles: (**a**) Pd, (**b**) Ag, (**c**) Pd-core/Ag-shell(1/1), (**d**) Pd-core/Ag-shell(1/2), (**e**) Pd-core/Ag-shell(1/3), and (**f**) Pd-core/Ag-shell(1/4) nanoparticles. D_{av} is the average diameter and σ is the standard deviation [26]

observed. This observation suggests that the Ag atoms, produced by reduction of Ag ions with hydride on the surface of the Pd seed, completely cover the Pd seed to form a Ag shell when the atomic ratio of Pd/Ag is less than 1/2. The presence of Ag atoms on the surface of the Pd/Ag bimetallic nanoparticles thus prepared was confirmed by FT-IR spectra of the adsorbed CO on the surface of the nanoparticles [28].

TEM images and size distribution histograms of Pd/Ag bimetallic nanoparticles are shown in Fig. 3.5. The bimetallic nanoparticles were spherical and small, and had a considerably uniform size. The average diameter of Pd/Ag (1/2 in atomic ratio) nanoparticles was 6.8 ± 1.8 nm, which is larger than that of seed Pd nanoparticles (an average diameter = 4.4 ± 1.2 nm) and smaller than that of Ag nanoparticles (10.6 ± 1.4 nm) that were prepared separately by glycol reduction. The average diameter of Pd/Ag increased with an increase in the Ag content in the bimetallic nanoparticles. The same tendency was observed in the case of Pd/Au [29] and Pd/Pt [21, 30] bimetallic nanoparticles prepared by the same method (a sacrificial hydrogen reduction method) [21].

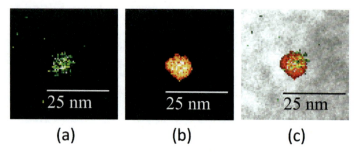

Fig. 3.6. EF-TEM images of Pd/Ag(1/2) bimetallic nanoparticles prepared by a sacrificial hydrogen reduction method: (**a**) Pd mapping, (**b**) Ag mapping, and (**c**) total mapping of Pd and Ag (*Pd* green, *Ag* red, *Pd + Ag* yellow) [26]

Detailed analyses of the Pd/Ag bimetallic nanoparticles were carried out using EF-TEM. Figure 3.6 shows the mapping of Pd and Ag in the Pd/Ag(1/2) bimetallic nanoparticles by using EELS. Green and red areas of the map show the coexistence of Pd and Ag. This mapping picture clearly indicates that the Ag surrounds Pd.

Based on the results of CO-FT-IR spectra, UV-Vis spectra, TEM images, and EF-TEM pictures, PVP-protected Pd/Ag(1/2) bimetallic nanoparticles prepared by a sacrificial hydrogen reduction method are considered to have a Pd-core/Ag-shell structure. This core/shell structure was also supported by comparing the catalytic activities of these nanoparticles toward the hydrogenation of methyl acrylate at 30°C under an atmospheric pressure of hydrogen. The Pd seed had a high catalytic activity (6.71 mol-H_2 mol-Pd^{-1} s^{-1}), while the Ag nanoparticles, prepared by reduction with ethylene glycol, showed almost no activity (0.049 mol-H_2 mol-Ag^{-1} s^{-1}). When the atomic ratio of Pd/Ag was varied, the bimetallic nanoparticles with Pd/Ag = 1/1 had a low activity (0.115 mol-H_2 mol-$(Pd+Ag)^{-1}$ s^{-1}), while those with Pd/Ag = 1/2 had almost no activity (0.037 mol-H_2 mol-$(Pd+Ag)^{-1}$ s^{-1}), similar to the case of Ag. This result supports that, when Pd/Ag = 1/2, the Pd core is completely covered by a Ag shell.

Now, the self-organization method was applied to fabricate the Pd/Ag/Rh trimetallic nanoparticles by physical mixture of PVP-protected Pd/Ag bimetallic nanoparticles with PVP-protected Rh nanoparticles. When a colloidal dispersion of PVP-protected Rh nanoparticles (an average diameter = 2.5 ± 0.83 nm), prepared by an alcohol reduction method, was mixed at room temperature with the colloidal dispersion of Pd/Ag(1/2) bimetallic nanoparticles (an average diameter = 6.8 ± 1.8 nm) prepared by a sacrificial hydrogen reduction method, the plasmon absorption at 380 nm gradually decreased in strength and disappeared completely after 20 min. This phenomenon suggests that the Ag surface of Pd/Ag bimetallic nanoparticles is covered by the Rh nanoparticles. Similar phenomena were observed in the mixture of colloidal dispersions of PVP-protected Ag nanoparticles and Rh

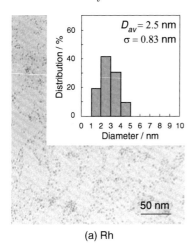 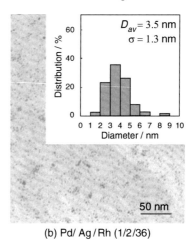

Fig. 3.7. TEM photographs of PVP-protected (**a**) Rh monometallic and (**b**) Pd/Ag/Rh(1/2/36) trimetallic nanoparticles: D_{av} is the average diameter, σ is the standard deviation [26]

nanoparticles, which we have previously reported [19]. During and after the mixing reaction of colloidal dispersions no aggregates or precipitates were observed.

In the CO-FT-IR spectra of the mixtures of Pd/Ag(1/2) bimetallic nanoparticles and Rh nanoparticles, the peaks characteristic to Rh rapidly increased in intensity with an increase in the amount of Rh nanoparticles added to the mixtures. This observation suggests the presence of Rh on the surface of particles resulting from the mixtures.

Figure 3.7 shows TEM images of Rh nanoparticles and the Pd/Ag/Rh(1/2/36) trimetallic nanoparticles, produced by the mixture of Pd/Ag(1/2) bimetallic nanoparticles and Rh nanoparticles at the atomic ratio of (Pd + Ag)/Rh = 1/12. The trimetallic nanoparticles were considerably uniform in size and had an average diameter of 3.5 ± 1.3 nm, although average diameters of Pd/Ag(1/2) and Rh nanoparticles were 6.8 and 2.5 nm, respectively. This suggests that the particles produced by the mixture are not composed of two independent kinds of particles, but of alloyed trimetallic nanoparticles. The HR-TEM image and a part of EF-TEM mapping are shown in Fig. 3.8 in black and white picture. Due to the EF-TEM mapping Rh atoms are located not in the center of the particles, but rather on the surface of the particles as Fig. 3.8b shows. The HR-TEM image shows that the particles are mostly single crystals and are composed of three elements on the basis of EDS spectra, although some polycrystalline structures are also observed. Since all of the bulk crystals of Pd, Ag, and Rh have a face-centered cubic structure, in the Pd/Ag/Rh trimetallic nanoparticles produced the Ag interlayer can grow

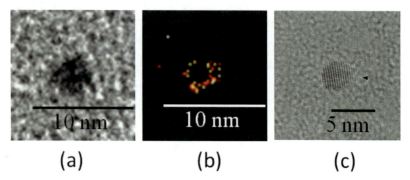

Fig. 3.8. EF-TEM images and HR-TEM photograph of Pd/Ag/Rh(1/2/36) trimetallic nanoparticles: (**a**) zero-loss image, (**b**) Rh mapping, and (**c**) HR-TEM photograph [26]

epitaxially on the Pd seed, and the Rh shell also can grow up epitaxially on the Ag interlayer, resulting in almost single crystal structure.

Although we have no strict evidence on the structure of the present PVP-protected Pd/Ag/Rh(1/2/36) trimetallic nanoparticles, the CO-FT-IR, TEM, HR-TEM, and EF-TEM results suggest the spontaneous formation of a three-layered core/shell structure or a Pd-core/Ag-interlayer/Rh-shell structure by self-organization. This structure is also supported by the high catalytic activity of these trimetallic nanoparticles, described in the next section.

3.4 Catalytic Properties of Pd/Ag/Rh Trimetallic Nanoparticles

The catalytic properties of Pd/Ag/Rh trimetallic nanoparticles, fabricated by the self-organization method described in the previous section starting from PVP-protected Pd-core/Ag-shell bimetallic nanoparticles and Rh nanoparticles, were examined. Here, we used Pd/Ag bimetallic nanoparticles only at the atomic ratio of 1/2 because the Pd seed core could be completely covered by Ag at an atomic ratio of 1/2. The trimetallic nanoparticles with various atomic ratios were used as catalysts for hydrogenation of methyl acrylate at 30°C under an atmosphere of hydrogen. The catalytic activity was calculated by dividing the initial rate of hydrogen uptake per second with total amount of used metals (Pd + Ag + Rh). The catalytic activity varied with the composition of trimetallic nanoparticles.

In the series of trimetallic nanoparticles prepared from Pd/Ag(1/2) bimetallic nanoparticles (an average diameter = 6.8±1.8 nm) and Rh nanoparticles (an average diameter = 2.5 ± 0.8 nm), the catalytic activity varied with the atomic ratio of Rh to (Pd + Ag). In other words, the highest catalytic activity was achieved by the trimetallic nanoparticles at the atomic ratio of 1/2/36 (Fig. 3.9a). Interestingly the catalytic activity had a sharp maximum

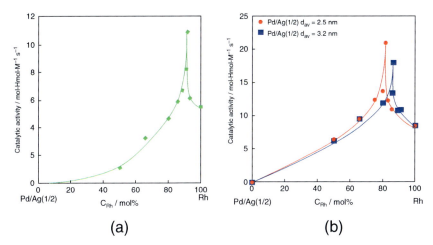

Fig. 3.9. The relationship between metal composition and catalytic activity of PVP-protected Pd/Ag/Rh trimetallic nanoparticles for hydrogenation of methyl acrylate. (**a**) The trimetallic nanoparticles were prepared by starting from Pd/Ag(1/2) bimetallic nanoparticles with an average diameter of 6.8 ± 1.8 nm. (**b**) The trimetallic nanoparticles were prepared by starting from Pd/Ag(1/2) bimetallic nanoparticles with an average diameter of (*filled circle*) 2.5 ± 0.4 nm, and (*filled square*) 3.2 ± 0.7 nm. The concentration of Rh C_{Rh} shows the % atomic ratio of Rh in total metal (Pd + Ag + Rh) [26]

peak at this point, where the average diameter was the smallest (3.5 ± 1.3 nm) among a series of trimetallic nanoparticles starting from the Pd/Ag(1/2) bimetallic nanoparticles (6.8 ± 1.8 nm) and Rh nanoparticles (2.5 ± 0.8 nm). Even when the catalytic activity was normalized by dividing the activity with total surface area of the catalyst, the same tendency was observed. Thus, the atomic ratio of 1/2/36 (Pd/Ag/Rh) could be a special ratio for the present system and could be determined naturally during the self-organization process. Note that these results were reproduced.

When the Pd/Ag(1/2) bimetallic nanoparticles with the smaller sizes than this case, i.e., those of an average diameter of 3.2 ± 0.7 and 2.5 ± 0.4 nm, were used instead of those of 6.8 ± 1.8 nm, the dependence of catalytic activity on the metal composition varied as shown in Fig. 3.9b. The highest catalytic activities were observed for atomic ratios of 1/2/21 (Pd/Ag/Rh) and 1/2/13.5, and the average diameters of trimetallic nanoparticles at these composition were 2.6 ± 0.6 and 2.2 ± 0.6 nm, respectively. These two kinds of Pd/Ag/Rh trimetallic nanoparticles showed the same tendency in structure analyses as those of trimetallic nanoparticles prepared from Pd/Ag(1/2) bimetallic nanoparticles (an average diameter = 6.8 ± 1.8 nm) and Rh nanoparticles. Again, the trimetallic nanoparticles with highest catalytic activity had the smallest average diameter. The high catalytic activity, however, is not due to a large surface area. In fact, even if the activity was normalized to the surface area, the trimetallic nanoparticles with the smallest size still had the highest catalytic activity.

We have to emphasize here that the average diameter of the Pd/Ag/Rh trimetallic nanoparticles at the atomic ratio of 1/2/13.5 thus fabricated starting from Pd/Ag(1/2) bimetallic nanoparticles (an average diameter = 2.5 ± 0.4 nm) and Rh nanoparticles (an average diameter = 2.5 ± 0.8 nm) was 2.2 ± 0.6 nm. In other words, the resulting trimetallic nanoparticles were smaller than the starting nanoparticles. This means that any kind of realignment should occur during the self-organization process. A similar consideration has also been discussed in the case of self-organization between Ag and Rh nanoparticles.

The catalytic activity of Pd/Ag/Rh trimetallic nanoparticles with a three-layered core/shell, i.e., a Pd-core/Ag-interlayer/Rh-shell structure prepared by self-organization was compared to those of the corresponding monometallic and bimetallic nanoparticles as well as trimetallic nanoparticles prepared by other methods. The results are shown in Fig. 3.10.

In Fig. 3.10, the catalytic activities of monometallic and bimetallic nanoparticles are also shown for comparison. The colloidal dispersions of PVP-protected Pd, Ag, and Rh monometallic nanoparticles used for these measurements were prepared by an alcohol reduction method in ethanol/water. The catalytic activity of Pd/Ag(1/2) bimetallic nanoparticles with a Pd-core/Ag-shell structure used for the preparation of the present trimetallic nanoparticles was also measured for comparison. In the case of other bimetallic nanoparticles, various kinds of reference nanoparticles were prepared with different atomic ratios by two methods, i.e., simultaneous reduction and physical mixing (self-organization). The catalytic activities of Ag/Rh(1/4) bimetallic nanoparticles prepared by physical mixture and of Pd/Rh(1/2) bimetallic ones prepared by simultaneous reduction are also shown in Fig. 3.10.

In a series of the Pd/Ag/Rh trimetallic nanoparticles prepared by mixing three kinds of Pd/Ag(1/2) bimetallic nanoparticles with Rh nanoparticles, the Pd/Ag/Rh(1/2/12.5) nanoparticles with an average diameter of 2.2 ± 0.6 nm had the highest catalytic activity (21.0 mol-H_2 mol-M^{-1} s^{-1}), while the Pd/Ag/Rh(1/2/36) nanoparticles with an average diameter of 3.5 ± 1.3 nm and Pd/Ag/Rh(1/2/21) nanoparticles with an average diameter of 2.6 ± 0.6 nm had activities of 11.0 and 18.0 mol-H_2 mol-M^{-1} s^{-1}, respectively. Thus, the smaller the particle size the higher the catalytic activity. This result is not due to the surface area. In fact, when the activity was normalized to the surface area, the trimetallic nanoparticles with the smallest size still had the highest catalytic activity.

Now, the catalytic activity of Pd/Ag/Rh(1/2/13.5) nanoparticles prepared by self-organization was compared with those of trimetallic nanoparticles prepared by other methods. At first, the simultaneous reduction method was used for preparation of trimetallic nanoparticles. Thus, a 1/9 (v/v) water/ethanol solution of Pd(OAc)$_2$, AgClO$_4$, and Rh(NO$_3$)$_3$ with a molar ratio of 1/2/13.5 was refluxed for 3 h under nitrogen. The obtained trimetallic nanoparticles

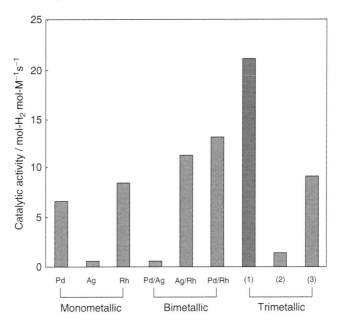

Fig. 3.10. Catalytic activity of Pd, Ag, and Rh monometallic nanoparticles, Pd/Ag(1/2), Ag/Rh(1/4), and Pd/Rh(1/2) bimetallic nanoparticles, and Pd/Ag/Rh(1/2/13.5) trimetallic nanoparticles: (1) The trimetallic nanoparticles prepared by self-organization starting from Pd/Ag(1/2) bimetallic and Rh nanoparticles (cf. Fig. 3.9b), (2) those prepared by simultaneous reduction, and (3) those prepared by self-organization by mixing of Pd, Ag, and Rh monometallic nanoparticles. The atomic composition Pd/Ag/Rh was kept constant 1/2/13.5 for all trimetallic nanoparticles. See the text for details of preparation of monometallic and bimetallic nanoparticles [26]

had an average diameter of 2.5 ± 0.7 nm. This was thought to be small enough for the particles to show a high catalytic activity, but the catalytic activity was, in fact, very small (0.59 mol-H$_2$ mol-M^{-1} s^{-1}). This is probably due to the structure of trimetallic nanoparticles prepared by simultaneous reduction, which may provide a structure different from Pd-core/Ag-interlayer/Rh-shell, like Pd-core/Rh-interlayer/Ag-shell. This structure is based on the formation mechanism proposed for simultaneous reduction [20] and is supported by the appearance of a plasmon resonance peak at 350 nm in the colloidal dispersion of these trimetallic nanoparticles.

The other preparation method was self-organization among the Pd, Ag, and Rh monometallic nanoparticles with average diameters of 2.3, 5.3, and 2.5 nm, respectively. PVP-protected Pd, Ag, and Rh nanoparticles separately prepared by alcohol reduction were mixed at room temperature, and the mixtures were kept at room temperature for 3 days. The resulting mixture had a wide size distribution with an average diameter of 3.3 ± 3.7 nm, and no plasmon absorption was observed. This suggests that the colloidal dispersions, prepared by self-organization, might be a mixture of Ag-core/Pd-shell and Ag-core/Rh-shell bimetallic nanoparticles, or alloyed bimetallic and/or trimetallic nanoparticles. In fact, the trimetallic nanoparticles had a reasonably high catalytic activity ($9.2 \, \text{mol-H}_2 \, \text{mol-M}^{-1} \text{s}^{-1}$). This catalytic activity data is roughly consistent with the structure speculated above. We also tried to synthesize PVP-protected Pd/Ag/Rh trimetallic nanoparticles using Pd/Ag bimetallic nanoparticles prepared by simultaneous reduction. However, the trimetallic nanoparticles with a three-layered core/shell structure could not be prepared by this method.

Based on the comparison among the catalytic activities of trimetallic nanoparticles shown in Fig. 3.10, the Pd/Ag/Rh(1/2/13.5) nanoparticles prepared by the self-organization from Pd-core/Ag-shell(1/2) bimetallic nanoparticles (an average diameter $= 2.5 \pm 0.4 \, \text{nm}$) and Rh nanoparticles had the highest catalytic activity. This can be understood by the successive electronic charge transfer from Rh-shell to Ag-interlayer and Ag-interlayer to Pd-core in three-layered core/shell-structured trimetallic nanoparticles. Other trimetallic nanoparticles did not have such an ideal structure for successive electronic effect [23], which could be the reason why the other trimetallic nanoparticles had lower activity.

3.5 Au/Pt/Rh Trimetallic Nanoparticles

As described in the previous sections, fabrications of bimetallic and trimetallic nanoparticles by self-organization were proceeded only between Ag and Rh atoms. In fact, molar enthalpy obtained by isothermal titration calorimetry (ITC) between Ag and Rh nanoparticles was as high as several hundreds $\text{kJ} \, \text{mol}^{-1}$ [22]. In a series of experiments, however, the molar enthalpies of the interactions between nanoparticles of noble metals were still in the order of a few hundreds $\text{kJ} \, \text{mol}^{-1}$ [22]. This result suggests that the self-organization method will be applied to the fabrication of the other kinds of trimetallic nanoparticles composed of noble metals without silver. From the viewpoint of sequential electron transfer [23], we chose the combination of Au, Pt, and Rh, the ionization potentials of which are known to be 9.225, 8.62, and 7.46 eV, respectively.

In order to construct the Au-core/Pt-interlayer/Rh-shell trimetallic nanoparticles, self-organization by physical mixture was carried out for the dispersions of PVP-protected Au-core/Pt-shell-structured bimetallic nanoparticles and those of PVP-protected Rh nanoparticles [31]. The Au-core/Pt-shell bimetallic nanoparticles can be easily prepared by simultaneous or coreduction of Au(III) ions and Pt(IV) ions in refluxing ethanol in the presence of PVP. The dispersion of PVP-protected Rh nanoparticles was separately prepared from RhCl_3 by alcohol reduction by the same way as that presented in the previous section.

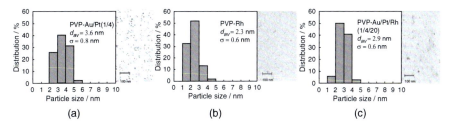

Fig. 3.11. TEM images and size distribution histograms of PVP-protected (**a**) Au/Pt bimetallic, (**b**) Rh monometallic, and (**c**) Au/Pt/Rh trimetallic nanoparticles [31]

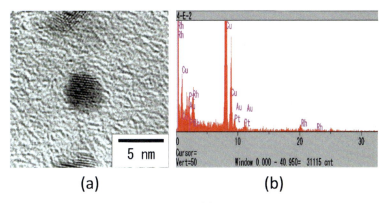

Fig. 3.12. (**a**) HR-TEM image and (**b**) EDS spectrum of PVP-protected Au/Pt/Rh(1/4/20) nanoparticles [31]

In the present self-organization method, the UV-Vis spectral change is too small to infer the occurrence of a reaction between the nanoparticles since Au/Pt bimetallic nanoparticles has no plasmon absorption. TEM photographs and size distribution histograms of these nanoparticles, shown in Fig. 3.11, however, suggest the self-organization because the size distribution histogram of the trimetallic nanoparticles is narrower than that expected by summation of the histograms of bimetallic and monometallic ones. HR-TEM image (Fig. 3.12a) of the trimetallic nanoparticles reveals that the nanoparticles form a single crystal, which means that mixture gives the spontaneous formation of single crystal. The energy dispersive X-ray spectrum (EDS) (Fig. 3.12) clearly shows the presence of Rh, Pt, and Au at an atomic ratio of 80.8:16.1:3.1, which is close to the fed ratio (80:16:4). Element mapping by TEM showed that the surrounding parts of the particles consist of Rh. This result may suggest that the trimetallic nanoparticles have a structure of a single crystal, but that Rh may possibly exist near the surface of the crystal, and Pt and Au may be located rather near the center of the particle.

XRD patterns of PVP-protected Au/Pt/Rh(1/4/20) trimetallic nanoparticles, and the related monometallic and bimetallic nanoparticles are shown

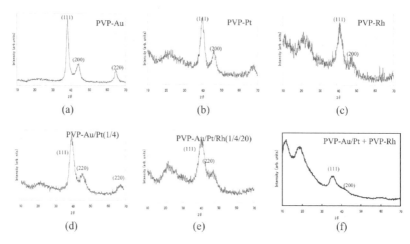

Fig. 3.13. XRD patterns of (a) PVP-Au, (b) PVP-Pt, (c) PVP-Rh, (d) PVP-Au/Pt(1/4), (e) PVP-Au/Pt/Rh, and (f) the 1:4 mixture of PVP-Au/Pt(1/4) and PVP-Ph nanoparticles [31]

Table 3.1. Spacing of lattice place (d) of bulk and nanoparticles of metals [31]

material	$d(\text{Å})$ at various hkl values			
	(111)	(200)	(220)	(311)
Au	2.36	2.04	1.44	1.23
PVP-Au	2.34	2.04	1.43	1.23
Pt	2.27	1.96	1.38	1.18
PVP-Pt	2.25	1.95	1.37	–
Rh	2.20	1.90	–	1.15
PVP-Rh	2.17	1.90	–	–
PVP-Au/Pt	2.28	1.98	1.37	–
PVP-Au/Pt/Rh	2.24	1.93	–	–
PVP-Au/Pt+PVP-Rh	2.27	–	–	–
	2.22	–	–	–

in Fig. 3.13. Table 3.1 summarizes the spacing of lattice plane (d) of metals in bulk and nanoparticles. These results may also suggest that the trimetallic nanoparticles are not mixtures of the related metal nanoparticles but new particles with a single crystal structure. The FT-IR spectrum of CO adsorbed on Au/Pt/Rh(1/4/20) trimetallic nanoparticles was rather similar to that of Rh nanoparticles, although they were not completely the same. This result again supports that the surface of trimetallic nanoparticles mainly consists of Rh metal.

The catalytic activity of the Au/Pt/Rh trimetallic nanoparticles was investigated for the hydrogenation of methyl acrylate in ethanol at 30°C under 1 atm of hydrogen pressure. Dependence of the catalytic activity upon the Rh content of Au/Pt/Rh trimetallic nanoparticles is shown in Fig. 3.14, where

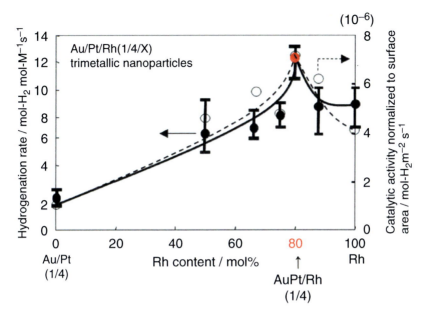

Fig. 3.14. Catalyst activity vs. Rh content of Au/Pt/Rh trimetallic nanoparticles. The *open circles and broken line* show the catalytic activity normalized to surface area [31]

the molar ratio of Au/Pt is kept constant at 1/4. Figure 3.14 reveals that the catalytic activity suddenly increases at the molar ratio of (Au + Pt) to Rh = 1/4, i.e., totally at the molar ratio of Au/Pt/Rh = 1/4/20. When the dependence of the average diameter of trimetallic nanoparticles on the Rh content was examined, the minimum diameter was observed at the content of Rh = 80% (Au/Pt/Rh = 1/4/20). This atomic ratio is completely the same with that observed at the highest catalytic activity. Thus, the catalytic activity is normalized to the surface area (open circles and broken line in Fig. 3.14). The highest normalized catalytic activity is still observed at the molar ratio of Au/Pt/Rh = 1/4/20. This suggests that this ratio 1/4 is special for the preparation of small particles, and thus, the high total surface area is convenient for the high catalytic acidity. However, note that, although monometallic Rh nanoparticles have the smaller average particle size than the Au/Pt/Rh(1/4/20) trimetallic nanoparticles, the catalytic activity of Rh is lower than that of the Au/Pt/Rh(1/4/20) trimetallic nanoparticles (about 57% of that of trimetallic ones). Thus, the Au/Pt/Rh(1/4/20) trimetallic nanoparticles have a special structure, which could be a reason for the high catalytic activity.

This idea was supported by comparing the catalytic activity of the trimetallic nanoparticles with those of other trimetallic ones having the same composition but prepared by different methods. Thus, other kinds

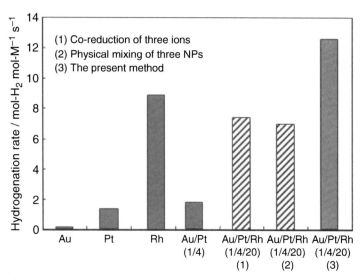

Fig. 3.15. Comparison of catalytic activities of various metal nanoparticles. See text for details of Au/Pt/Rh(1/4/20) (1), (2), and (3) [31]

of trimetallic nanoparticles were prepared by (1) coreduction of HAuCl$_4$, H$_2$PtCl$_6$, and RhCl$_3$ at the same ratio by refluxing ethanol/water and (2) physical mixture of Au, Pt, and Rh monometallic nanoparticles at the same ratio. The catalytic activities of these trimetallic nanoparticles as well as Au/Pt(1/4) bimetallic, and Au, Pt, and Rh monometallic nanoparticles are summarized in Fig. 3.15 for comparison. The Au/Pt/Rh trimetallic nanoparticles prepared by the present sophisticated method have the highest catalytic activity among the nanoparticles tested here. Again, the present trimetallic nanoparticles could have a special structure, which provided the high activity.

Here, again we have succeeded in fabricating the trimetallic nanoparticles having a three-layered core/shell structure, in which electron transfer occurs from the surface shell Rh atoms to the interlayer Pt atoms as well as from the interlayer Pt atoms to the central core Au atoms. This sequential electron transfer [23] could decrease the electron density of the surface Rh atoms, improving the catalytic activity. This idea can be confirmed by XPS data. The binding energies of PVP-protected Au/Pt(1/4) bimetallic nanoparticles and the related ones, as well as those of PVP-protected Au/Pt/Rh(1/4/20) trimetallic nanoparticle and the related ones, are shown in Tables 3.2 and 3.3, respectively.

In the case of Au/Pt bimetallic nanoparticles, the binding energy of Au 4f$_{7/2}$ in Au/Pt bimetallic nanoparticles shifts to lower energy in 0.15 eV, while that of Pt 4f$_{7/2}$ shifts to higher energy in 0.03 eV. Although these shifts were very small and comparable with an error bar, the measurements were carried out several times using different lots and the resulting data showed the

Table 3.2. Binding energy of PVP-protected Au/Pt bimetallic and the corresponding monometallic nanoparticles [31]

	Binding energy (eV)			Shift(eV)
	PVP-Au	PVP-Pt	PVP-Au/Pt	Δ
Au $4f_{7/2}$	84.25	–	84.10	−0.15
Pt $4f_{7/2}$	–	71.11	71.14	+0.03

Table 3.3. Binding energy of PVP-protected Au/Pt/Rh trimetallic and the corresponding monometallic nanoparticles [31]

	Binding energy (eV)				Shift(eV)
	PVP-Au	PVP-Pt	PVP-Rh	PVP-Au/Pt/Rh	Δ
Au $4f_{7/2}$	84.25	–	–	84.10	−0.15
Pt $4f_{7/2}$	–	71.11	–	71.29	+0.18
Rh $3d_{5/2}$	–	–	307.35	307.55	+0.20

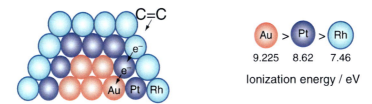

Fig. 3.16. Sequential electronic charge transfer in the Au-core/Pt-interlayer/Rh-shell trimetallic nanoparticles. *Cartoon* shows a cross section of a part of spherical particle as an example. Ionization energies of the corresponding bulk metals are also shown for comparison [31]

same tendency. Thus, these data may suggest the concept of electronic charge transfer from Pt in the surface to Au in the core. The similar electronic charge transfer may also be considered to occur from Rh to Pt and from Pt to Au in the Au/Pt/Rh(1/4/20) trimetallic nanoparticles, based on the data shown in Table 3.3. This sequential electronic charge transfer [23] can be illustrated as shown Fig. 3.16. This order of the charge transfer may consist of the order of the ionization potential of Au, Pt, and Rh in a bulk.

As described in the previous section, we have used a ITC technique to understand the driving force to form bimetallic nanoparticles by self-organization [22]. This ITC technique was applied to the present system to form trimetallic nanoparticles by the mixture of bimetallic nanoparticles and monometallic ones (self-organization). The results are shown in Table 3.4 as well as for the formation of Au/Pt bimetallic nanoparticles by the self-organization. The data clearly indicate the presence of the strong exothermic interaction between these particles. This strong exothermic interaction

Table 3.4. Molar enthalpy obtained with ITC for Au/Pt bimetallic and the corresponding monometallic nanoparticles [31]

Injection particle		Particle in cell	ΔH (kJ/mol)
PVP-Rh	$\rightarrow$	PVP-Au/Pt	-297.3
PVP-Au/Pt	$\rightarrow$	PVP-Rh	-47
PVP-Pt	$\rightarrow$	PVP-Au(L)	-534
PVP-Au(L)	$\rightarrow$	PVP-Pt	-78

could be a driving force for the self-organization to form the trimetallic nanoparticles.

Thus, again we succeeded in preparing colloidal dispersions of PVP-protected Au/Pt/Rh trimetallic nanoparticle by mixing the dispersions of PVP-protected Rh nanoparticles and those of PVP-protected Au/Pt(1/4) bimetallic nanoparticles with a Au-core/Pt-shell structure. The trimetallic nanoparticles thus fabricated may have a three-layered core/shell structure (Au-core/Pt-interlayer-/Rh-shell), which is suggested by HR-TEM, EF-TEM, and FT-IR-CO. The trimetallic nanoparticles with the composition of Au/Pt/Rh(1/4/20) and a three-layered core/shell structure have the highest catalytic activity for hydrogenation of methyl acrylate at 30°C in ethanol among the trimetallic nanoparticles with other compositions and other structures, and the corresponding bimetallic and monometallic nanoparticles. The high catalytic activity of trimetallic nanoparticles is probably due to the sequential electronic effect between elements in a particle [23].

3.6 Pt/Pd/Rh Trimetallic Nanoparticles

In the previous sections, we have succeeded in fabricating the three-layered trimetallic nanoparticles in a series of Pd/Ag/Rh and Au/Pt/Rh. They have the special atomic ratios, in which the highest catalytic activity and the smallest particle size were achieved. The atomic ratios were 1/2/13.5 (an average diameter $= 2.2 \pm 0.6$ nm) for Pd/Ag/Rh, and 1/4/20 (an average diameter $= 2.9 \pm 0.6$ nm) for Au/Pt/Rh, respectively.

Now, we have examined to fabricate trimetallic nanoparticles composed of Pt, Pd, and Rh. In this series, the ionization potentials of these metals are 8.62, 8.34, and 7.46 eV, respectively. Thus, spontaneous formation of the trimetallic nanoparticles having the Pt-core/Pd-interlayer/Rh-shell structure has been expected to proceed at room temperature starting from Pt-core/Pd-shell bimetallic nanoparticles and Rh nanoparticles.

Colloidal dispersions of PVP-protected Pt-core/Pd-shell bimetallic nanoparticles were prepared by alcohol reduction of H_2PtCl_6 and $PdCl_2$ in ethanol and water in the presence of PVP, as described previously [20]. The present Pt/Pd(1/4) bimetallic nanoparticles had a relatively uniform size with an average diameter $= 1.6 \pm 0.4$ nm (cf. Fig. 3.17). This was nearly the same

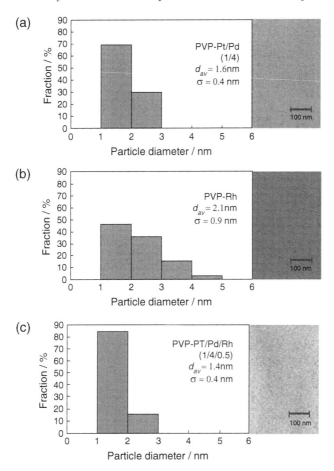

Fig. 3.17. TEM photographs and size distribution histograms of PVP-protected (**a**) Pt/Pd bimetallic, (**b**) Rh monometallic, and (**c**) Pt/Pd/Rh trimetallic nanoparticles

size and size distribution was as those reported previously [20]. Colloidal dispersions of PVP-protected Rh nanoparticles were also prepared by the alcohol reduction method. The Rh nanoparticles had an average diameter of 2.1 ± 0.9 nm (cf. Fig. 3.17).

Both dispersions of Pd/Pd bimetallic and Rh monometallic nanoparticles were mixed at various ratios at room temperature. The resulting dispersions were evaluated by the catalytic activity for hydrogenation of methyl acrylate at 25°C. The results are summarized in Fig. 3.18. The highest catalytic activity was observed at the molar ratio of Pt/Pd/Rh = 1/4/0.5. The results were quite different from those obtained in the cases of Pd/Ag/Rh and Au/Pt/Rh trimetallic nanoparticles.

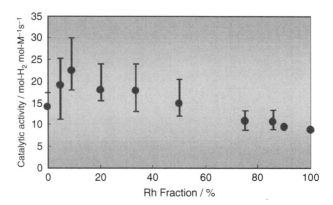

Fig. 3.18. Dependence of catalytic activity of PVP-protected Pt/Rd/Rh trimetallic nanoparticles upon the fraction of Rh. Pt/Pd ratio was kept constant at 1/4

Thus, in the cases of Pd/Ag/Rh and Au/Pt/Rh trimetallic nanoparticles the smallest particle sizes and the highest catalytic activities were observed at the special ratio of Pd/Pt/Rh = 1/2/13.5 (an average diameter = 2.2±0.6 nm) and Au/Pt/Rh = 1/4/20 (an average diameter = 2.9 ± 0.6 nm), respectively. At these ratios the structures of particles were spontaneously organized to give the thermodynamically stable particles. In the case of Pt/Pd/Rh trimetallic nanoparticles, however, the atomic ratio was 1/4/0.5, at which the number of Rh atoms on the shell was not enough to cover the Pt/Pd bimetallic core. In other words, both Rh and Pd atoms were located on the surface of Pt/Pd/Rh(1/4/0.5) trimetallic nanoparticles. In this particle, Pd atoms not only promote the catalytic activity of Rh atoms on the surface of the particle by locating adjacent to the Rh atoms (a ligand effect), but also directly interact with the substrate to improve the catalytic activity by an ensemble effect. This proposed mechanism can be illustrated in Fig. 3.19.

3.7 Concluding Remarks

Here, we presented the fabrication of three-layered core/shell-structured trimetallic nanoparticles by self-organization by the mixture of three kinds of core/-shell-structured bimetallic nanoparticles with Rh nanoparticles in solution at room temperature. The three-layered core/shell structures were suggested on the basis of HR-TEM, EF-TEM, FT-IR-CO, and the special high catalytic activities. In the case of Pd/Ag/Rh(1/2/13.5) and Au/Pt/Rh(1/4/20) nanoparticles Rh atoms completely cover the core bimetallic nanoparticles, i.e., all the surface atoms are Rh, while, in the case of Pt/Rd/Rh(1/4/0.5) nanoparticles, the surface atoms are composed of both Pd and Rh atoms. We do not have the exact reason why Rh atoms completely

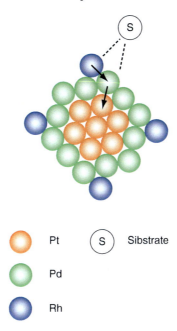

Fig. 3.19. Cross section of Pt/Pd/Rh(1/4/0.5) trimetallic nanoparticles

cover the surface of trimetallic nanoparticles in the former two cases, and do not completely cover the surface in the last case.

One possible explanation could be attributed to the chemical property of the surface atoms. The surface of a metal nanoparticle is covered by organic corona to stabilize the nanoparticle in dispersion. In the present case PVP molecules work as a protective agent surrounding the metal nanoparticle. In this case the PVP molecules interact with the metal atoms on the surface of the nanoparticle [32]. The strength of this interaction depends on the structure of protecting molecule and the kind of metal on the surface. In the three present cases, we should consider only the kind of metal on the surface because PVP is a common protecting agent.

In the former two cases, i.e., Pd/Ag/Rh and Au/Pt/Rh trimetallic nanoparticles, the stronger interaction of PVP with Rh than Ag (in the case of Pd/Ag/Rh trimetallic nanoparticles) or Pt (in the case of Au/Pt/Rh trimetallic nanoparticles) might result in the trimetallic nanoparticles being completely covered by Rh as the thermodynamically stable ones. In other words, the trimetallic nanoparticles, in which the surface atoms were both Rh and Ag or Pt, might not be so stable since the interaction of PVP with Ag or Pt is not so strong as with Rh. In the case of Pd/Pd/Rh trimetallic nanoparticles, in contrast, the strength of the interaction of PVP with Pd could be nearly in the same level as with Rh. Thus, the trimetallic nanoparticle covered completely by Rh might not be so stable and the thermodynamically most stable nanoparticles might have both Pd and Rh atoms on the surface.

78   N. Toshima

Since only a few examples have been reported on the three-layered core/-shell trimetallic nanoparticles or nanoclusters with enough small size, it is difficult to discuss the formation mechanism and the stability of the trimetallic nanoparticles. We wish more research will be carried out on this interesting system in the near future.

Acknowledgments

The author expresses his sincere thanks to all the coworkers and coauthors of cited references for their kind collaboration. This work was supported by a Grant-in-Aid for Scientific Research (B) (No. 15310078) from MEXT, Japan, and Core Research for Evolutional Science and Technology (CREST) from JST, Japan.

References

1. G. Schmid (ed.), *Clusters and Colloids from Theory to Application* (VCH, Weinheim, 1994)
2. B. Corain, G. Schmid, N. Toshima (eds.), *Metal Nanoclusters in Catalysis and Materials Science: The Issue of Size Control* (Elsevier, Amsterdam, 2008)
3. H. Hirai, N. Toshima, in *Taiored Metal Catalyst*, ed. by Y. Iwasawa (Reidel, Dordrecht, 1985), pp. 121–135
4. H. Bönnemann, W. Brijoux, R. Brinkmann, E. Dinjus, T. Joußen, B. Korall, Angew. Chem. Int. Ed. Engl. **30**, 1312 (1991)
5. C. Amiens, D. de Caro, B. Chaudret, J.S. Bradley, R. Mazel, C. Roucau, J. Am. Chem. Soc. **115**, 11638 (1993)
6. N. Toshima, Y. Shiraishi, T. Teranishi, M. Miyake, T. Tominaga, H. Watanabe, W. Brijoux, H. Bönnemann, G. Schmid, Appl. Organometal. Chem. **15**, 1 (2001)
7. H. Tsunoyama, H. Sakurai, Y. Negishi, T. Tsukuda, J. Am. Chem. Soc. **127**, 9374 (2005)
8. S. Sun, C.B. Murray, D. Weller, L. Folk, A. Moser, Science **287**, 1989 (2000)
9. M. Chen, J. Kim, J.P. Liu, H. Fan, S. Sun, J. Am. Chem. Soc. **128**, 7132 (2006)
10. M. Nakaya, M. Kanehara, T. Teranishi, Langmuir **22**, 3485 (2006)
11. T. Matsushita, J. Masuda, T. Iwamoto, N. Toshima, Chem. Lett. **36**, 1264 (2007)
12. T. Teranishi, M. Nishida, M. Kanehara, Chem. Lett. **34**, 1004 (2005)
13. (a) N. Watanabe, J. Kawamata, N. Toshima, Chem. Lett. **33**, 1368 (2004); (b) N. Watanabe, N. Toshima, Bull. Chem. Soc. Jpn. **80**, 2008 (2007)
14. S. Kobayashi, T. Miyama, N. Nishida, Y. Sakai, H. Shiraki, Y. Shiraishi, N. Toshima, J. Display Technol. **2**, 121 (2006)
15. Y. Shiraishi, N. Toshima, K. Maeda, H. Yoshikawa, J. Xu, S. Kobayashi, Appl. Phys. Lett. **81**, 2845 (2002)
16. H. Iida, T. Nakanishi, T. Osaka, Electrochim. Acta **51**, 855 (2005)
17. V. Ponec, Appl. Catal. A: General **222**, 31 (2001)
18. N. Toshima, K. Kushihashi, T. Yonezawa, H. Hirai, Chem. Lett. 1769 (1989)
19. K. Hirakawa, N. Toshima, Chem. Lett. **32**, 78 (2003)

3 Polymer-Assisted Composites of Trimetallic Nanoparticles 79

20. (a) N. Toshima, T. Yonezawa, K. Kushihashi, J. Chem. Soc. Faraday Trans. **191**, 2537 (1993); (b) T. Yonezawa, N. Toshima, J. Chem. Soc. Faraday Trans. **91**, 4111 (1985)
21. Y. Wang, N. Toshima, J. Phys. Chem. B **101**, 5301 (1997)
22. N. Toshima, M. Kanemaru, Y. Shiraishi, Y. Koga, J. Phys. Chem. B **109**, 16326 (2005)
23. N. Toshima, in *Macromolecular Complexes: Dynamic Interactions and Electronic Processes*, ed. by E. Tsuchida (Japan Scientific Societies Press, Tokyo, 1991), pp. 279–295
24. A. Henglein, J. Phys. Chem. B **104**, 6683 (2000)
25. N. Toshima, Y. Shiraishi, T. Matsushita, H. Mukai, K. Hirakawa, Int. J. Nanosci. **1**, 397 (2003)
26. T. Matsushita, Y. Shiraishi, S. Horiuchi, N. Toshima, Bull. Chem. Soc. Jpn. **80**, 1217 (2007)
27. Y. Shiraishi, N. Toshima, Colloids Surfaces A: Physicochem. Eng. Asp. **169**, 59 (2000)
28. J.S. Bradley, G.H. Via, L. Bonnevio, E.W. Hill, Chem. Mater. **8**, 1895 (1996)
29. Y. Shiraishi, D. Ikenaga, N. Toshima, Aust. J. Chem. **56**, 1025 (2003)
30. N. Toshima, Y. Shiraishi, A. Shiotsuki, D. Ikenaga, Y. Wang, Eur. Phys. J. D **16**, 209 (2001)
31. N. Toshima, R. Ito, T. Matsushita, Y. Shiraishi, Catal. Today **122**, 239 (2007)
32. H. Hirai, H. Chawanya, N. Toshima, Reactive Polym. **3**, 127 (1985)

4

Fabrication of Organic Nanocrystals and Novel Nanohybrid Materials

Tsunenobu Onodera, Hitoshi Kasai, Hidetoshi Oikawa,
and Hachiro Nakanishi

4.1 Introduction

Nonlinear optical (NLO) materials are required for the high-speed optical network systems dealing with a large amount of data. In particular, organic third-order NLO materials are powerful candidates for such optical devices because of the fast NLO response and easy processability in comparison with inorganics [1]. For example, polydiacetylene (PDA), classified as a one-dimensional π-conjugated polymer, possesses excellent third-order NLO response properties in the range of femtoseconds to subpicoseconds, which is attributed to delocalized π-conjugated electrons along the main chain backbone [2]. PDA crystals can be obtained through topochemical solid-state polymerization with phase transition from a diacetylene monomer single crystal to a polydiacetylene single crystal by UV-, γ-irradiation or heat treatment [3]. Thus far, the syntheses and optoelectronic properties of PDA derivatives have extensively been studied, especially the π-conjugation extended between main chain and side chains such as directly linked aromatic groups and ladder-type PDAs [2]. However, the magnitude of effective third-order NLO susceptibility $\chi^{(3)}(\omega)$ is still not sufficient for device applications.

On the other hand, hybridization and nanocomposition are considerably investigated as the most important subjects in materials science and technology. One of them is a research on electromagnetic-field enhancement in the hybridization between organics and novel metal nanoparticles. Upon photoexcitation of surface plasmon in the novel metal nanoparticles such as gold, silver, and copper, electromagnetic field will be concentrated near the metal–dielectric interface, exhibiting very strong amplification of local optical electric field. As a result, photoluminescence of fluorescent molecule located in optimum distance from the surface of noble metal nanoparticle are enhanced by interactions with surface plasmon resonance in the metal nanoparticle [4]. Furthermore, the enhancement of the local field has also been utilized to amplify the weak signals from NLO processes such as second-harmonic generation (SHG) [5] and surface-enhanced Raman scattering (SERS) [6]. Such

82   T. Onodera et al.

research on electromagnetic-field enhancement in the hybridized system has been attracting a growing interest, which will contribute to the development of novel hybridized materials on optical devices, energy conversion, and sensors.

Neeves and Birnboim [7] theoretically predicted that core–shell-type nanostructure consisting of a PDA core and a metal nanoshell exhibits remarkably large effective $\chi^{(3)}(\omega)$ values, which encourage us to develop a novel type of core–shell nanocrystal. In previous work on the hybridized core–shell nanoparticles, much attention was denoted to inorganic nanoparticles coated with organic materials, i.e., inorganic (core)/organic (shell) structure [8–10], because of improvement in redox properties and electrical conductivity of π-conjugated organic materials. Recently, we have also fabricated a metal (core)/PDA (shell) nanocrystal by the coprecipitation method, and discussed the influence of the difference in nanostructure on optical properties [11]. However, functional organic (core)/metal (shell) nanocrystals have scarcely been reported, because organic nanocrystals as a core have not been systematically studied for a long time. Although evaporation and melting methods are usually employed for preparing inorganic nanocrystals, these common methods are not applicable for organic nanocrystals due to instability at high temperatures, except for some thermal stable organic compounds. Furthermore, it was difficult to lower particle size to $1\,\mu m$ below by top-down type techniques.

Under such conditions, we have proposed the so-called reprecipitation method to nanocrystallize organic compounds such as thermally unstable diacetylene monomers [12]. This method is promising for fabricating PDA core for PDA nanocrystal (core)/metal (shell) nanocrystals. On the other hand, metal deposition methods have recently been developed for polycation-modified [13] and animo-terminated [14] submicroparticles. However, successful metal deposition by these methodologies has so far been limited to only polystyrene and silica submicroparticle templates. In addition, it was indispensable to chemically and/or physically modify the surface of cores in these methodologies. Therefore, we need to devise new metal-plating techniques to modify the surface of PDA nanocrystals without damaging π-conjugated main chains of PDA as a core.

This chapter provides an account of mild solution-phase methods that generate PDA nanocrystals and PDA (core)/silver (shell) hybridized nanocrystals with well-controlled shapes. First, we describe fabrication of PDA nanocrystals as a core and their size-dependent optical properties. Next, we demonstrate fabrication and optical properties of PDA nanocrystals deposited with silver nanoparticles by using two different schemes [15,16]: In scheme 1, PDA nanocrystal cores were coated with silver nanoparticles via surfactant binders. Silver seeds were deposited through in situ reduction of silver cations on the surface of the PDA nanocrystal (seed deposition process), the surface of which was previously modified with several types of surfactants. Afterwards, the coverage of silver on the PDA surface was increased by further reduction with hydroxylamine (seed growth process). In scheme 2, we developed a novel preparation technique, the visible-light-driven photocatalytic reduction method, for the purpose of fabricating PDA (core)/silver (shell) nanocrystals having higher coverage and strong interaction between the core and shell. Thus far, the photocatalytic action of π-conjugated polymers has scarcely been studied, apart from the photodeposition of platinum on polyaniline and polypyrrole films [17]. The present visible-light-driven photocatalytic reduction method provides some advantages that allow the fabrication of nanostructures with an organic core and a metallic shell (1) direct deposition of the metallic shell on the core without using a surfactant or a surface-modified core; (2) fewer byproducts because no reducing agent is used; and (3) the mild preparation condition of visible-light irradiation. It is very important to mildly modify the surface of nanocrystals without damaging π-conjugated main chains of PDA as a core. Finally, we conclude with some perspectives on areas for future work.

4.2 PDA Nanocrystals

4.2.1 Fabrication Technique: Reprecipitation Method

Organic nanoparticles can be fabricated by a convenient wet process, i.e., the reprecipitation method as shown in Fig. 4.1. The dilute solution of target compound dissolved in a good solvent is quickly injected into excess amount of vigorously stirred poor solvent. As a result, nanocrystals are obtained in the poor dispersed medium. The poor solvent for the target compound must be able to dilute the good solvent infinitely. Distilled water is usually used as a poor solvent. We emphasize that the reprecipitation method is powerful for obtaining size-controlled nanocrystals of various organic compounds. One typical example was demonstrated using DA derivatives [1,6-di-(N-carbazolyl)-2,4-hexadiene] [18]. Molecular structure and topochemical solid-state polymerization [3] of the DA monomer is displayed in Fig. 4.2. Typical reprecipitation procedure for preparing PDA nanocrystals is described as follows [19]. A 0.2 mL of 10^{-2} M acetone solution $(1\,\text{M} = 1\,\text{mol}\,\text{dm}^{-3})$ was injected into 10 mL of well-stirred water at room temperature. Within

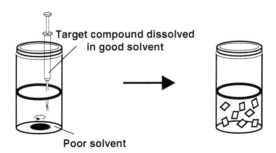

Fig. 4.1. Reprecipitation method

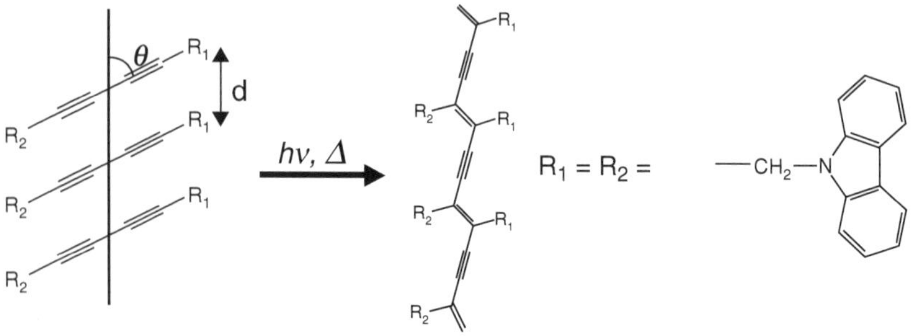

84   T. Onodera et al.

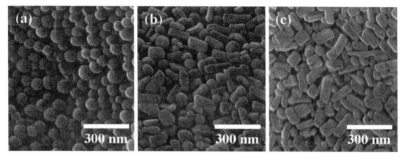

Fig. 4.2. Molecular structure and topochemical solid-state polymerization of diacetylene monomer. d and θ indicate stacking distance of DA monomers and angle between stacking axis and DA molecule, respectively

Fig. 4.3. SEM images of DA monomer nanoparticles with elapsed time of (**a**) 0 min, (**b**) 10 min, and (**c**) 20 min after the reprecipitation

ca. 20 min, DA monomers were nanocrystallized. By irradiating UV light or γ-ray, the slightly milky dispersion of DA monomer nanocrystals was changed to a blue dispersion of PDA nanocrystals. The size of PDA nanocrystals could be controlled by changing the combination between poor and good solvents, the concentration of the injected solutions, the speed of injection, temperature of the poor solvent, the mixing speed, and addition of surfactant. The typical crystal size was obtained from 10 nm to 5–10 μm in length.

Figure 4.3 shows scanning electron microscopy (SEM) images of DA monomer nanoparticles with time elapsed after the reprecipitation. The particle shape was roughly spherical just after reprecipitation procedure (Fig. 4.3a). At this stage, solid-state polymerization did not occur by UV-irradiation, and the resulting nanoparticle dispersion exhibited slightly pale-yellow color. On the contrary, by irradiating UV-light after the retention time of more than 10 min, DA nanoparticles were polymerized in solid state, providing blue dispersion liquid. This peak at around 650 nm arises from excitonic absorption in the π-conjugated chain of PDA nanocrystals [20]. It was noted that the absorbance gradually increased within the retention time of 0 to 10 min [21]. In addition, after the retention time of 20 min, the shape of DA monomer

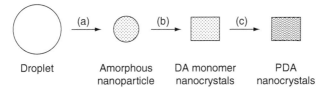

Fig. 4.4. Speculated mechanism of crystallization process of DA nanocrystals in the reprecipitation method: (**a**) removing of acetone from droplet and diffusion into surrounding water; (**b**) rearrangement of DA molecules in amorphous nanoparticles and subsequent crystallization; and (**c**) solid-state polymerization by UV-irradiation

nanoparticles was perfectly changed into a rectangle, similar to DA bulk crystals (Fig. 4.3c). These facts suggest that nanocrystallization in DA monomer nanoparticles is completed within the period of 10 min. Powder X-ray diffraction (XRD) measurements also support these discussions. Namely, while clear diffraction patterns of DA monomer nanocrystals, which were very similar to those of DA bulk crystals, were observed after retention times of more than 10 min, only hallo pattern was observed just after reprecipitation.

From the results described above, the crystallization process of DA in the reprecipitation method is speculated as displayed in Fig. 4.4 [22]. To deepen understanding of the reprecipitation method, crystallization process should be roughly divided into four stages (I) solution is injected into stirred poor solvent disperses and forms droplets; (II) good solvent diffuses from droplets and amorphous nanoparticles forms; (III) nucleation; and (IV) crystal growth. The stages (I) and (IV) seem to be common in various organic nanocrystals. However, the stages (II) and (III) are variable depending on a kind of compounds. In the case of DA monomer nanocrystals, the initial particles were evidently noncrystalline (and/or amorphous) ones in the supersaturated state just after the good solvent had diffused from droplets. Nucleation and crystal growth are considered to start in the amorphous nanoparticles via rearrangement of DA molecules. This is supported by the fact that the volume of amorphous nanoparticles (Fig. 4.3a) was almost the same as that of corresponding PDA nanocrystals (Fig. 4.3c), although the geometrical shapes were changed from sphere to rectangle. On the other hand, in perylene nanocrystals, the initial particles in the stages (II) and (III) seem to be superfine particles of perylene, and nucleation and crystal growth occur by thermal collision between superfine particles from the measurements of light scattering intensity [23, 24].

By clarifying such nanocrystallization mechanism, we can obtain important insight into the control of crystal size. In accordance with the scheme in Fig. 4.4, one should control the size of the droplets or of amorphous nanoparticles in the case of PDA nanocrystal. Namely, if the concentration of injected DA–acetone solutions is decreased, the crystal size will be reduced. Figure 4.5 shows the concentration and injected-amount dependence of light scattering intensity I_s. Based on Rayleigh scattering equation, I_s is proportional to NV^2;

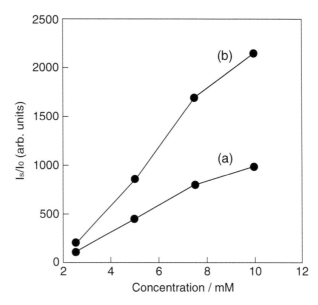

Fig. 4.5. Dependence of light scattering intensity I_s on concentration and injected amount. The injected amounts were (**a**) 0.1 mL and (**b**) 0.2 mL, respectively

N and V are the number and the volume of scatter, respectively [25]. Experimentally, I_s was proportional to the injected amount of DA–acetone solution at whole range of concentration, which implies the increase of scatters, i.e., the number of DA monomer nanoparticles [21]. On the other hand, I_s decreased with decreasing concentration of DA–acetone solution [21], which means the reduction of the volume of DA monomer nanoparticles under same condition on the injected amount. In fact, as shown in Fig. 4.6a,b, the crystal size was controlled to be ca. 100 nm when the concentration of DA–acetone solution was 10 mM, and was reduced to be ca. 50 nm at 2.5 mM. Furthermore, addition of an anionic surfactant SDS (sodium dodecylsulfate) into DA–acetone solution enabled to stabilize DA–acetone droplets and amorphous nanoparticles with a smaller size. As a result, the crystal size of PDA nanocrystals was reduced up to ca. 15 nm (Fig. 4.6c). However, the rate of crystallization was very slow, owing to protection of nanoparticle surface with SDS. The effect of added SDS is related to the negative ζ-potential of PDA nanocrystals. The detailed mechanism will be discussed later.

On the other hand, when DA monomer nanoparticles were prepared by adding SDS at 60°C, fibrous DA microcrystals grew with the elapsed time of 20 min (Fig. 4.6d) [26]. The diameter and length were 60 nm and 5–10 μm, respectively. Just after reprecipitation, the size of nanoparticles formed at 60°C was similar to that at room temperature. The nanocrystals formed at the earlier stage are considered to play the role of the nucleus for the growth of crystalline microfiber, which may be accelerated by the collision between

4 Fabrication of Organic Nanocrystals 87

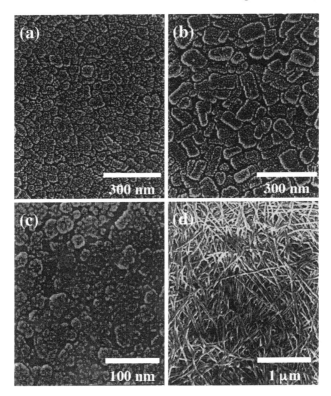

Fig. 4.6. SEM images of (**a**)–(**c**) PDA nanocrystals and (**d**) fibrous PDA microcrystals prepared under different conditions. The concentrations of DA–acetone solution were (**a, c, d**) 2.5 mM and (**b**) 10 mM, respectively. Water as a poor solvent was kept at (**a–c**) room temperature and (**d**) 60°C. (**c**) and (**d**) were prepared in the presence of SDS

DA amorphous nanoparticles and already grown fibrous crystals. Thus, one-dimensional crystal growth has proceeded by connecting DA amorphous nanoparticles to the only one preferential surface of already formed nanocrystals [26]. These fibrous DA microcrystals were also solid-state polymerized by UV-irradiation.

4.3 Optical Properties of PDA Nanocrystals

Figure 4.7 shows crystal size-dependent UV-visible extinction spectra of PDA nanocrystals dispersed in water [22, 26]. The excitonic absorption peak position (λ_{max}) was shifted to higher energy side with decreasing crystal size. Although the tendency is apparently similar to the well-known quantum size effect in semiconductor nanoparticles, the size region is quite different

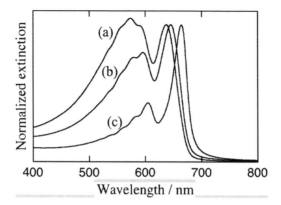

Fig. 4.7. UV-visible extinction spectra of PDA nanocrystals with different crystal sizes: (**a**) 50 nm; (**b**) 100 nm; and (**c**) 5–10 μm in length of PDA nanocrystals

from that of semiconductor nanoparticles. Namely, the size effect of organic nanocrystals was observed at one order of magnitude larger crystal size than that of semiconductors, and cannot be explained by means of conventional quantum size effect theory. The reasons for the present size effect are (1) a novel size effect related to a certain interaction between excitons and phonon in softened crystal lattice, (2) surface effect of nanocrystals, and (3) scattering effect. The size effect of PDA nanocrystals was investigated not only in their colloidal dispersion but also in a single nanocrystal level [27]. On the other hand, the size effect on photoluminescence was observed in various organic nanocrystals such as perylene nanocrystals [28, 29], which was also discussed on an exciton-phonon interaction.

4.4 Hybridized PDA Nanocrystals

4.4.1 Silver-Deposited PDA Nanocrystals Fabricated Using Surfactants as Binder

PDA nanocrystals deposited with silver nanoparticles have been fabricated for the first time by this authors' group. Here, we introduce the fabrication technique of hybridized PDA nanocrystals by using surfactant binders. In this method, some kinds of surfactants were used as a binder: anionic ones; sodium dodecylsulfate (SDS), sodium dodecylbenzenesulfonate (SDBS), heptacosa-10, 12-dinoic acid (14, 8-ADA), aerosol OT (AOT); and cationic one; cetyltrimethylammonium bromide (CTAB).

First, PDA nanocrystals, as a core, were fabricated by the reprecipitation method [21, 26] as follows. DA monomer acetone solution (1 mL, 5 mM) containing a surfactant was injected into vigorously stirred water (50 mL) at 45°C. Water used was purified up to 18.2 MW cm. After being maintained at room temperature for 30 min, a dark blue surfactant-modified PDA nanocrystals dispersion liquid with low scattering loss could be obtained by irradiating UV light with a handy lamp (254 nm, 4 W) for 1 h. After mixing silver nitrate aqueous solution (0.1 mL, 22.2 mM) with the PDA nanocrystals dispersion (5 mL), an aqueous solution of sodium borohydride (0.3 mL, 1.11 mM) was dropwise added to the mixture as a reducing agent. Through the reduction of silver cations, deposition of silver fine seeds on the surface of PDA nanocrystals could occur. The subsequent reduction of silver cations (0.4 mL, 22.2 mM) with hydroxylamine (4 mL, 2.22 mM) was performed in order to increase silver coverage on the PDA surface.

Dilute dispersion liquid of the resulting nanocrystals was cast on a silicon wafer, and then field emission SEM measurement was carried out without sputtering to evaluate the crystal size and morphology in detail. The UV-visible extinction spectra of the resulting nanocrystal aqueous dispersion were also measured.

Figure 4.8a shows the SEM image of SDS-modified PDA nanocrystals as a core. The image was not so clear, due to the charge-up effect of PDA nanocrystals without sputtering. However, it was recognized that the nanocrystal cores formed were cubic and ca. 70 nm in size. It was almost same with the value measured by the dynamic light scattering method. This fact means that isolated PDA nanocrystals were stably dispersed in water. As mentioned above, the nanocrystallization of PDA can proceed even in the presence of SDS in the reprecipitation method [26]. Just after injection of DA monomer–acetone solution into water, acetone droplets containing both DA monomer and SDS are first formed. Next, acetone would immediately diffuse to the water phase. Finally, the DA monomer amorphous nanoparticles would be surrounded and stabilized by SDS. Hydrophobic hydrocarbon chains of SDS could be in contact with the surface of DA monomer amorphous nanoparticles, and anionic moiety should be exposed to the water phase. The subsequent

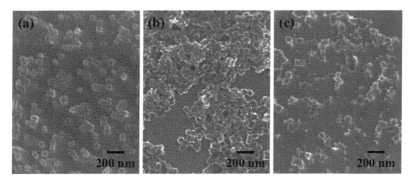

Fig. 4.8. SEM images of (**a**) SDS-modified PDA nanocrystal cores, (**b**) silver seed-deposited PDA nanocrystals, and (**c**) silver-deposited PDA nanocrystals and isolated silver seeds. These were all cast on a silicon wafer

nucleation and crystallization of the DA monomer in amorphous nanoparticles could thermally proceed. In addition, the morphologies of the resulting PDA nanocrystals was affected by water temperature. For example, SDS can protect and stabilize amorphous DA nanoparticles, and neither crystallization nor collision occurs below room temperature. On the other hand, both amorphous DA nanoparticles and DA nanocrystals may stably coexist at 60°C, owing to the presence of SDS. As a result, SDS was able to produce fibrous nanocrystals in Fig. 4.6d by inducing epitaxial growth of DA through collision of amorphous nanoparticles and a specific crystal face of already formed nanocrystals. However, by precisely controlling water temperature, only the cubic SDS-adsorbed PDA nanocrystals were successfully obtained at 45°C as shown in Fig. 4.8a. No epitaxial crystal growth of the DA monomer seems to occur at this temperature. The cubic PDA nanocrystals modified with SDS were useful for the fabrication of core–shell-type nanocrystals.

Other anionic surfactants were also adsorbed on the PDA core surface in the same manner as SDS. However, the ratio of adsorption to desorption of surfactant depends on the affinity between PDA and surfactant, critical micelle concentration (CMC), pH value, and so on. Above all, 14, 8-ADA, which was a DA derivative with carboxylic acid and served as a surfactant, was characterized by the control of pH because acid dissociation of carboxylic moiety greatly relates to hydrophile–lipophile balance (HLB value). 14, 8-ADA could be adsorbed on the surface of PDA at pH 9, when carboxylic moiety was fully dissociated.

When cationic surfactants were used, stable dispersion of surfactant-adsorbed PDA nanocrystals was not obtained because aggregation occurred due to electrostatic attraction between cationic surfactant and PDA nanocrystals with negative ζ-potential.

Table 4.1 summarizes the results of silver seed deposition. All the PDA nanocrystal dispersions turned into bright-greenish blue during in situ reduction of silver cations with sodium borohydride. However, it was found that metal deposition could occur only in the presence of anionic surfactants. These phenomena suggest that reduction of silver cations proceeds just on the surface of the PDA core. The hybridized nanomaterials are shown in Fig. 4.8b.

Table 4.1. Effect of added surfactant for silver seed deposition

Surfactant	Ionicity of surfactant	Amount of deposited silver seeds
None	–	None
SDS	Anionic	High
SDBS	Anionic	Middle
14,8-ADA	Anionic	Middle
AOT	Anionic	Middle
CTAB	Cationic	None

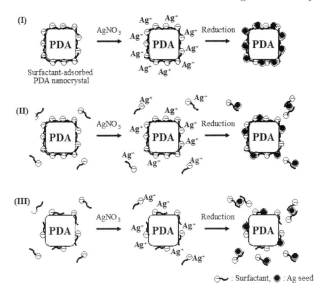

Fig. 4.9. Silver seed deposition on PDA nanocrystal core: (**I**) when surfactants added are almost adsorbed on surface of PDA nanocrystals; (**II**) in the presence of surfactant of excess amount; and (**III**) in the case of poor affinity between PDA core and surfactant

The fine silver seeds (bright dots) were observed on the PDA core surface, and the SEM image became clearer than that of Fig. 4.8a, due to the conductivity of silver seeds. Seed-deposited PDA nanocrystals on the silicon wafer seem to be aggregated during drying-up. In fact, dynamic light scattering analysis indicated that an aqueous dispersion of the isolated silver seed-deposited PDA-cores was once obtained. Among all the surfactants used, SDS successfully immobilized the largest number of silver seeds on the surface of PDA. From the perspective of the manufacture of NLO materials, the employment of 14, 8-ADA as a DA derivative also seems to be promising.

The difference in the amount of deposited silver seeds is due to not only the kind and the amount of surfactant used, but also affinity between PDA and surfactant. A schematic illustration of the proposed mechanism of silver seed deposition in the presence of surfactant is shown in Fig. 4.9. Case (I) indicates that added anionic surfactants are almost adsorbed onto the surface of PDA nanocrystals under an optimized concentration of the surfactant. Silver cations were captured by the adsorbed surfactants through electrostatic attraction, so that silver seeds were deposited on the PDA surface. In cases (II) and (III), however, isolated silver seeds are unexpectedly formed, owing to the excess amount of surfactant and poor affinity of surfactant for PDA, respectively. In fact, a number of isolated silver nanoparticles were formed by adding excess SDS as shown in Fig. 4.8c. Thus, it is significant to carefully control the ratios of the concentration of silver cations and surfactant to the

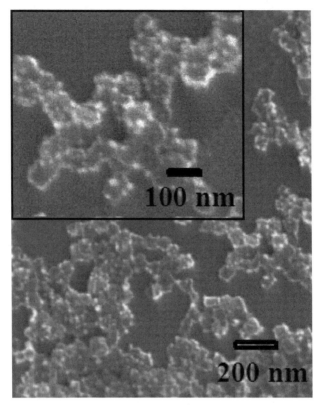

Fig. 4.10. SEM image of silver seed-grown PDA nanocrystals in the presence of sufficient SDS

number and average size of PDA nanocrystal cores. In addition, the number of deposited silver nanoparticles was also dependent upon the difference in affinity between the surfactant and PDA. In contrast, cationic surfactant did never trap silver cations on the PDA surface, due to the electrostatic repulsion. Finally, a subsequent reduction of silver cations with hydroxylamine was performed to obtain higher silver coverage on the surface of PDA nanocrystals. The color of the final dispersion liquid changed to deep green. Figure 4.10 shows SEM image of silver seed-grown PDA nanocrystals in the presence of sufficient SDS. The silver seeds evidently grew to be 10–20 nm in size on the surface of PDA nanocrystals. Silver metal itself generally has a self-catalyzing ability, and a weak reducing agent such as hydroxylamine plays the role of a surface-catalyzed reducing agent [30]. It is worth noting that no nucleation was generated, apart from silver seed deposited on the surface of the PDA core, during reduction with hydroxylamine. The average coverage with silver nanoparticles was around 60% from SEM images in the best case of the SDS system. However, external stimulus of ultrasound irradiation to hybridized

4 Fabrication of Organic Nanocrystals   93

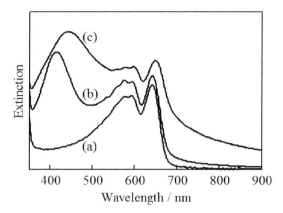

Fig. 4.11. UV-visible extinction spectra of (**a**) PDA nanocrystals as core, (**b**) silver seed-deposited PDA nanocrystals, and (**c**) silver seed-grown PDA nanocrystals dispersed in water. The extinction was normalized by the number of PDA nanocrystals

nanocrystal dispersion has unfortunately led to separation of PDA core and silver nanoparticles, owing to weak core–shell binding via a surfactant SDS.

Figure 4.11 shows the UV-visible extinction spectra of the PDA nanocrystals used as a core, the seed-deposited PDA nanocrystals, and the seed-grown PDA nanocrystals in the case of the SDS system. The peak at 642 nm in Fig. 4.11 is based on the excitonic absorption (EA) of the π-conjugated backbones of PDA nanocrystals [20]. While the localized surface plasmon (LSP) peak in the inverse core–shell-type structure [11] as described in the Introduction disappeared during solid-state polymerization of DA, the distinct LSP peak around 414 nm was still observed after deposition of silver seeds in the present case. We speculated that optoelectronic interaction at the PDA/Ag interface in the reverse nanostructure [11] might be significantly different from that in the present case. This difference is now under investigation [31]. Anyway, appearance of LSP peak is important to enhance effective $\chi^{(3)}$ value. On the other hand, the EA peak position was slightly redshifted as in the reverse core–shell-type structure [11].

After hydroxylamine was treated, the EA peak position was redshifted up to 649 nm. At the same time, the LSP peak was broadened and also redshifted to 442 nm, though the LSP peak of ordinary silver nanoparticles having the same size is usually sharp, and is observed at around 400 nm [32]. In this case, the broadening of the LSP peak is related to both an increase in the size of silver seeds and the dielectric constant of the surrounding medium [33, 34]. Furthermore, the size distribution of silver nanoparticles is also responsible for inhomogeneous broadening of the plasmon band. The redshift of the LSP peak may be based upon the increase in dipole interaction among silver nanoparticles on the same PDA nanocrystal because the average distance between silver nanoparticles became reduced with increasing volume of silver nanoparticles.

This interaction may also lead to the enhancement of the local electric field between silver nanoparticles [35]. On the other hand, the redshift of EA peak would be due to changes in dielectric properties around the PDA core. Anyway the silver deposition on the PDA nanocrystal can induce some changes of internal electric field in the PDA nanocrystal core, leading to the enhancement of the effective $\chi^{(3)}$ value in NLO properties.

4.4.2 Silver-Deposited PDA Nanocrystals Produced by Visible-Light-Driven Photocatalytic Reduction

Next, we introduce another fabrication method, i.e., visible-light-driven photocatalytic reduction to obtain higher coverage and strong interaction between a PDA core and a silver nanoparticle shell. Cubic PDA nanocrystals as a core were fabricated with no surfactants by the conventional reprecipitation method [21,22]. The crystal structure of the resulting PDA nanocrystals was recognized to be the same as that of the bulk crystal by measuring powder-XRD pattern. Next, aqueous solutions of silver nitrate ($AgNO_3$: 0.5 mL, 22.2 mM) and ammonia (NH_3: 0.2 mL, 111 mM) were added to the PDA nanocrystal dispersion liquid (5 mL). Afterwards, the mixture was allowed to stand under visible-light irradiation using a xenon lamp (150 W), equipped with an attenuator and a UV-cut filter ($\lambda > 420$ nm), for about 30 min at room temperature.

Figure 4.12a,b shows SEM images of bare PDA nanocrystals and silver-nanoparticle-deposited PDA nanocrystals formed by visible-light irradiation. The color of the resulting dispersion liquid was gradually changed into deep green from dark blue within tens minutes of visible-light irradiation time as displayed in the insets of Fig. 4.12a,b. By this photocatalytic reduction, as shown in Fig. 4.12b, a large number of silver nanoparticles were deposited in comparison with the previous method in Sect. 4.4.1 [15], and no isolated silver nanoparticles were not observed in the SEM image. This fact suggests that the reduction selectively occurred just on the surface of PDA nanocrystals. Moreover, even when ultrasound was irradiated, the present

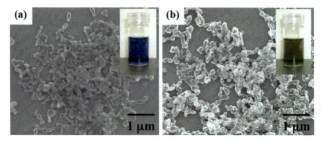

Fig. 4.12. SEM images of (**a**) as-prepared and (**b**) silver-deposited PDA nanocrystals. The *insets* are, respectively, the corresponding photographs of the dispersion liquids

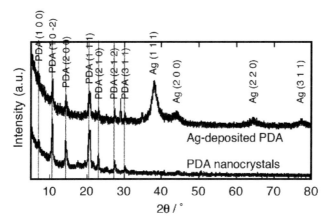

Fig. 4.13. Powder XRD patterns of as-prepared and silver-deposited PDA nanocrystals

hybridized nanocrystals were never separated into isolated PDA cores and silver nanoparticles. Figure 4.13 shows the powder-XRD pattern of the hybridized nanocrystals. It suggests that the resulting products consisted of PDA [36] and Ag metal, not containing silver oxide. The average size of deposited Ag nanoparticles was estimated to be 5.4 nm by the Scherrer equation [37] to the (1 1 1) diffraction peak. Figure 4.14a,b shows the TEM images of as-prepared PDA nanocrystals and silver-nanoparticle-deposited PDA nanocrystals produced by visible-light irradiation. Size of the deposited silver nanoparticles was around 5–10 nm in diameter, which is roughly in accord with the value estimated from the powder-XRD measurements. As shown in the inset of Fig. 4.14b, the electron diffraction pattern revealed clear Debye–Scherrer rings, i.e., diffractions from the (1 1 1), (2 0 0), (2 2 0), and (3 1 1) planes in the fcc lattice structure of the silver metal (JCPDS: 4-783).

This photoreduction of silver cation proceeded only under the coexistence of PDA nanocrystals as a core, silver cation (AgNO$_3$ aq.), and NH$_3$ aq. in the dispersion liquid under visible-light irradiation. That is to say, if one factor of aforementioned conditions was lacking, this reduction never takes place completely. In addition, the reduction on the surface of unpolymerized DA monomer nanocrystals was not observed. The DA monomer nanocrystals does not have band gap in the visible region. It is interesting that the silver cation was reduced to silver metal only through the "photocatalytic action" of PDA itself. To our knowledge, silver cation cannot be directly reduced to silver metal upon exposure to visible light. Figure 4.15a shows the band structure for PDA [38], the conduction band (CB) and valence band (VB) of which are −3.44 eV [−1.06 V vs. standard hydrogen electrode (SHE)] [39] and −5.77 eV (1.27 V vs. SHE) [39], respectively. The band gap of PDA is narrow, 2.33 eV in the visible region, and PDA provides a stronger reducing property than anatase-type TiO$_2$ (CB: −4.3 eV; VB: −7.5 eV) as a typical photocatalyst. On

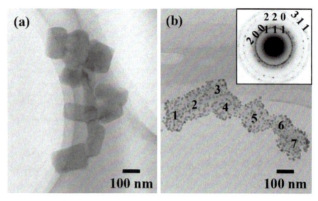

Fig. 4.14. TEM images of (**a**) as-prepared cubic PDA nanocrystals and (**b**) silver-deposited cubic PDA nanocrystals. The electron diffraction pattern in the *inset* indicates the Debye–Scherrer rings, corresponding to the (1 1 1), (2 0 0), (2 2 0), and (3 1 1) diffractions from fcc lattice structure of silver metal

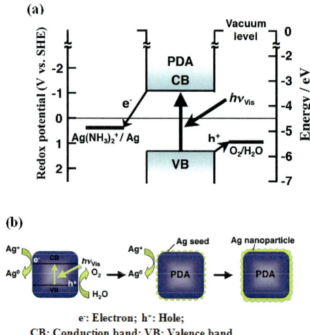

e⁻: Electron; h⁺: Hole;
CB: Conduction band; VB: Valence band

Fig. 4.15. Energy diagram and proposed scheme of photocatalytic reduction process. (**a**) The conduction band (CB) and valence band (VB) for PDA are, respectively, -3.44 and -5.77 eV vs. vacuum level, corresponding to -1.06 and 1.27 V vs. SHE. The redox potentials of $[Ag(NH_3)_2]^+$ as a major component of silver cation in the present system and of H_2O are 0.373 and 1.23 V vs. SHE, respectively. (**b**) Photocatalytic formation of silver nanoparticles as a nanoshell on the surface of PDA nanocrystal

the other hand, the redox potential of $[Ag(NH_3)_2]^+$ is 0.373 V vs. SHE and is located between the CB and the VB for PDA as shown in Fig. 4.15. Therefore, PDA is capable of reducing $[Ag(NH_3)_2]^+$ by the visible-light-excited electrons in the CB of PDA. The pH value of the dispersion liquid eventually changed from 9 to 6 through this reaction. This photocatalytic reduction also occurred similarly under the alkali condition of dilute NaOH aqueous solution, instead of by the addition of NH_3. When NaOH aqueous solution was used, silver nanoparticles were larger and deposited inhomogeneously, compared with those shown in the TEM image in Fig. 4.14b. These facts allow us to speculate on the possible evolution of H^+ by the following process:

$$2H_2O + 4h^+ \rightarrow O_2 + 4H^+ \tag{4.1}$$

Equation (4.1) means that the holes in the VB produced by photoexcitation are scavenged by water molecules and then a very small amount of oxygen is generated. The pH-dependent redox potential, E_{redox}, in (4.1) can be obtained from the Nernst equation [40]:

$$E_{redox} = 1.23 - 0.059 \times [pH] \quad (V \text{ vs. SHE}). \tag{4.2}$$

Equation (4.2) suggests that the oxidation of water does not proceed so much at low pH. In fact, the reduction did not proceed completely under an acidic condition. On the other hand, several oxidation mechanisms of NH_3 or NH_4^+ are reported by past literature [41]. However, it is now speculated that these are not the main hole-scavenging processes because of the aforementioned photocatalytic reduction in NaOH aqueous solution and the equivalent weight of NH_3 added to silver cation in order to form silver–amine complex, $[Ag(NH_3)_2]^+$, under the present experimental condition.

The present photoreaction may be attributed to the following factors (1) the electrostatic adsorption of $[Ag(NH_3)_2]^+$ on the surface of the PDA nanocrystals having negative zeta potential (ca. -40 mV) [21], (2) the self-catalyzed reduction [42], and (3) the suppression of electron–hole recombination due to rapid carrier separation [43] at the PDA/Ag interface as illustrated in Fig. 4.15b.

Figure 4.16 shows UV-visible extinction spectra of bare PDA nanocrystals and silver-deposited ones fabricated by visible-light-driven photocatalytic reduction method. EA peak of the bare PDA nanocrystals was around 653 nm. In the present hybridized nanocrystals (Fig. 4.16b), the EA peak position was redshifted by 12 nm, owing to changes in the dielectric properties around the PDA nanocrystal core. We consider that this fact is based on the higher coverage and the direct contact between PDA core and silver nanoparticles. On the other hand, LSP peak clearly appeared at around 448 nm when silver nanoparticles were directly contacted with PDA nanocrystal in the present case, which is also redshifted to the longer wavelength region in comparison with that in Sect. 4.4.1. The higher redshift of the LSP peak is mainly due to the increase in the dipole–dipole interaction among silver nanoparticles with

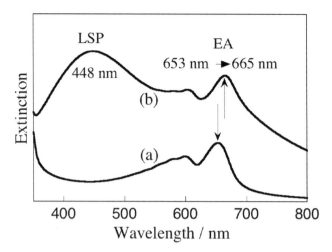

Fig. 4.16. UV-visible extinction spectra of (**a**) as-prepared cubic PDA nanocrystals and (**b**) silver-deposited cubic PDA nanocrystals. Both samples remain dispersed in water

higher coverage, because the average distance is shortened as silver nanoparticle deposition proceeds on the surface of the PDA nanocrystal core [44, 45]. It would be related to the effective dielectric constant around the deposited silver nanoparticles [36, 37], which seem to be influenced by certain interfacial Ag–PDA and/or Ag–Ag interactions. These spectral changes also support the fact that silver-deposited PDA nanocrystals were fabricated in the dispersion liquid.

4.5 Summary and Future Scopes

Through the systematic research on organic nanocrystals, we established for the first time a fabrication method for the PDA nanocrystals deposited with silver nanoparticles by using two different methods (1) in situ reduction using anionic surfactant binders and (2) visible-light-driven photocatalytic reduction. To the best of our knowledge, no mild metal-plating technique on a chemically inert surface of functional organic nanocrystals has ever been reported. In the in situ reduction, it is very important to optimize the type and amount of surfactant and the concentration ratios of the used surfactant to silver salt and PDA nanocrystals. A surfactant with high affinity for PDA core surface should be employed to further raise the coverage of silver. This process could be widely applied to various kinds of organic nanocrystal (core)/metal (nanoshell) system. On the other hand, in the visible-light-driven photocatalytic reduction, when the relationships among the band structure of the organic materials, the redox potential of the metal ions, and the hole scavengers were appropriately selected, it would be applied to various combinations of π-conjugated organic and polymer materials and noble metals, irrespective of the size, shape, and morphology of the organic materials. In addition, it was found that the extinction spectra of the PDA nanocrystals deposited with silver nanoparticles by the photo-driven technique exhibited a distinct LSP peak and changes in dielectric constant around the PDA cores. We expect that the optical phenomena enhance the effective $\chi^{(3)}(\omega)$ for the resulting PDA nanocrystals coated with silver nanoparticles. Furthermore, hybridized materials in this article will be useful for novel optoelectronic devices [46], organic/metal heterojunctions for integrated circuits in organic transistors, and new types of organic–metal hybridized systems such as metamaterials [47].

References

1. C. Bosshard, K. Sutter, P. Pröre, J. Hulliger, M. Flörsheimer, P. Kaatz, P. Güter, in *Organic Nonlinear Optical Materials (Advances in Nonlinear Optics)*, vol. 1, ed. by F. Kajzar, A.F. Garito (Gordon and Breach, Basel, 1995), Chap. 6, p. 89
2. A. Sarkar, S. Okada, H. Matsuzawa, H. Matsuda, H. Nakanishi, J. Mater. Chem. **10**, 819 (2000)
3. H. Bäsler, in *Polydiacetylenes (Advances in Polymer Science)*, vol. 63, ed. by H.-J. Cantow (Springer, Berlin, 1984), Chap. 1, p. 1
4. O.G. Tovmachenko, C. Graf, D.J. van den Heuvel, A. van Blaaderen, H.C. Gerritsen, Adv. Mater. **18**, 91 (2006)
5. Y. Shen, C.S. Friend, Y. Jiang, D. Jakubczyk, J. Swiatkiewicz, P.N. Prasad, J. Phys. Chem. B **104**, 7577 (2000)
6. S. Nie, S.R. Emony, Science **275**, 1102 (1997)
7. A.E. Neeves, M.H. Birnboim, J. Opt. Soc. Am. B **6**, 787 (1989)
8. S.T. Selvan, T. Hayakawa, M. Nogami, M. Möler, J. Phys. Chem. B **103**, 7441 (1999)
9. R. Shenhar, V.M. Rotello, Acc. Chem. Res. **36**, 549 (2003)
10. K.G. Thomas, P.V. Kamat, Acc. Chem. Res. **36**, 888 (2003)
11. A. Masuhara, H. Kasai, S. Okada, H. Oikawa, M. Terauchi, M. Tanaka, H. Nakanishi, Jpn. J. Appl. Phys. **40**, L1129 (2001)
12. H. Kasai, H.S. Nalwa, H. Oikawa, S. Okada, H. Matsuda, N. Minami, A. Kakuta, K. Ono, A. Mukoh, H. Nakanishi, Jpn. J. Appl. Phys. **31**, L1132 (1992)
13. F. Caruso, Adv. Mater. **13**, 11 (2001)
14. S.J. Oldenburg, R.D. Averitt, S.L. Westcott, N.J. Halas, Chem. Phys. Lett. **288**, 243 (1998)
15. T. Onodera, Z. Tan, A. Masuhara, H. Oikawa, H. Kasai, H. Nakanishi, T. Sekiguchi, Jpn. J. Appl. Phys. **45**, 379 (2006)
16. T. Onodera, H. Oikawa, A. Masuhara, H. Kasai, T. Sekiguchi, H. Nakanishi, Jpn. J. Appl. Phys. **46**, L336 (2007)
17. H. Nakano, Y. Tachibana, S. Kuwabata, Electrochim. Acta **50**, 749 (2004)
18. K.C. Yee, R.R. Chance, J. Polym. Sci., Polym. Phys. Ed. **16**, 431 (1978)
19. H. Nakanishi, H. Kasai, in *Handbook of Nanostructured Materials and Nanotechnology*, vol. 5, ed. by H.S. Nalwa (Academic, San Diego, 2000), Chap. 8

100   T. Onodera et al.

20. G. Weiser, Phys. Rev. B **45**, 14076 (1992)
21. H. Katagi, H. Kasai, S. Okada, H. Oikawa, K. Komatsu, H. Matsuda, Z. Liu, H. Nakanishi, Jpn. J. Appl. Phys. **35**, L1364 (1996)
22. H. Katagi, H. Kasai, S. Okada, H. Oikawa, H. Matsuda, H. Nakanishi, J. Macromol. Sci. Pure Appl. Chem. A **34**, 2013 (1997)
23. H. Kasai, H. Oikawa, S. Okada, H. Nakanishi, Bull. Chem. Soc. Jpn. **71**, 2597 (1998)
24. H.-R. Chung, E. Kwon, H. Oikawa, H. Kasai, H. Nakanishi, J. Cryst. Growth **294**, 459 (2006)
25. H.C. van de Hulst, *Light Scattering by Small Particles* (Dover, New York, 1981), Chap. 6
26. T. Onodera, T. Oshikiri, H. Katagi, H. Kasai, S. Okada, H. Oikawa, M. Terauchi, M. Tanaka, H. Nakanishi, J. Cryst. Growth **229**, 586 (2001)
27. V.V. Volkov, T. Asahi, H. Masuhara, A. Masuhara, H. Kasai, H. Oikawa, H. Nakanishi, J. Phys. Chem. B **108**, 7674 (2004)
28. T. Onodera, H. Kasai, S. Okada, H. Oikawa, K. Mizuno, M. Fujitsuka, O. Ito, H. Nakanishi, Opt. Mater. **21**, 595 (2002)
29. H. Oikawa, T. Mitsui, T. Onodera, H. Kasai, H. Nakanishi, T. Sekiguchi, Jpn. J. Appl. Phys. **42**, L111 (2003)
30. K.R. Brown, M.J. Natan, Langmuir **14**, 726 (1998)
31. H. Oikawa, A.M. Vlaicu, M. Kimura, H. Yoshikawa, S. Tanuma, A. Masuhara, H. Kasai, H. Nakanishi, Nonlinear Opt. Quantum Opt. **34**, 275 (2005)
32. M. Kerker, J. Colloid Interface Sci. **105**, 297 (1985)
33. H. Hvel, S. Fritz, A. Hilger, U. Kreibig, M. Vollmer, Phys. Rev. B **48**, 18178 (1993)
34. M.P. Pileni, A. Taleb, C. Petit, J. Dispersion Sci. Technol. **19**, 185 (1998)
35. H. Xu, E.J. Bjerneld, M. Käl, L. Böjesson, Phys. Rev. Lett. **83**, 4357 (1999)
36. P.A. Apgar, K.C. Yee, Acta Cryst. B **34**, 957 (1978)
37. H.P. Klug, L.E. Alexander, *X-Ray Diffraction Procedures for Polycrystalline and Amorphous Materials* (Wiley, New York, 1974), Chap. 9
38. W. Spannring, H. Bäsler, Chem. Phys. Lett. **84**, 54 (1981)
39. In general, the redox potential can be semiempirically correlated with the energy level. S.R. Morrison, *Electrochemistry at Semiconductor and Oxidized Metal Electrodes* (Plenum, New York, 1980), Chap. 1, p. 29
40. J.P. Hoare, in *Standard Potentials in Aqueous Solution*, ed. by A.J. Bard, R. Parsons, J. Jordan (Dekker, New York, 1985), Chap. 4, p. 49
41. J.T. Maloy, in *Standard Potentials in Aqueous Solution*, ed. by A.J. Bard, R. Parsons, J. Jordan (Dekker, New York, 1985), Chap. 7, p. 127
42. K.R. Brown, M.J. Natan, Langmuir **14**, 726 (1998)
43. T. Hirakawa, P.V. Kamat, J. Am. Chem. Soc. **127**, 3928 (2005)
44. S.L. Westcott, S.J. Oldenburg, T.R. Lee, N. Halas, J. Chem. Phys. Lett. **300**, 651 (1999)
45. K.E. Peceros, X. Xu, S.R. Bulcock, M.B. Cortie, J. Phys. Chem. B **109**, 21516 (2005)
46. A.E. Neeves, M.H. Birnboim, J. Opt. Soc. Am. B **6**, 787 (1989)
47. A. Ishikawa, T. Tanaka, Opt. Commun. **258**, 300 (2006)

Part II

Nanohybridized Thin Films

5

Polymer Nanoassemblies and Their Nanohybridization with Metallic Nanoparticles

Masaya Mitsuishi, Jun Matsui, and Tokuji Miyashita

5.1 Introduction

Since the last two decades, hybrid nanoassemblies have been the subject of considerable interest. Bottom-up approaches such as layer-by-layer (LbL) adsorption [1,2], self-assembled monolayer (SAM) [3], and Langmuir–Blodgett (LB) technique [4,5] make it possible to develop myriad nanoarchitectures on solid substrates. These approaches and their combinations offer a promising route to polymer nanoarchitectures for a wide variety of functional materials and surfaces. Using these nanostructures, the surface properties can be controlled uniformly, even on a nanoscale. These compelling qualities of ultrathin polymer films have been expanded to broad scientific fields. In fact, the research related with these topics now leads to an exponentially increasing number of publications. Needless to say, it is important to intensify the interdisciplinary interactions, for example, between organic (especially, polymer) materials and inorganic nanoparticles in basic and application-oriented work.

We have investigated functionalization of polymer LB films (Fig. 5.1) consisting of poly(alkylacrylamide)s and poly(alkylmethacrylamide)s [4,6–9]. Poly(N-dodecylacrylamide) (pDDA) proved to have an excellent property for LB assembly [6]; the material provides a highly oriented and densely packed monolayer at the air–water interface. Figure 5.2 shows an example of surface pressure-area isotherms of amphiphilic acrylamide polymers. The surface pressure rises steeply and shows high collapse pressure, indicating that the polymers take a stable monolayer formation. Varying the side chain length, the film thickness can be tuned at the nanometer length scale (1–2 nm in length) [10,11]. Besides, widely various functional molecules were incorporated and distributed uniformly in a two-dimensional (2D) field. The key factor is the existence of a 2D hydrogen-bonding network based on acrylamide groups between polymer backbones, which enhance the monolayer stability at the air–water interface, leading to quantitative deposition onto solid substrates via vertical dipping. We recently designate polymer monolayer and the LB films

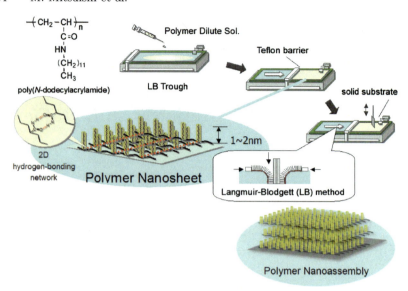

Fig. 5.1. Polymer nanosheets

consisting of alkyl acrylamide polymer as polymer nanosheets. Using these advantages, directional (vectorial) electron transfer between several redox components was achieved in multilayered polymer nanosheets [12], extending to logic gate operations [13,14]. We have demonstrated photochemically patterned polymer nanosheets for three-dimensional (3D) nanoarchitecture [15]. Compared with low-molecular-weight fatty acids and other polymer LB films, the pDDA nanosheets give simplicity and versatility in preparation and usage. In fact, preparation of pDDA monolayers requires no addition of ions to the water subphase to stabilize the monolayer formation. The pDDA monolayer can be deposited onto numerous substrates, such as quartz, glass, ITO glass, metals, plastic fiber, and spin-coated films.

Recently, metal nanoparticles have received much attention because they show fascinating features such as quantum size effect [16], catalytic activity [17], and localized surface plasmon [18–20]. We have investigated the preparation of metal nanoparticle arrays templated with polymer nanosheets through electrostatic interaction. Polymer nanosheets, which consist of cationic comonomer and N-dodecylacrylamide capable of forming LB films, serve as a good template for metal nanoparticle immobilization. We assembled functional polymer nanosheets with metal nanoparticle monolayers to utilize a localized surface plasmon as an excitation source [7]. The hybrid polymer nanoassemblies show strong extinction bands in the visible light wavelength because of the localized surface plasmon resonance of the metal nanoparticle arrays. They are expected to be available for opto/electronic device application based on their surface plasmon photonics. Quite recently, we have successfully assembled core–shell-type metallic nanoparticle covered

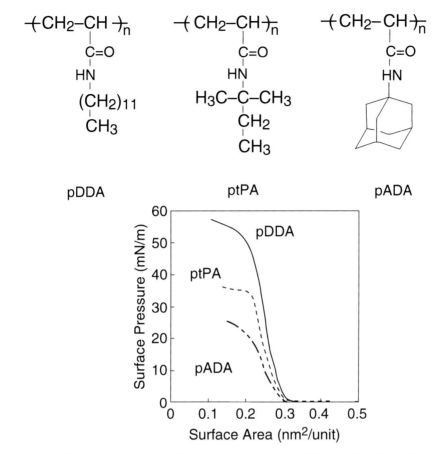

Fig. 5.2. Chemical structures of representative amphiphilic acrylamide polymers and surface pressure-area isotherms at 15°C

with amphiphilic polymer brushes [21, 22]. This approach also offers the potential for nanoassembly construction. In this chapter, despite the considerable variety of contributions, we specifically address recent topics of hybrid polymer nanoassemblies via polymer nanosheets. We will focus on (a) fabrication of hybrid polymer nanoassemblies via a bottom-up approach based on the LB technique, (b) spectroscopic characterization of hybrid polymer nanoassemblies, and finally (c) synthesis and preparation of hybrid polymer nanoassemblies with magnetic nanoparticle covered with polymer brushes. We will limit the present review to hybrid polymer nanoassemblies via polymer nanosheets; some of the other topics such as free-standing polymer nanosheets and application-oriented works have been described in recent reviews [6, 7].

5.2 Hybrid Nanoassemblies via Polymer Nanosheets

The experimental procedure for preparation of hybrid nanoassemblies is outlined in Fig. 5.3 [23]. Bilayers of poly(*N*-dodecylacrylamide-*co*-4-vinylpyridine) (p(DDA/VPy)) or poly(*N*-dodecylacrylamide-*co*-*N*-2-(2-(2-aminoethoxy)ethoxy)ethylacrylamide) (p(DDA/DONH)) nanosheets were transferred onto silicon substrates by the vertical dipping method of the LB technique. Monodispersed gold nanoparticles were prepared by reduction of $HAuCl_4$ with trisodium citrate dihydrate in an aqueous solution under reflux [24]. Silver nanoparticles were synthesized from $AgNO_3$ and trisodium citrate dihydrate in a similar manner. The diameter determined by SEM observation was ca. 30 nm for gold nanoparticles and ca. 80 nm for silver nanoparticles. Substrates were immersed in metal nanoparticle solution (pH 6.0) for several hours. After immersion, the substrates were rinsed in pure water and then finally dried with nitrogen gas. Figure 5.4 shows time dependence of reflectivity under surface plasmon resonance (SPR) measurement. Briefly, gold nanoparticles were adsorbed onto p(DDA/VPy) nanosheets (transferred on the gold film (50 nm)). In situ observation of the reflectivity from the substrate coupled with a high-refractive index prism was carried out with a He–Ne laser. As shown clearly in Fig. 5.4, the reflectivity at the resonance angle dramatically increased after flowing gold nanoparticle solution, indicating that gold nanoparticles were adsorbed onto p(DDA/VPy) nanosheets, with the exception of pDDA nanosheets. From the SEM observation, they were found to take a uniformly distributed monolayer over a large area [23]. As an advantage of polymer nanosheets, gold nanoparticles were uniformly immobilized onto the cationic polymer nanosheets. Interestingly, the amount of immobilized gold nanoparticles strongly depends on VPy contents (Fig. 5.5).

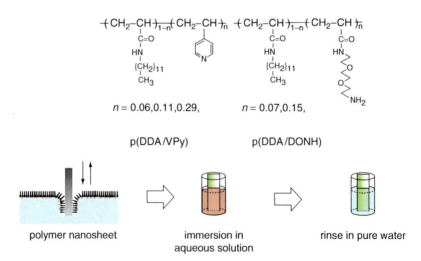

Fig. 5.3. Chemical structure of cationic polymer nanosheets and schematic illustration of the preparation process of hybrid polymer nanoassemblies

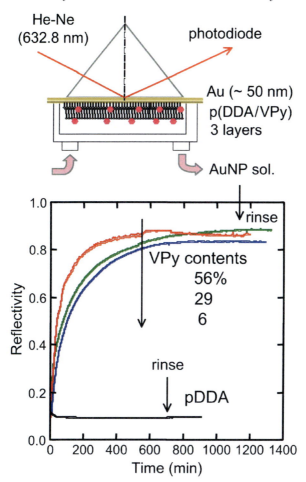

Fig. 5.4. Time course of SPR reflectivity measured at the resonance angle

The surface charge density can be controlled by varying VPy contents. The nanoparticles form a monoparticle monolayer due to electrostatic repulsion among the nanoparticles. It took more than six hours to reach the saturation of metal nanoparticle immobilization on polymer nanosheets. Similar to the gold nanoparticle immobilization, the silver nanoparticles can be randomly distributed on p(DDA/DONH) nanosheets [25]. The surface coverage of the silver nanoparticles immobilized on the p(DDA/DONH) nanosheet increased with DONH content. In other words, the amount of the silver nanoparticles can be also tuned by changing DONH contents.

The pDDA nanosheets were transferred on p(DDA/VPy29) nanosheets, as a function of the number of layers, to examine the effect of cationic polymer nanosheets on metal nanoparticle immobilization. Figure 5.6 gives the clear

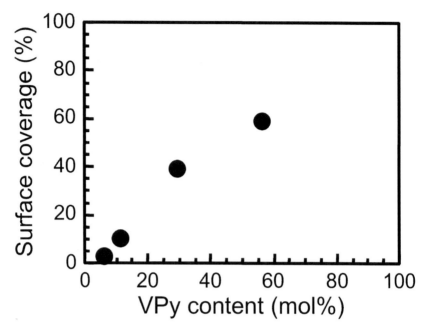

Fig. 5.5. Plots of the surface coverage of gold nanoparticles immobilized on p(DDA/VPy) nanosheets as a function of VPy contents

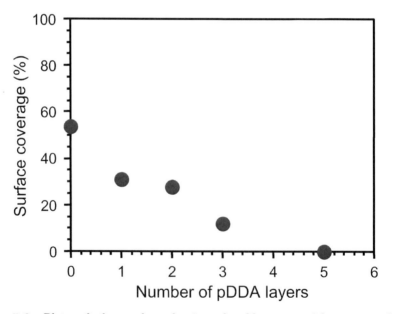

Fig. 5.6. Plots of the surface density of gold nanoparticles on two-layer p(DDA/VPy29) nanosheets as a function of the number of pDDA spacer layers

evidence that the amount of immobilized gold nanoparticles depends strongly on the spacer length. The amount decreases as the number of deposited pDDA spacer layers increases. Eventually, no adsorption of gold nanoparticles (apart from those immobilized through nonspecific physisorption) was observed over five-layer pDDA nanosheets. The findings imply that a long-range electrostatic interaction exists between gold nanoparticle and p(DDA/VPy) nanosheets, and that this interaction is a predominant driving force for gold nanoparticle immobilization. As for the effect of cationic polymer nanosheets, the surface charge density is also controllable by varying the number of deposited cationic polymer nanosheets (Fig. 5.7); the amount of immobilized gold nanoparticles was saturated above two-layer cationic polymer nanosheets.

The polymer nanosheets, p(DDA/VPy) LB films, have another useful feature: the photopatterns can be drawn [26–28]. As reported previously, deep UV irradiation through a photomask gives a fine photopattern of p(DDA/VPy) nanosheets. Three-layer p(DDA/VPy) nanosheets were irradiated under air atmosphere, and photodecomposition occurred selectively at the irradiated portion (Fig. 5.8). Patterned p(DDA/VPy) nanosheets were immersed in a gold nanoparticle solution. Gold nanoparticles were directed

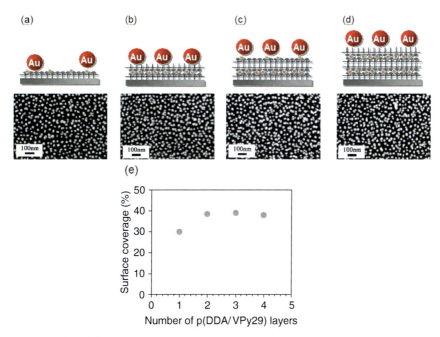

Fig. 5.7. (a)–(d) Schematic illustration and SEM images of gold nanoparticle monolayer formation as a function of the number of underlying p(DDA/VPy29) nanosheets: (a) monolayer, (b) two-layer, (c) three-layer, and (d) four-layer p(DDA/VPy29) nanosheets. (e) Plots of the surface coverage obtained from (a) to (d)

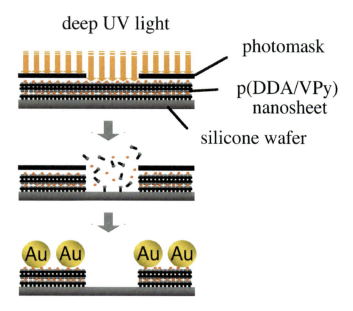

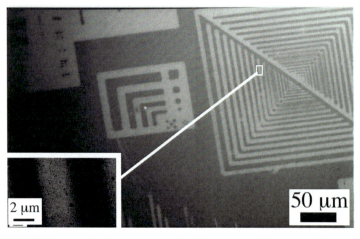

Fig. 5.8. SEM image of photopatterned gold nanoparticle arrays. Reprinted with permission from [23]. Copyright 2003, American Chemical Society

onto the unirradiated portion of p(DDA/VPy) nanosheets, producing clearly configured gold nanoparticle monolayer patterns with 2.0 μm resolution. That is, p(DDA/VPy) nanosheets can act as a good template for nanoparticle ordering.

5.3 Spectroscopic Properties of Hybrid Assemblies

As mentioned in Sect. 5.1, localized surface plasmon resonance (LSPR) is one of the fascinating features of noble metal nanoparticles, and metal nanoparticle ordering in hybrid nanoassemblies leads to effective utilization of LSPR. Spectroscopic characterization of metal nanoparticles on a solid substrate is sometimes difficult because of aggregate formation of metal nanoparticles. In solution, the molar extinction coefficient of metal nanoparticles is determined quantitatively using Mie theory. A deep understanding of spectroscopic properties of metal nanoparticle will help us find new functionality that is related to photonic and electronic device application. We examined spectroscopic properties of isolated metal nanoparticles on solid substrates by preparing uniformly distributed metal nanoparticle monolayers. Figure 5.9 shows an environmental scanning electron microscope (ESEM) image of gold nanoparticles immobilized on two-layer p(DDA/VPy29) nanosheets on a transparent glass substrate after 4-h immersion. In this condition, the gold nanoparticles do not take a notable aggregate formation on p(DDA/VPy) nanosheets: almost all the gold nanoparticles are isolated from their neighbors. The

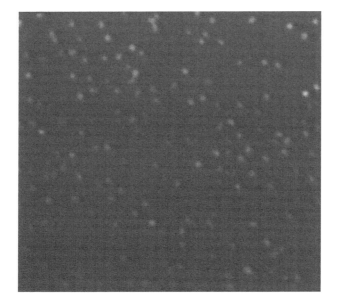

Fig. 5.9. ESEM image of gold nanoparticles immobilized onto two-layer p(DDA/VPy29) nanosheets

UV-Vis extinction spectrum of the substrate exhibits a strong extinction band at 525 nm because of dipole-like SPR, that is, collective oscillation of free electron gas in the gold nanoparticles coupled with incident light. The ESEM image shows that the surface density of the gold nanoparticle was determined as $51 \pm 3\,\mu m^{-2}$. We can determine the molar extinction coefficient of the gold nanoparticle on substrates, assuming that the reflection loss from the gold nanoparticle is negligible in UV-Vis extinction measurement: $7.1 \times 10^{12}\,M^{-1}\,cm^{-1}$ for gold nanoparticles (30 nm ϕ). It is noteworthy that the molar extinction coefficient includes the contribution from both absorption and scattering caused by LSPR. Considering the metal nanoparticle size, the contribution from absorption seems to be predominant for smaller gold nanoparticles (30 nm ϕ). Similarly, the molar extinction coefficient of silver nanoparticles ((averaged) 80 nm ϕ) was determined as roughly $9.5 \times 10^{12}\,M^{-1}\,cm^{-1}$. These large molar extinction coefficient values imply that metal nanoparticles serve as an effective light energy acceptor in photoexcited energy transfer processes.

Gold nanoparticles used in craftwork have shown beautiful colors, with their colors maintained exactly as they were several hundred years ago. Their optical properties were then exploited for coloration of glass, ceramics, China, and pottery. These optical properties arise from individual localized surface plasmons of gold nanoparticles; the various hues of the nanoparticles have been ascribed to various nanoparticle sizes, nanoparticle shape, stages of agglomeration, and overall composition. Recently, surface plasmon coupling, also known as particle–particle coupling, has received considerable attention. As the interparticle distance decreases, the plasmon coupling wavelength shifts to lower energy (redshift), which implies that color tuning as a function of interparticle spacing is possible with a well-defined nanostructure. Control of the surface plasmon coupling at the nanometer length scale is achieved by varying the interlayer distance between gold nanoparticle monolayers with the number of inserting pDDA. The gold nanoparticle shows scarlet color in water. The same color appears in a gold nanoparticle monolayer immobilized on two-layer p(DDA/DONH) nanosheets after 4-h immersion in an aqueous gold nanoparticle solution. The gold nanoparticle monolayer comprises uncoupled and isolated nanoparticles, which implies that no interparticle coupling occurred between adjacent gold nanoparticles in the monolayer because of its well-separated configuration. After repeating bilayer deposition of p(DDA/DONH) nanosheet and gold nanoparticle adsorption with an n layer (n = 0, 4, 8, 12, 16, 20, 24) pDDA spacer four times, UV-Vis extinction spectral measurement was implemented (Fig. 5.10a). The extinction spectra in the extinction peaks at 538 and 637 nm are assigned, respectively, to isolated particle and interparticle surface plasmon coupling. It is noteworthy that the extinction peak for isolated gold nanoparticles shows a redshift because the environment moves from air to pDDA nanosheets. For n = 0, no pDDA nanosheet was inserted between gold nanoparticle monolayers. Also, dipole-like surface plasmon coupling between two adjacent gold nanoparticle monolayers was the strongest,

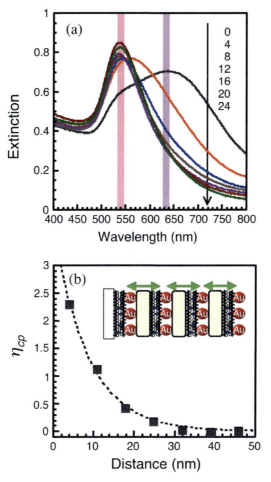

Fig. 5.10. (a) UV-Vis extinction spectra of gold nanoparticle multilayers (after four deposition cycles) as a function of the number of pDDA spacer layers. The values show the numbers of pDDA spacer layers. (b) Plots of the surface plasmon coupling efficiency as a function of spacer distance between adjacent gold nanoparticle monolayers. The *inset* shows the layer structure. Reprinted with permission from [7]. Copyright 2007, The Society of Polymer Science, Japan

resulting in the redshift and broadening of the extinction band. The extinction band shape resembles that of gold nanoparticle in solution when the number of pDDA layers increases. Using the two optical extinction values at 538 and 637 nm, the efficiency of surface plasmon coupling is estimated as [7]

$$\eta_{\rm cp} = \frac{(A_{637}/A_{538})_n - (A_{637}/A_{538})_{24}}{(A_{637}/A_{538})_{24}}, \quad (5.1)$$

where A_{538} and A_{637} represent optical extinction at 538 and 637 nm, respectively. Figure 5.10b illustrates the distance-dependent behavior for interlayer surface plasmon coupling: the η_{cp} value decreases exponentially as the distance between gold nanoparticle monolayers increases. Assuming that the monolayer thicknesses of p(DDA/DONH) and pDDA are 2.0 and 1.7 nm, the separation distance between gold nanoparticle monolayers is determined. The experimental data were fitted with a single exponential function (Fig. 5.10b, solid line). Consequently the decay distance $(d_{1/e})$, of which the magnitude falls to $1/e$, was determined as $d_{1/e} = 8.7$ nm, indicating that effective interlayer surface plasmon coupling occurs within the spacer separation distance of 8.7 nm for a 30 nm ϕ gold nanoparticle.

5.4 Hybrid Polymer Nanosheet Fabricated from Core–Shell Nanoparticles

The other strategy to fabricate hybrid nanosheet is directly fabricating monolayer of a core–shell nanoparticle, which composed of inorganic core and polymer shell. Several methods, such as emulsion polymerization [29], layer-by-layer assembly [30], and grafting polymerization [31] have been reported to coat inorganic nanoparticle core with polymer shell. Several groups fabricated hybrid monolayer films using inorganic core–polymer shell nanoparticles. For example, Fukuda et al. [32] reported the fabrication of monolayer of gold nanoparticle which was coated by high-density polymer brushes using the air–water interface. Recent developments of living radical polymerization, mainly in atom transfer radical polymerization (ATRP), enable to synthesize high-density polymer brushes onto a nanoparticle surface [31]. Because the polymerization proceeded in a living fashion, it is easy to control the polymerization degree. Using the polymer brushes as a shell, the distance between the nanoparticles is controlled by changing the polymerization degree of the polymer. Yoshinaga et al. [33] also studied the assembly of polymer-grafted nanoparticle at the air–water interface. They studied the effect of not only the grafted polymers' molecular weights but also the type of polymer to be grafted.

As mentioned in the previous sections, the polymer nanosheets can be prepared using poly(alkylacrylamide)s and poly(alkylmethacrylamide)s (p(alkylMA)s). Therefore, we employ the acrylamide as a polymer shell to fabricate hybrid nanosheet from the core–shell-type nanoparticles. In the following section, synthesis of the core–shell nanoparticle with a magnetic core and polyalkylmethacrylamide shell was described [21, 22]. The magnetic nanoparticle was selected as an inorganic core because magnetic nanoparticles are useful in widely various applications including electromagnetic devices, high-density storage media, ferrofluids, magnetic resonance imaging, drug delivery, etc. Moreover, development of nanoparticles' organization techniques is important to apply its unique properties for devices.

5.4.1 Synthesis of Magnetic Nanoparticle Covered with Poly(*N*-Alkylmethacrylamide)s

The alkylmethacrylamide polymers were selected as a shell for their ability to form a polymer nanosheet at the air–water interface. The synthesis of the magnetic nanoparticle covered with p(alkylMA)s was shown in Scheme 5.1 [21]. Usually ATRP of acrylamide derivatives is difficult because amide group coordinates with copper catalyst and/or their propagating radical undergo S_N2 substitution, both of which result in polymerization termination [34]. Therefore, only a few example of an ATRP of acrylamide derivative have been reported [35, 36]. In the present case, two-step route was applied to synthesized Fe_3O_4–p(alkylMA)s nanoparticle to avoid difficulty in the synthesis of poly(methacrylamide)s covered particle. First, reactive

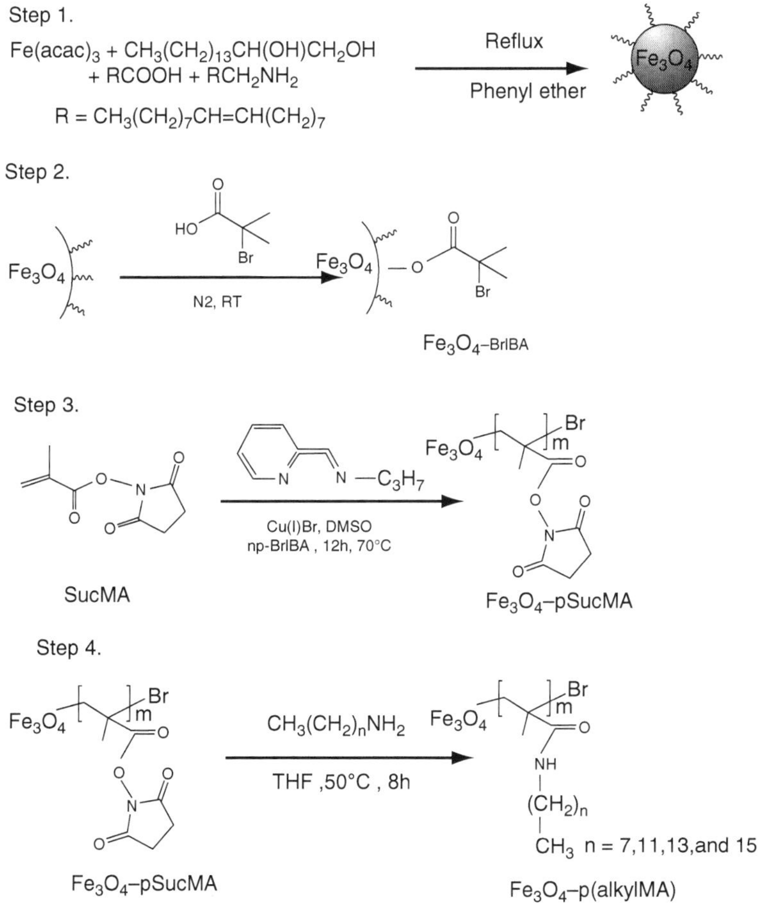

Scheme 5.1.

polymer, N-hydroxysuccinimide methacrylate (SucMA) was grafted to Fe_3O_4 nanoparticle surface using surface-initiated ATRP to synthesize polysuccinimide methacrylate-coated iron oxide nanoparticles (Fe_3O_4–pSucMA). pSucMA grafted onto the nanoparticle surface can act as a reactive polymer shell. These reactive polymer shells can react readily with various alkyl amine derivatives to produce Fe_3O_4–p(alkylMA)s via substitution reaction between pSucMA and corresponding alkylamines.

Magnetic nanoparticle was synthesized by high-temperature solution phase reaction (Scheme 5.1, step 1). Diameter of the particle was determined to be 4.7 ± 1.0 nm from TEM images. The particle was first modified with an ATRP initiator, 2-methyl-2-bromo-isobutyric acid (BrIBA) through ligand exchange reaction [37].

Fe_3O_4–BrIBA was used as a macroinitiator to polymerize SucMA on the nanoparticle surface (Scheme 5.1, step 3). After performing the ATRP reaction of SucMA, the FT-IR spectrum shows a new peak related to imide vibration band at 1,816 and 1,789 cm^{-1} and ester C=O stretching band at 1,747 cm^{-1}, which indicate that SucMA is polymerized at the nanoparticle surface (Fig. 5.11a). The pSucMA on the nanoparticle surfaces was reacted with various alkyl amines. Octylamine, dodecylamine, tetradecylamine, and hexadecylamine, respectively, were used to fabricate p(alkylMA)s-coated iron oxide nanoparticles of Fe_3O_4–pC$_8$MA, Fe_3O_4–pC$_{12}$MA, Fe_3O_4–pC$_{14}$MA, and Fe_3O_4–pC$_{16}$MA (Scheme 5.1, step 4). The substitution reactions were followed by FT-IR spectra. As an example, Fig. 5.11b shows the FT-IR spectra of Fe_3O_4–pC$_{14}$MA after the substitution reaction. After the substitution reaction, both the imide and ester peaks of pSucMA almost disappear; the characteristic amide band for NH stretching at 3,307 cm^{-1} and the C=O vibration band at 1,641 cm^{-1} appear. The FT-IR spectra indicate that pSucMA is converted into pC$_{14}$MA by a substitution reaction. Moreover, ^{1}H NMR spectrum of pC$_{14}$MA cleaved from Fe_3O_4 nanoparticle shows no peak related to pSucMA. These results indicate that the substitution reaction occurred completely. TEM observation shows core–shell structures of the Fe_3O_4–pC$_{14}$MA particles, which confirm the presence of the polymer thin shell around the nanoparticle surface (Fig. 5.12). Molecular weights of the grafting polymers were measured by cleaving the polymer shell from the nanoparticle surface. The number-average molecular weights (M_n) of p(alkylMA)s were ca. 3,000–4,000 with small polydispersity ($M_w/M_n = 1.13$).

5.4.2 Monolayer Behavior of p(alkylMA)s-Coated Iron Oxide Nanoparticles at the Air–Water Interface

The behaviors of monolayer composed of the nanoparticles coated with various p(alkylMA)s on the water surface were studied using π-A isotherm measurements. A chloroform solution containing the nanoparticles was spread onto the water surface to measure the π-A isotherm. The π-A isotherms with various alkyl chain lengths in Fe_3O_4–p(alkylMA)s show a sharp rise in surface

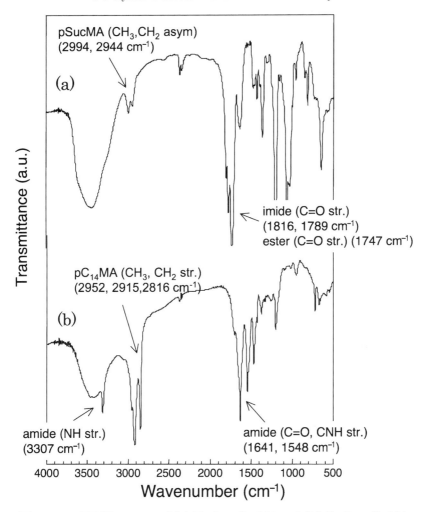

Fig. 5.11. FT-IR spectra of (a) Fe$_3$O$_4$–pSucMA and (b) Fe$_3$O$_4$–pC$_{14}$MA

pressure with a high collapse pressure (Fig. 5.13). The pressure curve stands up more sharply with increasing alkyl chain length from Fe$_3$O$_4$–pC$_8$MA to Fe$_3$O$_4$–pC$_{14}$MA and the collapse pressure also increases, whereas the Fe$_3$O$_4$–pC$_{16}$MA shows a decrease in collapse pressure. The phenomenon is similar to the monolayer behavior of N-alkylmethacrylamide polymer itself [8]. Monolayers of N-alkylmethacrylamide polymers with various alkyl chain lengths show high collapse pressure with increasing alkyl chain lengths from pC$_8$MA to pC$_{14}$MA. However, after a certain chain length, the collapse pressure of pC$_{16}$MA decreases due to strong hydrophobic interaction of the long alkyl chains. The π-A isotherm result indicates that the nanoparticle coated by the polymer having an excellent property for monolayer formation yields a stable

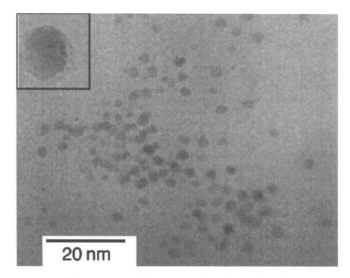

Fig. 5.12. TEM image of Fe_3O_4–$pC_{14}MA$

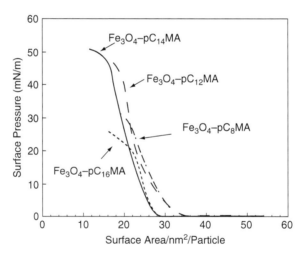

Fig. 5.13. Surface pressure-area isotherms of Fe_3O_4–p(alkylMA)s at the air–water interface

monolayer at the water surface. Fe_3O_4–$pC_{14}MA$ shows the steepest rise in surface pressure and highest collapse pressure. The limiting surface area per particle is determined to be 25 nm^2 by extrapolating the linear portion of the steep rise in π-A isotherm to zero surface pressure. The value is closed to the cross sectional area calculated from the nanoparticle diameter, assuming that all particles are spherical with 5 nm diameter (20 nm^2). The π-A isotherm results demonstrate successful monolayer formation of Fe_3O_4–$pC_{14}MA$ at the air–water interface.

5.4.3 Surface Morphology of Nanoparticle Monolayer at Different Deposition Pressures

Fe$_3$O$_4$–pC$_{14}$MA can be transferred onto a solid substrate using the LB technique. The film morphology of the deposited film depends on deposition pressure (π) and stabilization time (t). For example, the film deposited on a substrate at $\pi = 30\,\mathrm{mN\,m^{-1}}$ and $t = 30\,\mathrm{min}$ shows loosely packed spider web like structure (Fig. 5.14a). On the other hand, when the stabilization time increased to $t = 120\,\mathrm{min}$, a closed-packed monolayer was transferred onto a solid substrate (Fig. 5.14b). The thickness of both films is 4–5 nm, which indicates that the film is composed of monolayer of Fe$_3$O$_4$–pC$_{14}$MA regardless of the packing density. Moreover, the monolayer can be closely packed by increasing the deposition pressure. Figure 5.14c shows AFM images of

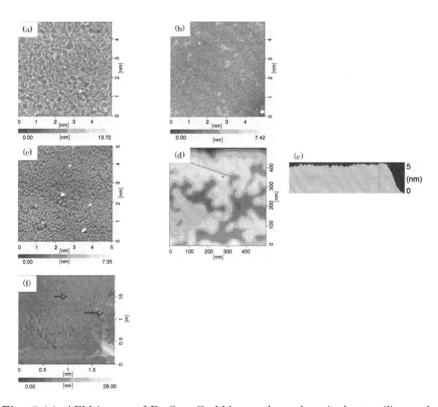

Fig. 5.14. AFM images of Fe$_3$O$_4$–pC$_{14}$MA monolayer deposited onto silicon substrate at different deposition pressure: (**a**) deposited at $30\,\mathrm{mN\,m^{-1}}$ with stabilization time of 30 min; (**b**) deposited at $30\,\mathrm{mN\,m^{-1}}$ with stabilization time of 2 h; (**c**) deposited at $40\,\mathrm{mN\,m^{-1}}$ with stabilization time of 30 min; (**d**) small area image of (**c**); (**e**) height profile across the sample surface of (**d**); and (**f**) deposited at $45\,\mathrm{mN\,m^{-1}}$. *Arrows* in (**f**) indicate overlapping of particles with small aggregation

Fe$_3$O$_4$–pC$_{14}$MA deposited at surface pressure of 40 mN m^{-1} with 30 min of stabilization time. The nanoparticle assemblies become more closely packed than in the monolayer transferred at $\pi = 30$ mN m^{-1} with same stabilization time. The magnified image shows a 4–5 nm high flat domain, which indicates that the domains comprise monolayers of Fe$_3$O$_4$–pC$_{14}$MA (Fig. 5.14d,e). Figure 5.14f shows the Fe$_3$O$_4$–pC$_{14}$MA monolayer that is deposited close to the collapse pressure ($\pi = 45$ mN m^{-1}, stabilization time of 30 min). The image shows that the domains are mostly in contact; it also reveals a densely packed monolayer with a small overlapping region at the edge of the domain.

5.4.4 Multilayer Formation

A multilayer film of Fe$_3$O$_4$–pC$_{14}$MA was fabricated by sequential deposition of the monolayer using the LB technique. The UV-Vis spectra of Fe$_3$O$_4$–pC$_{14}$MA nanoparticles LB film deposited on a quartz substrate at $\pi = 30$ mN m^{-1} with different numbers of layers are shown in Fig. 5.15a. The multilayer deposition was carried out with stabilization time of 30 min; therefore, the transfer ratio of first few layers was not in unity because of the low packing of the monolayer and the optical density of the transferred film is low. After 5–6 layers deposition, the monolayer becomes stable, and the spectra show a linear increase of optical absorption density at 360 nm with an increasing number of the layers of Fe$_3$O$_4$–pC$_{14}$MA. The linear relationship between the absorbance and the number of layers suggests that regular deposition takes place over the particle monolayer (Fig. 5.15b). The magnetic property measurement of the 30 layers of the prepared Fe$_3$O$_4$–pC$_{14}$MA film shows superparamagnetic behavior at room temperature, which is a characteristic feature of size-dependent magnetic properties (Fig. 5.16).

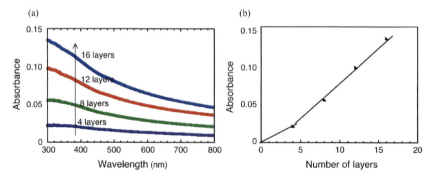

Fig. 5.15. (a) UV-Vis spectra of Fe$_3$O$_4$–pC$_{14}$MA particles LB thin film with 4, 8, 12, and 16 layers deposited onto a quartz substrate. (b) Absorption intensity at 360 nm as a function of the number of layers. The line is a guide for eye

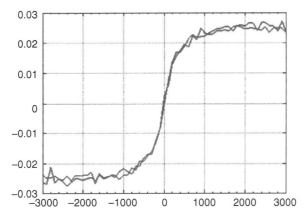

Fig. 5.16. Magnetic hysteresis loop of 30 layers of Fe_3O_4–$pC_{14}MA$ LB film measured at room temperature

5.5 Conclusion

We described fascinating aspects of hybrid polymer nanoassemblies consisting of polymer nanosheets and noble metal nanoparticle arrays. The advantage of this approach lies in precise, nanometer-scale positioning of noble metal nanoparticles and functional molecules along with the film surface normal. By varying the functional group's molar contents and immersion time in an aqueous metal nanoparticle solution, reproducible and precise control of metal nanoparticle immobilization is achieved. This approach provides a suitable configuration for a deep understanding of the relationship between the materials' spectroscopic properties and localized surface plasmon generated from metal nanoparticles: a well-separated metal nanoparticle monolayer was stacked on the functional polymer nanosheets. This approach also enables fruitful characterization of the interaction between metal nanoparticles and functional molecules in hybrid polymer nanoassemblies. An understanding of the relationship between material properties and optical/spectroscopic properties of functional molecules will open up new avenues for future nanodevice applications. In addition, the strong extinction band from metal nanoparticles implies that the metal nanoparticle serves as an effective quencher when the nanoparticle is coupled with a suitable luminophore. In particular, this tendency is observed for smaller metal nanoparticles. In this sense, detailed investigation of luminescence or SHG properties of hybrid polymer nanoassemblies will provide a deep understanding of localized surface plasmon coupling with functional molecules [38–40]. Due to the page's limitation, we skip fabrication of free-standing hybrid nanoassemblies [41]. Moreover, another technique to fabricate hybrid polymer nanosheet, which used inorganic core–polymer shell nanoparticle, was described. Poly(alkylMA)s were coated onto an inorganic nanoparticle by two-step reaction. Monolayer properties of the nanoparticles are highly dependent on shells polymer side chain structure. $pC_{14}MA$ is concluded to be the best polymer to induce the nanoparticles to form a stable monolayer at the air–water interface. The monolayer can be transferred onto a solid substrate to fabricate hybrid polymer nanosheets. The technique of pSucMA coating is versatile and applicable to other inorganic nanoparticle. Moreover, due to high reactivity of SucMA, other functional molecules can easily be attached onto shell surface.

Acknowledgments

The work was supported by Grants-in-Aid for Scientific Research ((S) No. 17105006), and Priority Areas (No. 446, No. 19022003, "Super-Hierarchical Structures"). J.M. thanks JST-PRESTO for their financial support.

References

1. V.V. Tsukruk, Prog. Polym. Sci. **22**, 247 (1997)
2. G. Decher, J.B. Schlenoff, *Multilayer Thin Films: Sequential Assembly of Nanocomposite Materials* (Wiley-VCH, New York, 2002)
3. A. Ulman, *Ultrathin Organic Films* (Academic, New York, 1991)
4. T. Miyashita, Prog. Polym. Sci. **18**, 263 (1993)
5. G. Wegner, Macromol. Chem. Phys. **204**, 347 (2003)
6. M. Mitsuishi, J. Matsui, T. Miyashita, Polym. J. **38**, 877 (2006)
7. M. Mitsuishi, M. Ishifuji, H. Endo, H. Tanaka, T. Miyashita, Polym. J. **39**, 411 (2007)
8. Y.Z. Guo, F. Feng, T. Miyashita, Macromolecules **32**, 1115 (1999)
9. T. Miyashita, Y. Mizuta, M. Matsuda, Br. Polym. J. **22**, 327 (1990)
10. T. Taniguchi, Y. Yokoyama, T. Miyashita, Macromolecules **30**, 3646 (1997)
11. F. Feng, A. Aoki, T. Miyashita, Chem. Lett. **27**, 205 (1998)
12. A. Aoki, Y. Abe, T. Miyashita, Langmuir **15**, 1463 (1999)
13. J. Matsui, M. Mitsuishi, A. Aoki, T. Miyashita, Angew. Chem. Int. Ed. **42**, 2272 (2003)
14. J. Matsui, M. Mitsuishi, A. Aoki, T. Miyashita, J. Am. Chem. Soc. **126**, 3708 (2004)
15. Y. Kado, M. Mitsuishi, T. Miyashita, Adv. Mater. **17**, 1857 (2005)
16. J. Zheng, P.R. Nicovich, R.M. Dickson, Annu. Rev. Phys. Chem. **58**, 409 (2007)
17. M.C. Daniel, D. Astruc, Chem. Rev. **104**, 293 (2004)
18. K.A. Willets, R.P. Van Duyne, Annu. Rev. Phys. Chem. **58**, 267 (2007)
19. J.R. Lakowicz, Anal. Biochem. **337**, 171 (2005)
20. E. Hutter, J.H. Fendler, Adv. Mater. **16**, 1685 (2004)
21. S. Parvin, E. Sato, J. Matsui, T. Miyashita, Polym. J. **38**, 1283 (2006)
22. S. Parvin, J. Matsui, E. Sato, T. Miyashita, J. Colloid Interface Sci. **313**, 128 (2007)
23. H. Tanaka, M. Mitsuishi, T. Miyashita, Langmuir **19**, 3103 (2003)
24. K.C. Grabar, R.G. Freeman, M.B. Hommer, M.J. Natan, Anal. Chem. **67**, 735 (1995)

5 Polymer Nanoassemblies and Their Nanohybridization 123

25. M. Mitsuishi, H. Tanaka, T. Miyashita, Trans. Mater. Res. Jpn. **30**, 639 (2005)
26. A. Aoki, M. Nakaya, T. Miyashita, Macromolecules **31**, 7321 (1998)
27. T.S. Li, J.F. Chen, M. Mitsuishi, T. Miyashita, J. Mater. Chem. **13**, 1565 (2003)
28. A. Aoki, T. Miyashita, Polymer **42**, 7307 (2001)
29. J.M. Asua, Prog. Polym. Sci. **27**, 1283 (2002)
30. F. Caruso, G. Sukhorukov, *Coated Colloids: Preparation, Characterization, Assembly and Utilization* (Wiley-VCH, Weinheim, 2003)
31. Y. Tsujii, K. Ohno, S. Yamamoto, A. Goto, T. Fukuda, in *Surface-Initiated Polymerization I*, vol. 197, ed. by R. Jordau, R. Advmcula, M.R. Buchmeiser (Springer, Berlin, 2006), p. 1
32. K. Ohno, K. Koh, Y. Tsujii, T. Fukuda, Angew. Chem. Int. Ed. **42**, 2751 (2003)
33. E. Mouri, Y. Okazaki, S. Tatsuno, H. Karakawa, K. Yoshinaga, J. Polym. Sci. B **44**, 2789 (2006)
34. M. Teodorescu, K. Matyjaszewski, Macromolecules **32**, 4826 (1999)
35. M. Teodorescu, K. Matyjaszewski, Macromol. Rapid Commun. **21**, 190 (2000)
36. Y. Xia, X.C. Yin, N.A.D. Burke, H.D.H. Stover, Macromolecules **38**, 5937 (2005)
37. Y. Wang, X. Teng, J.S. Wang, H. Yang, Nano Lett. **3**, 789 (2003)
38. M. Ishifuji, M. Mitsuishi, T. Miyashita, Appl. Phys. Lett. **89**, 3 (2006)
39. H. Tanaka, M. Mitsuishi, T. Miyashita, Chem. Lett. **34**, 1246 (2005)
40. S.L. Pan, L.J. Rothberg, J. Am. Chem. Soc. **127**, 6087 (2005)
41. H. Endo, Y. Kado, M. Mitsuishi, T. Miyashita, Macromolecules **39**, 5559 (2006)

6

Single Molecular Film for Recognizing Biological Molecular Interaction: DNA–Protein Interaction and Enzyme Reaction

Kazue Kurihara

6.1 Introduction

Protein–protein and protein–substrate interactions play essential roles in biological functions. Surface forces measurement and atomic force microscopy, which directly measure the interaction forces as a function of the surface separation, enable us to quantitatively evaluate these interactions [1–3]. We have employed the surface forces measurement [4] and colloidal probe atomic force microscopy [5] to study interactions involved in specific molecular recognition of DNA–protein and enzyme–substrate reaction. Studied are interactions between nucleic acid bases (adenine and thymine) [6], Spo0A–DB (the DNA-binding site of a transcription factor Spo0A), and DNA [7,8], those between subunits I and II of heptaprenyl diphosphate (HepPP) synthase in the presence of a substrate ((E, E)-farnesyl diphosphate, FPP) and a cofactor (Mg^{2+}) [9–11], and the selectivity of the substrates in this enzymatic reaction [12]. Keys of our approach are the preparation of well-defined samples and the appropriate analysis. We have modified the substrate surfaces with these proteins using the Langmuir–Blodgett (LB) method. This chapter reviews the LB modification method and subsequent demonstrations of biological specific interactions employing this approach.

6.2 Surface Forces Measurement

Surface forces measurement directly determines interaction forces between two surfaces as a function of the surface separation (D) using a simple spring balance. Instruments employed are a surface forces apparatus (SFA), developed by Israelachvili and Tabor [4], and a colloidal probe atomic force microscope introduced by Ducker and Sendan [5] (Fig. 6.1). The former utilizes the crossed cylinder geometry and the latter uses the sphere-plate geometry. For both geometries, the measured force (F) normalized by the mean radius (R) of cylinders or a sphere, F/R, is known to be proportional to

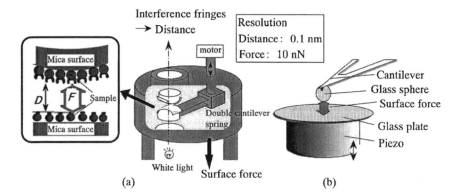

Fig. 6.1. Surface forces apparatus (**a**) and the colloidal probe atomic force microscope (**b**)

the interaction energy, G_f, between flat plates (Derjaguin approximation), $F/R = 2\pi G_f$ [1]. This enables us to quantitatively evaluate the measured forces, e.g., by comparing them with a theoretical model.

Sample surfaces are atomically smooth surfaces of cleaved mica sheets for SFA, and various colloidal spheres and plates for a colloidal probe AFM. These surfaces can be modified using various chemical modification techniques such as Langmuir–Blodgett deposition [13, 14] and silanization reactions [15, 16]. For more detailed information, see the original papers and texts.

6.3 Interaction Between Nucleic Acid Bases

Elucidation of the mechanism that governs the complementary nucleic acid base pairing is of fundamental importance in biology and is highly relevant to our understanding of molecular recognition. Surface forces measurement has been shown to provide a wealth of information on the interaction forces between apposed molecular layers. We have, therefore, undertaken systematic surface force measurements between apposed amphiphilic monolayers composed of adenine and thymine, adenine and adenine, and thymine and thymine at various pH values from 5.6 to 9.3; and reported for the first time that interaction between the complementary adenine–thymine monolayers is always attractive and stable independently of the surface separation and pH.

Nucleic acid base monolayers were formed on these mica surfaces by the LB deposition of amphiphiles **1** and **2** (Fig. 6.2), which had ammonium groups and nucleic acid bases at their opposite terminals. The cationic ammonium groups of **1** and **2** anchor themselves on the negatively charged mica surface, thereby forming nucleic acid base monolayers.

Force profiles measured between adenine–thymine monolayers (complementary pair) are shown in Fig. 6.3. The long-range attraction appears at a

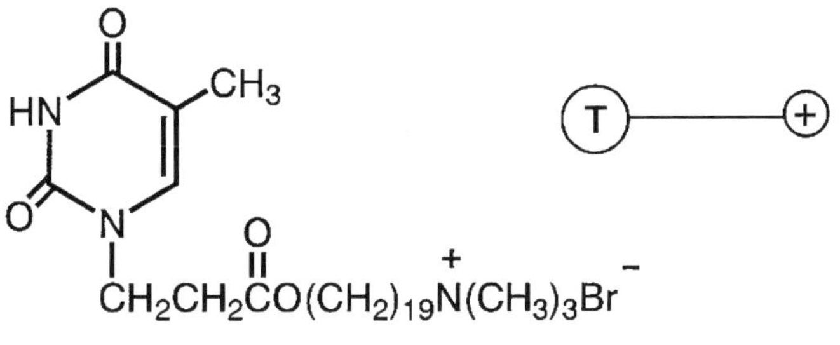

$$\text{Thymine derivative} \qquad \mathbf{1}$$

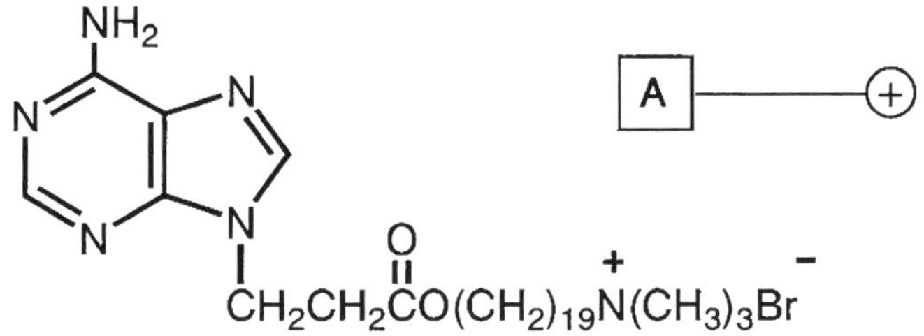

Thymine derivative   **1**

Adenine derivative   **2**

Fig. 6.2. Amphiphiles bearing a nucleic acid base, thymine (**1**), and adenine (**2**), and the ammonium group at each terminals

separation of 150 nm and increases with decreasing the distance. These attractions can be accounted in terms of the sum of the long-range hydrophobic attraction [17] and the attractive double-layer force, although we do not know the exact fraction of each component yet. Apparently, nature has effectively manipulated pK_as and hydrophobicities of nucleic acid bases.

6.4 Surface Immobilization of Protein Modified with His-Tag

Poly(histidine)-tagged (His-tag) proteins are immobilized on a substrate modified with an LB film of a chelate amphiphile. We have first confirmed that LB modification is also feasible for the colloidal probe AFM using LB forming lipid functionalized by the iminodiacetic acid group (DSIDA, Fig. 6.4) [18]. Interactions between DSIDA monolayers have been studied for different ionization states of the iminodiacetate group and for the chelate complex formation with Cu^{2+} ion. This monolayer is then used for immobilizing His-tag proteins to form two-dimensionally organized layers of proteins [19, 20]. Importance

128   K. Kurihara

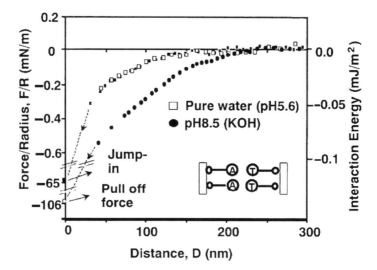

Fig. 6.3. Force profiles measured between a complementary pair of adenine–thymine (**1–2**) surfaces in pure water (pH 5.6) (*open square and filled square*) and in KOH (pH 8.5) (*filled circle*)

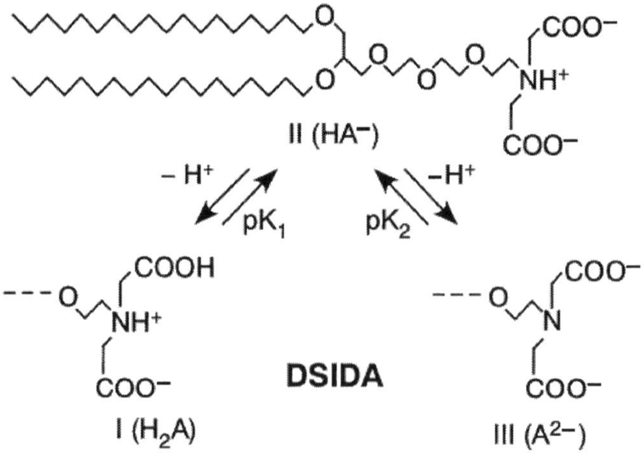

Fig. 6.4. Structural formula of N-(8-(1,2-di(octadecyloxy)propanoxy)-3,6-dioxyaoctyl) iminodiacetic (DSIDA): I, II, and III represent different protonation states of the iminodiacetate group

of preparing oriented protein layers is apparent for investigating specific interactions of proteins by the surface forces measurement. (Fig. 6.1).

6.4.1 pH Dependence of DSIDA Interactions

The electric double-layer force between DSIDA monolayers changes depending on pH of the aqueous phase (10^{-4} M NaBr) as shown in Fig. 6.5a. Here, the pH value is adjusted by adding appropriate amount of NaOH or HNO_3. Under pH 6.0, no appreciable repulsion is observed, whereas the electric double-layer repulsion emerges at pH 6.5 and increases with increasing pH. No pull-off force is observed at pH 9.5. Pull-off forces at pH 7.5 and 5.5 are obtained to be 1.0 ± 0.5 and 4 ± 3 mN m^{-1}, respectively.

To examine the pH dependence in detail, the surface charge density of the DSIDA monolayer surface, estimated from a measured force profile by fitting to a theoretical one assuming the constant potential [9], is plotted as a function of pH of the aqueous bathing solution. The charge density changed stepwise with rising pH (Fig. 6.5b): it increases sharply till pH 7.5, and exhibited plateau between 7.5 and 8.5, then rose again beyond 8.5. The iminodiacetic acid group is known to form various deprotonation species: species I, II, and III as shown in Fig. 6.4 [21]. Therefore, the multiple step change of the surface charge shown in Fig. 6.5b can be attributable to dissociation of I to II below pH 7.5, stable formation of II between pH 7.5 and 8.5, and further dissociation

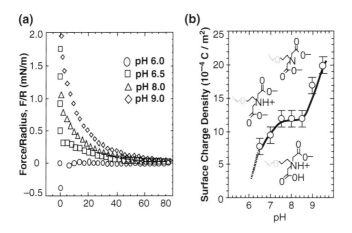

Fig. 6.5. Force profiles between DSIDA monolayer surfaces in 10^{-4} M NaBr aqueous solutions at various pHs adjusted by addition of NaOH or HNO(3) (**a**); the surface charge density of the DSIDA surface as a function of pH (**b**). An *arrow* represents the position where the surface jumps into contact. The charge density was calculated from the surface potential obtained by fitting the force profile. The marks of I, II, and III indicate formation of each dissociation state DSIDA shown in Fig. 6.3, and *solid lines* are drawn to guide the eyes

130   K. Kurihara

of II to III beyond 8.5. The pK_1 of DSIDA monolayer in 10^{-4} M NaBr likely lays around 6.5 and pK_2 is around 9. The pK_1 value of iminodiacetic acid in water has been reported to be 3.5 [21]. Such considerable shift of pKs of monolayers from those in solutions has been reported to occur in many systems [22]. The species II of iminodiacetic acid has been known to be stable at pH 4.5–8.5 in water [18], and is stable at pH 7.5–8.5 in the case of DSIDA monolayers.

6.4.2 Interactions Between DSIDA Monolayers in the Presence of Cu^{2+}

The iminodiacetate group is known to form a chelate complex with Cu^{2+} ion. Force profiles between DSIDA monolayers have been measured in the presence of $CuCl_2$ at pH 7.5 [18]. The double-layer force decreases with increasing $CuCl_2$ concentration from 10^{-7} to 10^{-5} M owing to chelation of the iminodiacetate group with Cu^{2+}. Pull-off force remains practically constant, $0.9 \pm 0.6\,mN\,m^{-1}$. The most pronounced drop of the repulsion has been detected between 10^{-7} and 10^{-6} M $CuCl_2$, which is consistent with the concentration range of $CuCl_2$ found for clear change in the π-A isotherm of DSIDA. The force profiles between DSIDA monolayers in 10^{-5} M $CuCl_2$ are maintained even after the water phase is replaced by pure water, demonstrating that the Cu^{2+} ion binding is irreversible.

6.4.3 AFM Imaging of Immobilized His-Tag Sigma A Protein

Two-dimensionally organized layers of His-tag proteins can be formed on DSIDA–Cu^{2+} monolayers. Sigma A has been immobilized on the DSIDA–Cu^{2+} monolayer using LB method. Figure 6.6 shows an atomic force microscope image contact mode of Sigma A adsorbed on the glass surface modified with DSIDA–Cu^{2+} monolayer. Regular and dense immobilization of the Sigma A proteins has been found. The distance between the proteins was nearly constant, ca. 10 nm, which is twice the size of Sigma A, 4.6 nm, calculated from its molecular weight assuming a spherical shape. The double-layer repulsion between charged proteins may have prevented. Sigma A proteins form the closest packing on the surface.

6.5 Specific Interactions Between Enzyme–Substrate Complexes: Heptaprenyl Diphosphate Synthase

Prenyl diphosphate synthases catalyze the sequential head-to-tail condensation of isopentenyl diphosphate (IPP) with allylic substrates to give linear prenyl diphosphates in the biosynthetic pathway of isoprenoid compounds. Heptaprenyl diphosphate (HepPP) synthase from *Bacillus subtilis*, which

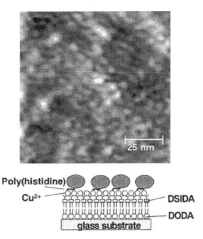

Fig. 6.6. Atomic force microscope images of Sigma A immobilized on glass surface modified with DSIDA–Cu^{2+} monolayer, and a schematic drawing of the surface modification using the poly(histidine)-tag. DODA is dioctadecyldimethyl ammonium used to render the surface hydrophobic

forms the HepPP with a chain length of C$_{35}$, is composed of two nonidentical protein subunits, neither of which has catalytic activity alone. The gel column chromatography has demonstrated that the subunits I and II most likely associate in the presence of the allylic substrate (E,E)-farnesyl diphosphate (FPP) and Mg^{2+} (a cofactor) to form a catalytically active complex, which represents an intermediary state during the catalysis (Fig. 6.7) [9]. However, there has been no direct evidence to support the complex formation between the two subunits.

It should be interesting to employ the direct interaction measurement for studying this enzymatic process. We immobilized the subunits I and II of HepPP synthase on the glass surfaces and measured the interactions between them in various conditions using the colloidal probe AFM [10, 11].

6.5.1 Interaction Between Subunit I and Subunit II upon Approach

Figure 6.8 shows the interaction forces between subunits I and II *upon approach*. Only repulsive forces are observed under all the solution conditions studied. We compress the layers till the force of 4.0 mN m^{-1}. Under this condition, these forces are reproducible after repeating compression. The isoelectric points of subunits I and II are known to be 5.1 and 5.2, respectively. Therefore, each subunit must be negatively charged in the solution at pH 8.3, giving rise to the repulsive double-layer force. Indeed, the observed repulsive force is described by an exponential function as expected for the double-layer

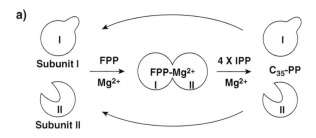

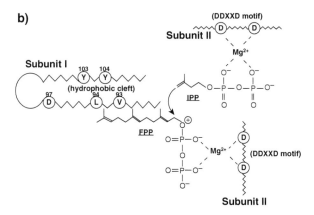

Fig. 6.7. (a) A hypothetical mechanism of the catalytically active complex of HepPP synthase of *B. subtilis* [9]. (b) A hypothetical scheme for the binding of FPP and IPP between the two subunits of *B. subtilis* HepPP synthase [9, 23]

force, and its decay length of 9.2 ± 0.4 nm is in good agreement with the Debye length of 9.2 nm calculated for the corresponding salt concentration of 1.1 mM [1].

6.5.2 Interaction Between Subunit I and Subunit II upon Separation

Figure 6.9 presents the surface forces between subunits I and II *upon separation* at pH 8.3. In the solutions containing only Mg^{2+}, FPP, or IPP, the interaction is always repulsive, and reversible both on approach and on separation. However, in the solution containing both Mg^{2+} and FPP, adhesive forces are observed upon separation though the repulsive force was observed upon approach. The intensity of the apparent adhesive force taken as shown in Fig. 6.8 is 0.20 ± 0.06 mN m^{-1}. When FPP is replaced by IPP, the interaction is only repulsive again. One may note that the interactions between identical proteins (subunits I and I, or II and II) are always repulsive under all solution conditions (data not shown). This confirmed that (1) the proteins are charged

6 Single Molecular Film   133

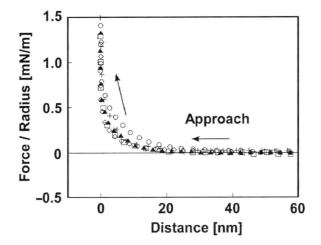

Fig. 6.8. Force profiles of interactions between subunit I and subunit II upon approach under various conditions. *Open circle* Tris buffer solution, *open square* Tris buffer solution containing Mg^{2+}, *open diamond* Tris buffer solution containing FPP, *times symbol* Tris buffer solution containing IPP, *filled triangle* Tris buffer solution containing $Mg2^{2+}$ and FPP, *plus symbol* Tris buffer solution containing Mg^{+2} and IPP

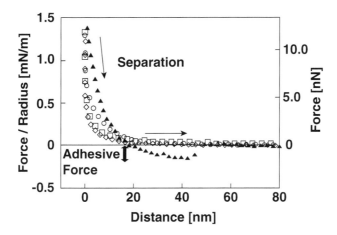

Fig. 6.9. Force profiles of interactions between subunit I and subunit II upon separation under various conditions. *Open circle* Tris buffer solution, *open square* Tris buffer solution containing Mg^{2+}, *open diamond* Tris buffer solution containing FPP, *filled triangle* Tris buffer solution containing $Mg2^{2+}$ and FPP, *plus symbol* Tris buffer solution containing Mg^{+2} and IPP

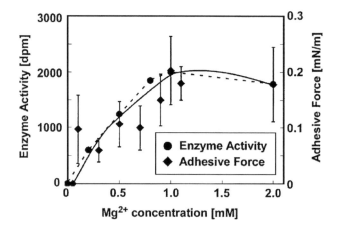

Fig. 6.10. The average adhesive force and the enzyme activity between subunits I and II vs. the Mg^{2+} concentration. *Filled circle* enzyme activity in the 25 mM Tris buffer solution containing 25 mM NH_4Cl, 10 mM 2-mercaptoethanol, 0–2.0 mM Mg^{2+}, 15 μM FPP, and 0.3 μM (1-^{14}C) IPP (1.95 TBq mol^{-1}) at pH 8.5. *Filled diamond* the average adhesive force in the 0.1 mM Tris buffer solution containing 1.0 mM NaCl, 0–2.0 mM Mg^{2+}, and 15 μM FPP at pH 8.3

and (2) there is no specific and nonspecific interaction between the identical proteins. Therefore, the observed adhesive force can be attributed to the specific interaction between subunits I and II which are associated by Mg^{2+} and FPP. Our observation well agreed with the report that a catalytically active complex is formed only in the presence of FPP and Mg^{2+} [9]. Thus, it is likely that we can form the intermediate complex by bringing two subunits into contact by AFM, and detect the adhesive force which possibly bridged the subunits by FPP, which and Mg^{2+}. This has been confirmed by the Mg^{2+} concentration dependence on adhesion force between subunit I and subunit II in the presence of FPP is similar to one for the catalytic activity of this enzyme (Fig. 6.10) [11], and by the influence of the reactant IPP and FPP.

This study has demonstrated the new approach of the force measurement for studying the elementary steps of enzyme reaction. It is possible to demonstrate the selectivity of the substrate chain length in the specific interaction between subunits I and II as described in the following section [12].

6.5.3 Selectivity in Substrate–Enzyme Complexation

The selectivity of substrate in substrate–enzyme complexation of heptaprenyl diphosphate synthase has been directly investigated using colloidal probe atomic force microscopy (AFM). The substrates studied are pyrophosphate (PPi), isopentenyl diphosphate (IPP), geranyl diphosphate (GPP), farnesyl monophosphate (FP), and farnesyl geranyl diphosphate (FGPP)

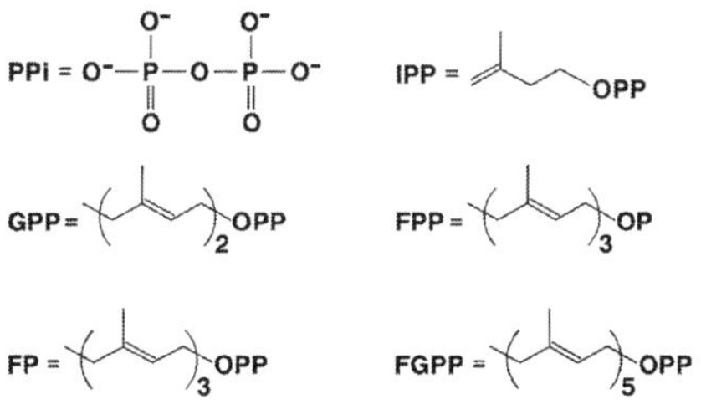

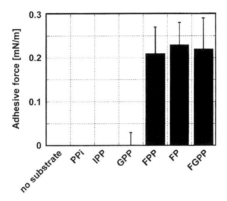

Fig. 6.11. Substrates used to examine specificity of the substrate–enzyme interaction

Fig. 6.12. Histogram of adhesion force between subunit I and subunit II for various substrates

(Fig. 6.10) [12]. No adhesion has been observed in the case of PPi, IPP, and GPP. On the other hand, the significant adhesion has been observed for phosphate derivatives which bear prenyl units longer than 3. This is in good agreement with the selectivity of the substrates by this enzyme which catalyzes the condensation reaction of four IPP molecules with FPP to give heptaprenyl (C_{35}) diphosphates [23]. Figure 6.11 shows the histogram of adhesion force between subunit I and subunit II, for various substrates. We conclude that more than 15 carbons are required for complexation of subunit I and subunit II. The strength of adhesion forces observed is almost the same as summarized in Fig. 6.12.

The selectivity observed in this study is well agreed with the function of this enzyme, which catalyzes the polymerization reaction of FPP with four IPP. During the catalytic reaction, the complex of C_{25}FGPP with subunits I and II should be formed by Mg^{2+} ion bridging and the hydrophobic interaction. It will be interesting to test how heptaprenyl diphosphate interacts with subunits. However, unfortunately, enough amount of heptaprenyl diphosphate has not been available for our experiments. It is also interesting that the

136   K. Kurihara

monophosphate is enough to form the substrate–enzyme complexes, addressing a question for a future study about an as yet unidentified role of the phosphate group.

6.6 Sequence-Dependent Interaction Between a Transcription Factor and DNA

Specific interactions between a protein, a DNA-binding site of a transcription factor, Spo0A-DB, and DNA have been studied [8]. The protein is modified with His-tag to form two-dimensionally organized layers through binding of His-tag to DSIDA–Cu^{2+}. A biotin-terminated DNA bearing a specific recognition sequence for Spo0A-DB, 5′-(TGTCGAA)$_4$-3′-biotin (DNA1), is immobilized on the surface modified with an avidin bound to a biotin-lipid monolayer. Several DNAs of which sequences are different from DNA1 in various manners have been used as references. Surface force profiles and pull-off forces between Spo0A-DB/Spo0A-DB, DNA/DNA, and Spo0A-DB/DNA layers are measured in aqueous solutions (0.1 mM Tris–HCl, 1.0 mM NaCl) at various pHs. Spo0A-DB and DNA1 exhibit the specific attraction and adhesion that are weakened when the DNA sequence is modified. It is possible to detect changes in the interaction caused by modification of one nucleic acid base in DNA1.

6.7 Conclusion

We have employed the colloidal probe atomic force microscopy for studying interactions involved in specific functions (reactions) of proteins and other biomolecules. The studies well demonstrate that the surface forces measurement is useful for examining various elementary steps of these functions, many of which are difficult to monitor as an individual process. For example, it is possible, for the first time, to detect the interactions involved in the complex formation of enzyme subunits, a cofactor and a substrate. These studies demonstrate a novel approach for studying complicated biological functions and/or reactions at the molecular level.

References

1. J.N. Israelachvili, *Intermolecular and Surface Forces*, 2nd edn. (Academic, London, 1991)
2. V.T. Moy, E.-L. Florin, H.E. Gaub, Science **266**, 257 (1994)
3. D.E. Leckband, J.N. Israelachvili, F.-J. Schmitt, W. Knoll, Science **255**, 1419 (1992)
4. (a) D. Tabor, R.H.S. Winterton, Proc. R. Soc. A **312**, 435 (1969); (b) J.N. Israelachvili, G.E. Adams, J. Chem. Soc.: Faraday Trans. I **74**, 975 (1978)

5. W.A. Ducker, T.J. Senden, R.M. Pashley, Langmuir **8**, 1831 (1992)
6. K. Kurihara, T. Abe, N. Nakashima, Langmuir **12**, 4053 (1996)
7. M. Fujita, Y. Sadaie, J. Biochem. **124**, 89 (1998)
8. Y. Shimizu, M. Fujita, R. Koitabashi, R. Ishiguro, T. Suzuki, D.Y. Sasaki, K. Kurihara (in preparation)
9. Y.-W. Zhang, T. Koyama, D.M. Marecak, G.D. Prestwich, Y. Maki, K. Ogura, Biochemistry **37**, 13411 (1998)
10. T. Suzuki, Y.-W. Zhang, T. Koyama, D.Y. Sasaki, K. Kurihara, Chem. Lett. **33**, 536 (2004)
11. T. Suzuki, Y.-W. Zhang, T. Koyama, D.Y. Sasaki, K. Kurihara, J. Am. Chem. Soc. **128**, 15209 (2006)
12. T. Suzuki, T. Koyama, K. Kurihara, Colloid Polym. Sci. **286**, 107 (2008)
13. K. Kurihara, Adv. Colloid Interface Sci. **71–72**, 243 (1997)
14. K. Kurihara, T. Kunitake, N. Higashi, M. Niwa, Thin Solid Films **210/211**, 681 (1992)
15. H. Okusa, K. Kurihara, T. Kunitake, Langmuir **10**, 3577 (1994)
16. M. Mizukami, K. Kurihara, Progr. Colloid Polym. Sci. **106**, 266 (1997)
17. K. Kurihara, T. Kunitake, J. Am. Chem. Soc. **115**, 10927 (1992)
18. R. Ishiguro, D.Y. Sasaki, C. Pacheco, K. Kurihara, Colloids Surfaces A **146**, 329 (1999)
19. D.R. Shnek, D.W. Pack, D.Y. Sasaki, F.H. Arnold, Langmuir **10**, 2382 (1994)
20. K. Ng, D.W. Pack, D.Y. Sasaki, F.H. Arnold, Langmuir **11**, 4048 (1995)
21. K. Ueno, *Chelate Chemistry* (Nankodo, Tokyo, 1969)
22. K. Kurihara, T. Kunitake, N. Higashi, M. Niwa, Langmuir **8**, 2087 (1992)
23. Y.-W. Zhang, X.-Y. Li, H. Sugawara, T. Koyama, Biochemistry **38**, 14638 (1999)

Part III

Nanohybridized Fine Porous Materials

7

Organic–Inorganic Hybrid Mesoporous Silica

Satoru Fujita, Mahendra P. Kapoor, and Shinji Inagaki

7.1 Introduction

Surfactant-mediated self-assembly approach has played a important role in materials science since the discovery of the well-ordered mesoporous silica, M41S [1,2], and FSM-16 [3,4]. Numerous researchers reported the successful synthesis of mesoporous silicas with hexagonal and cubic symmetries by the use of surfactant micellar structures as template or organizing agent. These materials exhibit a large specific surface area and narrow pore size distribution ranging from 2 to 10 nm. Today, a large number of mesoporous materials of varying composition, pore size, and wall thickness are available, derived using anionic, cationic, neutral, Gemini, triblock copolymers and oligomers under a wide range of acidic and basic conditions. Mesoporous materials are attractive as covalent scaffolding hosts for many applications and provide an excellent opportunity for organosilicon chemistry by incorporating organic components into mesoporous silica. The integration of organic functionality within these mesoporous solids has greatly expanded and advanced their applications such as adsorption, catalysis, chromatography, gas storage, optical and electrical devices, and so on. Mesoporous organic–inorganic hybrid materials are synthesized in three ways: by grafting organic unit on the silica surface in the channel, by direct synthesis from a mixture of monosilylated organic precursor [R–Si(OR′)$_3$] and pure silica precursor [Si(OR′)$_4$], and by direct synthesis from polysilylated organic precursor such as bridged organosilane [(R′O)$_3$Si–R–Si(OR′)$_3$]. A polysilylated organic compound is an excellent precursor because 100% of polysilylated precursor produces highly ordered mesoporous material, while monosilylated precursor needs to be mixed with large amount of pure silica precursor (>75%) to produce ordered mesoporous material. Moreover, a polysilylated organic compound has a strong nature of self-organization which is a great advantage for the formation of mesoporous organosilica with crystal-like pore wall structure.

This chapter describes the preparation, characterization, properties, and potential applications of mesoporous organic–inorganic hybrid materials.

Especially, the emphasis is on the preparation and functionalization of PMOs. Furthermore, the main focus will be on PMOs with molecular-level periodicity and its applications.

7.2 Postsynthetic Grafting Methods

Grafting generally refers to postsynthesis of a prefabricated mesoporous silica support by attachment of functional organic moieties to the surface of the mesopores (Fig. 7.1a). Mesoporous silica posses surface silanol (SiOH) which can be presented in high concentration and act as anchoring points for organic functionalization. Surface modification with organic units is most commonly employed by silylation. During silylation, free (SiOH) and geminal (Si(OH)$_2$) silanol groups are modified; the original structure of the mesoporous support is generally maintained.

Commonly, grafting which only is referred as silylation can be performed using silane coupling agents such as alkoxysilanes, chlorosilanes, silylamines, and disilazanes [5–11]. The primary modification via postsynthesis grafting usually involves the reaction of an organoalkoxysilane with the silanol functions of the pore walls.

Furthermore, Jaroniec and coworkers [12] used trimethylsilyl, butyldimethylsilyl, octadimethylsilyl, 3-aminopropylsilyl, and octylsilyl compounds for surface modification of MCM-41 (pore size ~ 5 nm). Lim and Stein [13] revealed the postsynthesis modification with vinyl ligand. Later, the disilazane compounds were reported for higher degree of silylation of MCM-41 materials. Shephard and coworkers [14] adopted the two stages silylation where they were able to selectively functionalize the internal surface with 3-aminopropylsilane. Tatsumi's group [15] adopted an interesting approach leading to methyl group covalently bonded to surface silicon species. Their strategy involved the esterification of the surface hydroxyl groups with butanol (Si–OBt) followed by

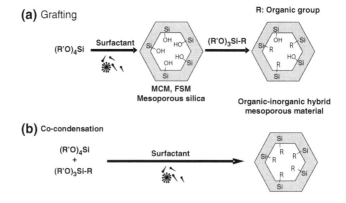

Fig. 7.1. Procedures of grafting and cocondensation syntheses

the Grignard reaction to form Si–CH$_3$ species. Jaenicke et al. [16, 17] prepared NH$_2$ group functionalized nanoporous silicas via direct grafting of 3-aminopropyltriethoxysilane precursor, while Choudary and coworkers [18] used trimethoxysilylpropylenediamine into MCM-41. Lin et al. [19] revealed that surfactant-silyl exchange process is greatly favored when applied to mesoporous silica derived under acidic conditions. It was explained due to the weaker interaction between the surfactant and the silica walls compared to nanoporous materials obtained under basic conditions.

The other approaches are the secondary surface modification that is the transformation of already exist functionalities by additional treatments. For example, Cauvel et al. [20] also incorporated the basic functional groups via primary and secondary modifications in two stages wherein they first grafted the chloropropyltriethoxysilane and then in refluxing in the presence of piperidine in toluene. The residual OH group was neutralized by reaction with hexamethyldisilazane either before or after contacting with the secondary amine. Likewise, Rao et al. [21] disclosed the two-step modifications of internal surface of MCM-41 using 3-trimethoxysilylpropoxymethyloxirane followed by the reaction of an epoxy group with 1,5,7-triazabicyclo{4,4,0}-dec-5-ene under very mild conditions. In other modifications, the MCM-41 materials were also functionalized with acidic species. The most studied example is the grafting of Brönsted propylsulfonic acid species on MCM-41 silica surface. The route adopted was the surface coating using 3-mercaptopropyltriethoxysilane followed by the controlled oxidation in the presence of hydrogen peroxide. Recently, numerous studies about photochemistry using mesoporous silica have been reported. For instance, Mal et al. [22, 23] successfully created the photochemically controlled system for compound uptake and release by anchoring coumarin to the pore openings of MCM-41 silica. Tourné-Péteilh et al. [24] constructed another potential active compound transport system by anchoring ibuprofen to the MCM-41. Rodman et al. [25] developed an optical sensor based on mesoporous silica for the quantitative analysis of Cu^{2+} ions in aqueous solutions. This sensor relies on the formation of a copper tetraamine complex on diffusion of Cu^{2+} into the pores. More recently, Zhu and Fujiwara [26] reported the fully controlled storage and release system by installing two photomechanical units which behave as stirrer and gate functions into coumarin-azobenzene-modified MCM-41.

7.3 Direct Synthesis from Monosilylated Organic Precursors

The above grafting procedures tend to create mesoporous organosilicas with nonuniform distribution of organic units and result in low organic loading, particularly when the higher loading of the organic is in demand. However, cocondensation directed templated synthesis, one-pot synthesis, results in uniform distribution of organic moieties in the pore walls (Fig. 7.1b).

General preparation of cocondensation uses monosilylated organic precursor, (R'O)$_3$SiR, with tetraalkoxysilanes, (R'O)$_4$Si, in the presence of directed template leading to mesoporous materials with organic residues anchored covalently to the pore walls. Cocondensation procedures in template-assisted synthesis were first introduced by Burkett et al. [27]. They prepared phenyl containing MCM-41 materials. Later, Macquarrie [28] detailed the cocondensation of siloxane and organosiloxane precursor by combining the sol-gel technique and the supramolecular templating technique to generate ordered mesoporous silica-based nanocomposites in a single step. Mesoporous silica using the nonionic amine route modified with cyanoethyl species was reported. The incorporation of the phenyl, allyl, aminopropyl, and mercaptopropyl functional groups was also attempted. The hexagonal phases synthesized showed the incorporation of up to 20 mol% of organic functionality into the order silica framework without any disruption of the long-range mesoscopic order. Pinnavaia's group [29–32] applied the one-step preparation procedure of surface-functionalized mesoporous organosilicas to self-assembly of hexagonal mesoporous silica using nonionic alkylamine surfactant and different organotriethoxysilanes such as 3-mercaptopropyl-, octyl-, phenyl-, butyl-, propyl-, and ethyltriethoxysilane. Two methods were described for syntheses: first, the substitution of structure-directing agent involving the partial replacement of equal molars, x of moles of TEOS and alkylamine template with x mole of organosilane precursor. This method was found to be useful for incorporating organic species whose molecular size is comparable to the alkylamine template. The second method was the direct addition of organosilane precursor into alkylamine template and TEOS mixture while the same ratio was maintained necessary to synthesize unfunctionalized mesoporous silicas. The later method was found to be suitable for the inclusion of organic moieties smaller than the alkylamine template.

Especially, there were numerous studies about thiol functional groups anchored mesoporous organosilica [33–42]. A thiol functional group anchored mesoporous organosilica was first prepared via cocondensation of TEOS or TMOS and 3-mercaptopropyltrimethoxysilane (MPTMS) precursors. The anchored mercaptopropyl groups were then oxidized to sulfonic acid functionality using nitric acid or hydrogen peroxide (H$_2$O$_2$) (Fig. 7.2) [33–36].

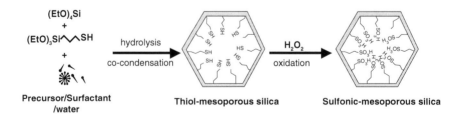

Fig. 7.2. Synthesis of sulfonic acid-functionalized mesoporous organosilicas

7 Organic–Inorganic Hybrid Mesoporous Silica   145

Diaz et al. [37, 38] summarized the one-pot synthesis functionalization of MCM-41 mesoporous materials using 3-mercaptopropyltrimethoxysilane under wide range of synthesis parameters. They also presented the details of characterization of the resultant functionalized material and found that the average distance between the organic group and the surface was about 5–6 Å, whereas the average density of the organic functional groups was estimated to be 3–4 organic moieties per $100\,Å^2$. Stucky et al. [39] combined the cocondensation step and subsequent oxidation step to produce sulfonic acid-functionalized organosilica. They used 3-mercaptopropyltrimethoxysilane in the presence of hydrogen peroxide and prepare the sulfonic acid bearing SBA-15-type mesoporous silica. Later, they have applied the similar strategy for simultaneous incorporation of multiple surface functionalities. In one of the variation, they used the mixture of 3-mercaptopropyltrimethoxysilane and benzyltrimethoxysilane precursors in the presence of TEOS, pluronic P123 triblock copolymer, and hydrogen peroxide to produce SBA-15 mesoporous materials with two functional groups. Later in other development, they have used methyltriethoxysilane instead of benzyltrimethoxysilane for variations. From the results, it was also concluded that in situ oxidation of thiol functional group to sulfonic acid functionality by hydrogen peroxide was more efficient compared to the postsynthesis oxidation. The efficiency was increased to 100% compared to 30–77% in the later case. In addition, the materials showed the considerable higher acid exchange capacity $(H^+ g^{-1})$ of $\sim 2.0\,mmol\,g^{-1}$ of silica.

Therefore, the direct one-pot synthesis route provides a more regular distribution of organic groups inside the channel pores than by grafting and allows the control of the organic content. However, the extent of loading is limited as the presence of the cocondensate disturbs the periodicity and regularity of the forming pores. In this context, several organofunctionalized mesostructures have been prepared through direct assembly pathways, as well as through grafting reactions of preassembled frameworks using different organosilane precursor [35–42]. In another precursor, Lim et al. [43,44] reported the vinyl-functionalized organosilicas via one-step condensation and used the bromination reaction to confirm the occurrence and accessibility of vinyl functional groups. Furthermore, functionalized mesoporous silicas with more complex organic groups have been obtained by cocondensation reactions [45–51]. Corriu et al. [45] anchored chelating cyclam molecules by substitution of the chlorine atoms on previously synthesized 3-chloropropyl-functionalized silicas, and showed that almost all cyclam units were localized on the pore surface and were freely accessible to complexation by Cu^{2+} and Co^{2+} ions. Jia et al. [46] synthesized functionalized mesoporous silica with the chelate ligand 3-(2-pyridyl)-1-pyrazolylacetamide. Huq et al. [47] were successful in functionalizing mesoporous silica by coupling the cyclodextrin units to 3-aminopropyltriethoxysilane and cocondensed with TEOS. Liu et al. [48] synthesized functionalized silica with calixarene amide. Mann et al. [49,50] synthesized 3-(2,4-dinitrophenylamino)propyl-modified MCM-41.

Brinker's group [51] synthesized nanocomposite films bearing photosensitive azobenzene units of 4-(3-triethoxysilylpropylureido)azobenzene, obtained by the coupling of triethoxysilylpropylisocyanate with 4-phenylazoaniline in the cocondensation reaction.

7.4 Direct Synthesis from Polysilylated Organic Precursors

The modification of mesoporous silica materials with organic moieties can be achieved in aforementioned two methods: grafting and direct synthesis by cocondensation. However, a major drawback of these two techniques is that only a fraction of the inorganic matrix can be organically modified. Typically, not more than 25% (R'O$_3$)Si–R can be used in a cocondensation process, whereas the rest of the material network is formed with Si(OR')$_4$ as the source, resulting in inorganic SiO$_2$. Furthermore, monosilylated organic precursors have often yielded materials with a lower degree of structural ordering because of sharing of the same confined channels by organic functional moieties that tend to perturb the micellar template structure. However, the recent development of the direct synthesis of periodic mesoporous organosilicas (PMOs) from polysilylated organic precursors seems to be a solution to the above these problems. PMOs are, in particular, obtained by using organosilanes containing bridged functional groups, and have a highly ordered mesostructure with well-defined hexagonal and cubic structures, uniform pore size distribution, and high surface area. The bis-silylated organic precursors yield mesoporous materials, in which the organic components are distributed homogeneously in the framework at a molecular level (Fig. 7.3).

In the present section, we review the synthesis of various PMO materials, and functionalized PMO. The enormous choice of polysilylated molecular precursors each having a different organic linker "R" provides a broad range of capabilities to precisely control the unique surface properties without disturbing the periodic pore structure. A diversity of organic spacers have been successfully employed in the synthesis of mesoporous organosilicas using a variety of surfactants in basic, acidic, or neutral media, and various chemical, electrical, and optical functionalities have been integrated within the

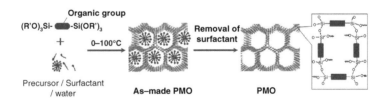

Fig. 7.3. Synthesis of PMO from organic-bridged silsesquioxane precursor

mesoporous network. Hexagonal, cubic, and wormhole framework structures have been reported.

7.4.1 Synthesis of PMOs from 100% Polysilylated Organic Precursors

PMOs with Ethane Linker

The first studies on PMOs were conducted by three independent groups in 1999 [52–55]. The first of these was by Inagaki et al. [52], who used 1,2-bis (trimethoxysilyl)ethane as the framework precursor and octadecyltrimethylammonium chloride as surfactant under basic conditions. They obtained the highly ordered 2D hexagonal phase and 3D hexagonal mesosphase depending on the synthesis temperature and alkyl chain length of the surfactants (Fig. 7.4a,b).

Nitrogen physisorption measurements revealed specific inner surface areas of 750 (2D hexagonal) and 1,170 m^2 g^{-1} (3D hexagonal) and pore diameters of 3.1 (2D hexagonal) and 2.7 nm (3D hexagonal). ^{29}Si-NMR measurements showed that the Si–C bond was not cleaved during the synthesis. Stein et al. [53] reported PMOs with ethane and ethylene groups synthesized using

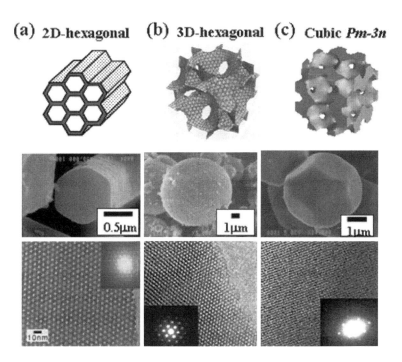

Fig. 7.4. SEM and TEM images of ethane-silica hybrid materials with (a) 2D hexagonal, (b) 3D hexagonal, and (c) cubic $Pm - 3n$

1,2-bis(triethoxysilyl)ethane and 1,2-bis(triethoxysilyl)ethylene organosilane precursors in the presence of the surfactant hexadecyltrimethylammonium. The resulting materials had a comparatively low long-range order, which was wormhole-like, rather than strictly parallel 2D hexagonally arranged pores. Further, Ozin et al. [54,55] reported the synthesis of PMOs using an ethylene-bridged precursor mixed in different ratios with TEOS in the presence of the surfactant hexadecylammonium bromide in basic media. The ethene-bridged PMO material exhibited a 2D hexagonally ordered pore system with a specific surface area of $600 \, m^2 \, g^{-1}$ and a pore diameter of 3.9 nm. That the C=C bonds were incorporated into the silica framework was confirmed by the bromination reaction [53]. Elemental analysis showed a degree of bromination (10%) relative to the C=C bond content. Further, an ethane-bridged PMO with cubic symmetry $(Pm - 3n)$, which is analogous to the SBA-1 pure silica phase, has been synthesized by Guan et al. [56] and Sayari et al. [57]. The crystal-like external morphology of these hybrid materials, as revealed by scanning electron microscopy (SEM), was described as a decaoctahedron comprising 6 squares and 12 hexagons (Fig. 7.4c). Kapoor and Inagaki [58] presented a unique route for the synthesis of ethane-bridged mesoporous silica with cubic symmetry using 1,2-bis(trimethoxysilyl)ethane and binary surfactant mixtures of octadecyltrimethylammonium chloride and the oligomeric Brij-30 surfactants $C_{12}H_{25}(EO)_4OH$ as structure-directing agents under basic conditions. Highly ordered ethane silicas with cubic $(Pm - 3n)$ symmetry akin to SBA-1 were obtained. The pore size and specific surface area were 2.8 nm and $740 \, m^2 \, g^{-1}$, respectively. The influence of the chain length of the surfactant on the synthesis of ethane-bridged PMOs was researched by Hamoudi et al. [59]. The length of the alkyl chain in the surfactant varied between C_{10} and C_{18}. The pore size increased with increasing chain length of surfactant. Liang et al. first described the synthesis of ethane-bridged organosilica using a binary mixture of Gemini (C_{18-3-1}) and cationic (C_{16}) surfactants as template. The resultant materials exhibited a relatively low structural order and poor long-range order with a broad low-angle Bragg reflection [60]. The synthesis of ethane-bridged PMOs could also be achieved under acidic conditions. The reaction pathway was different in this case. The "$S^+X^-I^+$"-route was adopted in acidic conditions in contrast to the usual "S^+I^-"-route under basic conditions. However, the acid-based approach produced poorly ordered mesoporous materials. Ren et al. [61] used 1,2-bis(triethoxysilyl)ethane as precursor in the presence of cetylpyridinium bromide as structure director. Others have described in detail the synthesis of average pore ethane-linked mesoporous organosilicas under acidic conditions in the presence of biodegradable Brij-56 [$(EO)_{10}C_{16}H_{33}$] and Brij-76 [$(EO)_{10}C_{18}H_{37}$] oligomers as structure directors [62,63]. The material showed a long-range ordered pore system with hexagonal symmetry and a pore diameter of 4.5 nm. Burleigh et al. [64–66] showed the synthesis of ethane-bridged organosilica using both Brij-56 and Brij-76 surfactants under a wide range of acid concentrations. The structural ordering of the resultant ethane mesoporous organosilica was higher with Brij-76 (pore diameter 4.2–4.5 nm) than with Brij-56 (pore diameter 3.6–3.9 nm), while synthesis performed in the presence of oligomeric surfactants with larger head groups such as Brij-58 and Brij-78 resulted in disordered materials with wide pore size distributions.

Fröba et al. [67] used a neutral pluronic P123 triblock copolymer as structure-directing agent and 1,2-bis(trimethoxysilyl)ethane as organosilica source under acidic conditions. Their material exhibited hexagonal symmetry akin to SBA-15 silica with large pores (6.5 nm). A well-ordered PMO with large, cage-like pores (12 nm) was obtained using the triblock polymer surfactant B50-6600 ($EO_{39}BO_{47}EO_{39}$) and 1,2-bis(triethoxysilyl)ethane as precursor under acidic conditions [68]. Guo et al. [69] found that the addition of a salt such as NaCl improves the degree of ordering of the materials. Cho et al. [70] achieved the synthesis of large-pore and thick pore wall (4.2 nm) ethane-bridged organosilicas with 2D hexagonal symmetry using the triblock copolymer ($EO_{16}(L_{28}G_3)EO_{16}$). The material showed enhanced thermal and hydrothermal stability and low microporosity. Meanwhile, Zhu et al. [71] synthesized enlarged-pore ethane-bridged organosilica using pre-existing lyotropic liquid crystal phases as template in the binary pluronic P123/water system to obtain monolithical mesoporous material with a specific surface area and pore size of $950 \, m^2 \, g^{-1}$ and 7.7 nm, respectively. Recently, Guo et al. [72] reported that the addition of a large amount of K_2SO_4 to a synthesis mixture containing 1,2-bis(trimethoxysilyl)ethane and pluronic F127 ($EO_{106}PO_{70}EO_{106}$) under acidic conditions affords a high-quality cubic ($Im - 3m$) ethane silica mesophase with large cavities 9.8 nm in diameter and a specific surface area of $980 \, m^2 \, g^{-1}$. Furthermore, Cho et al. [73] attempted the synthesis of large-pore ethane-bridged organosilica akin to SBA-16 by a cocondensation process using an ethane-bridged precursor (less than 10 mol%) and TEOS in the presence of the triblock copolymer F127. More recently, Djojoputro et al. [74] successfully synthesized ethane-bridged PMO hollow spheres, with diameters in the 100–1,000 nm range and a tunable wall thickness, by a dual templating approach using a fluorocarbon surfactant and cetyltrimethylammonium bromide.

PMOs with Various Organic Linkers

The most commonly studied typical organic linker group is ethane ($-CH_2-CH_2-$) because 1,2-bis(trialkoxysilyl)ethane is a widely available silsesquioxane precursor. However, the ethane linker group has limited chemical functionality. Thus, other linker groups have been used to form bis-silyl hybrid mesophases: methylene ($-CH_2-$), ethenylene ($-CH=CH-$), butylene ($-CH_2CH_2CH_2CH_2-$), phenylene ($-C_6H_4-$), thiophenylene ($-C_4H_2S-$), biphenylylene ($-C_6H_4-C_6H_4-$), tolyl, xylyl, dimethoxyphenyl, and vinyl. An overview of the structures of silsesquioxane precursors that were successfully used to create PMOs is presented in Scheme 7.1.

For example, Asefa et al. [75] reported the synthesis of the shortest organic-bonded periodic mesoporous materials with 2D hexagonal mesophases

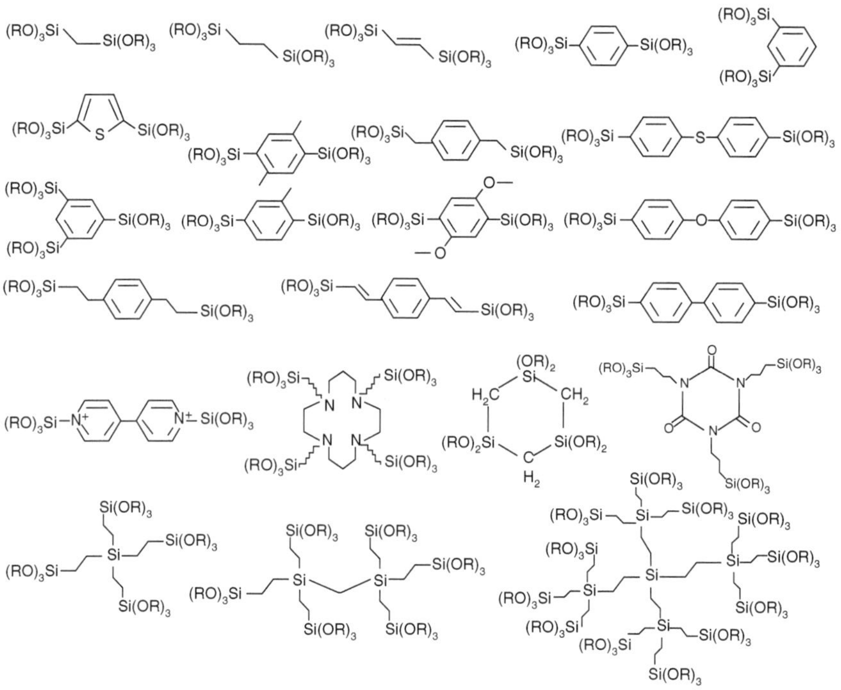

150   S. Fujita et al.

Scheme 7.1. Representative polysilsesquioxane precursors used for the synthesis of PMOs

containing methylene linker, which is isoelectronic with the oxygen atoms in MCM-41 materials. The synthesis of ethenylene-bridged mesoporous organosilica was also reported, showing the wide range of opportunities for further surface modifications based on olefin chemistry. Melde et al. [53] synthesized mesoporous ethylene silica with 2.4 nm wormhole-like channels, and pure ethylene silica mesoporous solids having 4 nm pores with hexagonal symmetry were synthesized in basic conditions. Wang et al. [76] prepared high-quality ethenylene silica mesophases under acidic conditions with varying pore sizes using oligomeric (4–5 nm) and triblock copolymer (8–9 nm) surfactant templates. They have found that addition of butanol to the polymeric reaction mixture improved the structural order and narrowed the pore size distributions. Burleigh et al. [66] reported a synthesis using a ethenylene-bridged precursor and Brij-76 surfactant under acidic conditions, while Nakajima et al. [77–79] detailed the synthesis of ethenylene-bridged mesoporous silicas with large pores using P123. Recently, Zhou et al. [80] successfully synthesized highly ordered ethylene-bridged PMO (face-centered cubic $Fm-3m$) with an ultra-large pore size of up to 14.7 nm using triblock copolymer F127 as template and 1,3,5-trimethylbenzene as swelling agent.

A new structural class of macromolecules, dendritic polymers, has attracted the attention of the scientific community. These nanometer-sized, polymeric systems are hyperbranched materials having compact hydrodynamic volumes in solution and high surface functional group content. This unique combination of properties makes them ideal candidates for nanotechnology applications in both biological and materials sciences. Landskron and Ozin [81] reported dendritic mesoporous organosilica compositions from the self-assembly of various dendrimer building blocks under ionic and nonionic surfactant routes using octadecyltrimethylammonium chloride or triblock copolymer surfactant templates. This was clearly an application of the structural Aufbau principle to materials science technology; the presence of intact Si–C$_4$ building blocks was confirmed by ^{29}Si MAS NMR.

In an analogous PMO group, a variety of nonsilica-based mesoporous materials have been successfully synthesized using phosphonic acid in the presence of surfactants. Surface functionalization techniques using organophosphorus coupling molecules have been established by Kimura et al. [82–84] and Mutin et al. [85]. Ordered mesoporous aluminum organophosphonates (AOPs) were obtained using aluminum chloride, methylene, ethylene, and propylene phosphonates, and the triblock copolymers P123, F68, and F127 as surfactant.

Further, several attempts have been made for the synthesis of aromatic-bridged PMO to introduce an enhanced functionality in mesoporous materials. Ozin et al. [55] described the synthesis of phenylene-bridged mesophases using cetylpyridinium chloride as a structure director and 1,4-bis(triethoxysilyl) benzene precursor in an acidic medium. The pore diameter was 2.0 nm with a very high surface area $(1,360\,\mathrm{m^2\,g^{-1}})$, although Si–C bond cleavage could not be completely avoided. They also reported mesophases from 1,4-bis(triethoxysilyl)-2-methyl-benzene, 1,4-bis(triethoxysilyl)-2,5-dimethyl-benzene, and 1,4-bis(triethoxysilyl)-2,5-dimethoxy-benzene precursors using cetylpyridinium chloride as surfactant under acidic conditions and also by adding ammonium fluoride as catalyst after neutralization [86]. Further, their group reported the synthesis of mesoporous 1,3,5-phenylene-silica with three-point attachment [87]. Burleigh et al. [65] and Sayari et al. [62, 88] reported a similar synthesis using benzene-bridged precursors in the presence of Brij-type surfactants. Hunks and Ozin [89] synthesized 4-phenylether and 4-phenylsulfide-bridged periodic mesoporous silicas using bis-4-(triethoxysilyl) phenylether and bis-4-(triethoxysilyl)phenylsulfide precursors in the presence of the oligomeric surfactant Brij-76. The addition of a small amount of NaCl salt assisted the interaction between hydrophilic head groups of the oligomeric template and the inorganic species. The materials showed a single low-angle peak in the XRD patterns. When this was combined with the TEM results, it was concluded that the material had a wormhole-type structure. The channel wall thickness was approximately 3.0 nm, while the pore diameter ranged from 2 to 3 nm. The specific surface area was $630\,\mathrm{m^2\,g^{-1}}$. Similarly, periodic mesoporous structures with aryl-methylene bridging groups could be obtained from the corresponding 1,4-(triethoxysilyl)(CH$_2$)$_n$C$_6$H$_4$ ($n = 1$ or 2) precursors in the presence of the surfactant Brij-56 under acid conditions.

In addition to the aforementioned aromatic-bridged mesoporous silicas, the synthesis of thiophene-bridged periodic mesoporous silicas was also reported. Ozin's group [55] described the synthesis using the precursor 2,5-bis (triethoxysilyl)thiophene in the presence of cetyltrimethylammonium bromide as surfactant template. The Si–C bonding was found to cleave under basic conditions, while mild acidic conditions were shown to promote stability. Recently, Morell et al. [90] synthesized very high-order thiophene-bridged periodic mesoporous materials using a 2,5-bis(triethoxysilyl)thiophene precursor in the presence of the triblock copolymer surfactant pluronic P123. The pore size ranged between 5 and 6 nm with a specific surface area of up to 550 m^2 g^{-1}. ^{29}Si NMR and Raman spectroscopy confirmed that under the conditions used (even though a highly acidic medium was used), only less than 4% of the Si–C bond was cleaved. Furthermore, Morell et al. [91] successfully synthesized highly ordered bifunctional PMOs containing different amounts of aromatic thiophene and benzene groups using the triblock copolymer pluronic P123 and the oligomeric surfactant Brij-76 under acidic conditions. They employed 2,5-bis(triethoxysilyl)thiophene and 1,4-bis(triethoxysilyl)benzene as organosilica precursors. The bifunctional aromatic PMO materials exhibited a 2D hexagonal mesostructure with a pore size of 5.4 nm. Furthermore, Fröba's group [92] determined the vibrational spectroscopic features of ethane-, thiophene-, benzene-, biphenyl-bridged PMO by IR and Raman spectra.

7.4.2 Synthesis of PMOs by Cocondensation

A variety of functionalized PMOs have been obtained in the last decade. In contrast to the grafting and cocondensation reactions, the organic units in PMOs are two point attached within the silica matrix through covalent bonds and are completely homogeneously distributed. PMO materials exhibit ordered pore arrangements accompanied by sharp pore size distributions since structure-directing agents are employed for their synthesis. By varying the organic moieties of the organosilica precursors, the chemical and physical properties of PMOs can be tailored to specific needs (Scheme 7.2). Because of their ordered and uniform porosity, functionalized PMOs are attractive for applications, such as catalysis, adsorption, photonics, chromatography, electrochemiluminescence, gas storage, and host–guest chemistry.

Multifunctionalized PMOs

Ozin's group [93] reported bifunctionalized PMOs with bridging ethyl groups in their walls and vinyl groups protruding into their channels. Burleigh et al. [94] obtained bifunctionalized PMOs having both ethane and benzene bridging groups via cocondensation of the corresponding ethane- and

7 Organic–Inorganic Hybrid Mesoporous Silica 153

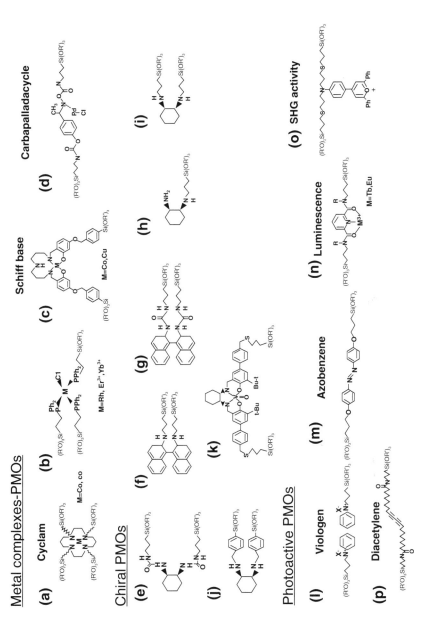

Scheme 7.2. Precursors of functionalized PMOs

benzene-bridged monomer precursors in the presence of the surfactant Brij-76. Depending on the molar ratio of the two monomer precursors in the initial reaction gel, a well-ordered mesoporous material with hexagonal symmetry could be formed. Adjustment of the hydrolysis and condensation rates of individual precursors was rather difficult; however, elemental analysis revealed that the amount of benzene moieties was higher than that of ethane moieties in the resultant mesoporous organosilicas. Furthermore, the possibility of the formation of multifunctional PMOs having four bridging moieties such as methane, ethane, ethylene, and benzene was explored, and a material with a somewhat low order was obtained. In addition, materials combining the two aromatic bridging moieties were synthesized using phenylene and thiophene bridging monomer precursors. Corriu et al. [95] synthesized a bifunctional mesoporous material with an acidic ($-SO_3H$) framework and basic ($-NH_2$) channel pores by cocondensation of a ternary mixture of bistrimethoxysilyl-4,5-dithiooctane $(MeO)_3-(CH_2)_3-S-S-(CH_2)_3-Si(OMe)_3$, TEOS, and 3-tertbutyloxycarbonyl(aminopropyl)trimethoxysilane $(MeO)_3Si-(CH_2)_3-NHBoc$ in the presence of P123. The material exhibited a worm-like structure with a specific surface area of $560\,m^2\,g^{-1}$ and a pore size of 8.9 nm. Furthermore, they reported the direct synthesis of bifunctional mesoporous organosilicas containing chelating groups in the framework and reactive functional groups ($-CN$, $-SH$, $-Cl$) in the channel pores. The bridged organosilicas used 1,4,8,11-tetra-kis(triethoxysilylpropyl)-1,4,8,11-tetra-aza-cyclotetradecane and the corresponding copper(II) chloride complex as chelating precursors [96].

Metal Complex: PMOs

Corriu et al. [96, 97] incorporated large chelating agents into mesoporous silica (Scheme 7.2a). They induced a cyclam derivative into the framework of silica materials through the neutral synthetic route with P123 and F127. The material obtained exhibited only a low mesoscopic order; however, it was able to bind a large concentration of the metal ions Cu^{2+} and Co^{2+}. Corriu et al. [98] also reported functionalization of mesoporous organosilicas by rare-earth complexes. A mesoporous material containing free phosphine oxide ligands in the pore walls was prepared. The phosphine oxide ligands were proved to operate as templates for Er^{3+} and Yb^{3+} ion complexation (Scheme 7.2b). Similarly, Dufaud et al. [99] synthesized SBA-3 incorporating phosphine-ligated transition metal complexes. In another attempt, Corriu [100] synthesized hybrid materials containing metal Schiff base complexes (Scheme 7.2c). Corma et al. [101] reported PMO containing a carbapalladacycle complex as heterogeneous catalyst for Suzuki cross-coupling (Scheme 7.2d). Zhu et al. [102] reported the synthesis of propylethylenediamine bridging group functionalized ethane-bridged organosilicas via cocondensation of the ethane-bridged precursor [1,2-bis(triethoxysilyl)ethane] and

Cu^{2+}-complexed precursor bis[(3-trimethoxysilyl)propyl]ethylenediamine in the presence of the triblock copolymer surfactant pluronic P123. The pore size of the resultant functionalized materials could be varied from 11 to 21 nm by changing the initial mole ratios of the two precursors (complex/ethane) from 0.1 to 0.3. The materials also allowed reversible change or substitution of Cu^{2+} ions by other ions, such as Zn^{2+}. Furthermore, functionalized PMOs have found application as absorbents. Zhang et al. [103] prepared a tetrasulfide bridging mesoporous silica, which showed a high affinity for Hg^{2+} cations as well as some affinity for other cations such as Cd^{2+}, Zn^{2+}, Pb^{2+}, and Cu^{2+}. Meanwhile, Jaroniec and Olkhovyk [104] integrated the isocyanurate group into periodic mesoporous organosilicas for high affinity to Hg^{2+} (1.8 g Hg^{2+} g^{-1} adsorbent).

Chiral PMOs

There are a few examples of PMOs with chiral moieties either in the mesopore or in the mesoporous framework prepared by the cocondensation method [105–109]. Álvaro et al. [105] synthesized chiral organic bridges into MCM-41 analogous hybrid materials by using mixtures of bis-silylated binaphthyl or cyclohexadiyl precursors (Scheme 7.2e–g) and TEOS. Li et al. [106] synthesized mesoporous ethane silicas with $trans$-$(1R, 2R)$-diaminocyclohexane in the mesopores by the cocondensation of 1,2-bis(trimethoxysilyl)ethane and chiral precursor (Scheme 7.2h). The resulting materials had an ordered mesoporous structure with a pore diameter of ~ 3.3 nm, a specific surface area of ~ 890 m^2 g^{-1} and a ligand loading of up to 0.57 mmol g^{-1}. The chiral diamino groups were located in the mesopores of PMOs, and were attached on the pore walls through the propyl linkages. These PMOs, after complexing with (Rh(cod)Cl)$_2$, exhibited $\sim 96\%$ conversion and ee values around 23% ee for the asymmetric transfer hydrogenation of acetophenone. Owing to the ethane bridges in the PMOs, the hydrophobic surface enhanced the catalytic activity. Li et al. also synthesized PMOs that incorporated the chiral diaminocyclohexane moiety by cocondensation of chiral precursors (Scheme 7.2i,j) with TMOS. These materials have highly ordered mesostructures with a pore size of ~ 2.3 nm, a specific surface area of $\sim 1,070$ m^2 g^{-1}, and ligand loading amount of up to 1.02 mmol g^{-1}. The catalyst (Rh(cod)Cl)$_2$ incorporated by the chiral diamino groups with phenyl groups as linker in the framework of the PMO exhibited 97% conversion and an ee value of $\sim 30\%$ for the asymmetric transfer hydrogenation of ketones. Baleizão et al. [109] reported that PMOs with chiral vanadyl Schiff base complex in the framework showed enantioselectivity for the asymmetric catalytic cyanosilylation of benzaldehyde (Scheme 7.2k). Recently, Polarz et al. [110] and Ide et al. [111] synthesized functionalized mesoporous ethylene silicas from 100% chiral borated ethylene-bridged disilane precursors prepared by enatioselective hydroboration.

156   S. Fujita et al.

Photoactive PMOs

Several studies have dealt with photochemical PMOs (Scheme 7.2l–p) [112–116]. García et al. [112] synthesized a PMO containing electron acceptor viologen units using the precursor 4,4′-bipyridinium (Scheme 7.2l). The ability of bipyridinium units to act as electron acceptor termini has been demonstrated by observation of the radical cation in the photochemical and thermal activation of as-made materials. Corriu et al. [113] reported a photoresponsive PMO introducing a bridged azobenzene moiety using 4,4′-[(triisopropoxysilyl)propyloxy]azobenzene as precursor (Scheme 7.2m). Minoofar et al. [114] reported the synthesis of a PMO containing luminescent molecules and verified their locations by luminescence lifetime measurement (Scheme 7.2n). Furthermore, Álvaro et al. [115] and Peng et al. [116] synthesized PMOs containing triphenylenepyrylium and diacetylene moieties, respectively (Scheme 7.2o,p). For instance, Peng et al. [116] produced a unique PMO embedded with polydiacetylene via cooperative assembly of cetyltrimethylammonium bromide and diacetylene-bridged silsesquioxane. The diacetylenic molecules spontaneously organized around the surfactant liquid crystalline structure, forming a mesoscopically ordered composite with molecularly aligned diacetylenic units. Surfactant removal followed by polymerization yielded a polydiacetylene–PMO with a 2D hexagonal mesostructure ($P6\,mm$) and a pore size of 5 nm. The reversible chromatic rapid responses to external stimuli were demonstrated by subjecting this material to thermal cycles between 20 and 103°C.

Applications of PMOs

A numerous studies of catalysis using functionalized PMO have been reported. For instance, Yang et al. [117] showed the benefits of periodic pore surface structures in which the ethane linker served to enhance selectivity and activity during esterification. Recently, the structural properties of hydrothermally stable sulfonic acid-functionalized ethane-bridged mesoporous organosilicas derived from a triblock copolymer surfactant (pluronic P123) template were presented. The positive influence of ethane bridging groups on the catalytic behavior of the materials was described [118]. In another report, the hydrolysis of sugars on ethane- and phenylene-bridged mesoporous hybrid materials was described [119]. Kapoor et al. [120] have shown that sulfonic acid-functionalized derivatives of three-dimensional ($Pm - 3n$) cubic phenylene-bridged hybrid mesoporous silica materials derived from the precursor 1,4-bis(triallyl)phenylene are effective in Friedel–Craft acylation reactions. Nakajima et al. [77–79] showed the esterification and pinacol–pinacolone rearrangement reactions using a new type of functionalized PMO, in which the ethenylene sites on the surface were modified to phenylene sulfonic acid via a Diels–Alder reaction.

Furthermore, isomorphous substitution of heteroatoms in the organosilica framework would be useful to enhance the catalytic properties. For instance,

Kapoor et al. [121–123] synthesized a new class of titanium containing hybrid silsesquioxane mesophases with integral ethane organic functionality from a single organosilane source, in which the titanium is necessarily incorporated, along with the ethane constituents, into channel wall as molecularly dispersed bridging ligands. Gold nanoparticles supported on highly hydrophobic ethane-bridged Ti-incorporated mesoporous material were reported for enhanced vapor phase epoxidation of propene using H_2 and O_2 [122]. Hughes et al. [124] recently presented the synthesis of ethane-bridged mesoporous silicas with incorporated aluminum, which could offer new prospects for the application of such hybrid silicas to acid catalysis. However, Yang et al. [125] presented the synthesis of aluminum-containing mesoporous phenylene silicas with crystal-like pore walls and demonstrated their application in alkylation of 2,4-di-tert-butylphenol with cinnamyl alcohol. Fukuoka et al. [126] showed the improved catalysis of PMO-embedded Pd nanowires for measuring the reaction rate of CO oxidation in the presence of excess O_2.

Recently, Hudson et al. [127] reported that chloroperoxidase was successfully immobilized on PMO. Chloroperoxidase is a versatile heme peroxidase that exhibits halogenase and peroxidative activity. The chloroperoxidase on PMO was formed using bis(3-(trimethoxysilyl)propyl)amine, trimethylbenzene, F127, monochlorodimedone, hydrogen peroxide, and so on. These materials showed significant losses in activity over 20 cycles in a low ionic strength assay.

Fukuoka et al. [128] successfully synthesized pure Pt and Rh and mixed (Pt/Rh, Pt/Pd) necklace-shaped nanowires inside an ethane-bridged PMO and confirmed their results by HRTEM and EDX. It was concluded that the mixed metallic nanowires consisted of uniform Pt/Rh alloy phases, while the monometallic Pt and Rh had a mean length of about 120 and 48 nm, respectively. Fukuoka et al. [129] also synthesized Pt nanowires, which they isolated by removing the mesoporous silica framework using a dilute aqueous HF solution. The mixed (Pt/Rh, Pt/Pd) nanowires in periodic mesoporous silica exhibit very interesting magnetic properties. The magnetic susceptibility below 90 K was approximately three times higher than expected for the simple sum of the value for bulk Pt and Pd. This was related to the low dimensionality of the metal topology.

7.5 PMOs with Crystal-Like Pore Walls

The bridge organic component in the PMOs described above is randomly arranged within the pore walls, and the chemical functionality of pore wall surfaces is limited. By contrast, the inherent potentials of electronic, optical, and sensors are enhanced in the case of molecular periodicity in the walls. In this section, several research studies about PMO materials combining an ordered mesoporous structure and molecular periodicity within pore walls will be discussed.

7.5.1 Synthesis of Various PMOs with Crystal-Like Pore Walls

The first report on PMOs with crystal-like pore walls came from Inagaki et al. [130]. They reported the first surfactant-mediated synthesis of ordered benzene-bridged hybrid mesoporous organosilicas using 1,4-bis(triethoxysilyl) benzene and alkyltrimethylammonium chloride as surfactant (Fig. 7.5).

The benzene-bridged hybrid mesoporous organosilicas showed a hexagonal array of mesoporous and crystal-like pore walls, observed as molecular periodicity in the walls along the channel directions due to the π–π stacking of benzene bridging groups. The resulting material had a hexagonal array of mesopores with a lattice constant of 52.5 Å and exhibited an atomic-scale periodicity with a spacing of 7.6 Å along the channel direction within all pore walls. In addition to three peaks in the small-angle scattering regime ($2\theta < 10°$) with d-spacings of 45.5, 26.0, and 22.9 Å, the material also showed four sharp diffraction peaks at d-spacings of 7.6, 3.8, 2.5, and 1.9 Å in the medium scattering angle region ($2\theta = 10$–$50°$). The higher-order reflections with a spacing of 7.6 Å, the TEM images of stacked lattice fringes with a uniform basal spacing of 7.6 Å, along with the corresponding electron diffraction spot confirmed the atomic-scale periodicity in the pore walls (Fig. 7.6a). The BET-specific surface area and pore size were $818 \text{ m}^2 \text{ g}^{-1}$ and 3.8 nm, respectively. Figure 7.6b shows the pore surface structure of mesoporous phenylene-silica with crystal-like pore walls.

Benzene rings are aligned in a circle around the pore, fixed at both sides by silicate chains. The silicate chains are terminated by silanol (Si–OH) at the surface. Hydrophobic benzene layers and hydrophilic silicate layers are arranged alternately at an interval of 7.6 Å along the channel direction. The periodically arranged hydrophobic–hydrophilic surface is highly suitable for use as catalyst and host material for inclusion chemistry because it could enable structural orientation of guest molecules or clusters enclosed in the pores. Later, Bion et al. [131] reported the synthesis of benzene-bridged hybrid mesoporous solids having crystal-like pore walls with varying pore diameters by using C_{14} to C_{18} trimethylammonium halide surfactants under basic conditions. The pore diameters ranged from 2.3 to 2.9 nm depending on the alkyl chain length of the respective surfactant. Onida et al. [132, 133] conducted

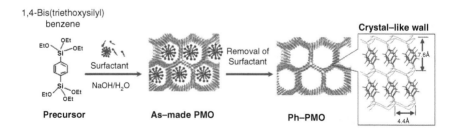

Fig 7.5. Synthesis of mesoporous organosilica with crystal-like pore wall structure

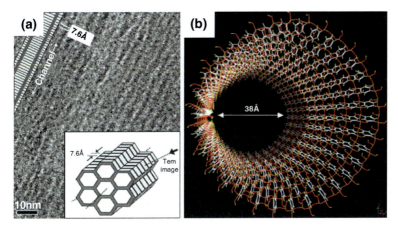

Fig 7.6. (a) TEM image of molecular-scale periodicity in the pore walls. (b) Simulated image of the crystal-like pore walls structure of surfactant-free mesoporous phenylene-silica with crystal-like pore walls

an infrared and ab initio molecular modeling study to deduce the surface properties of mesoporous phenylene-silica hybrid materials with crystal-like pore walls. Goto et al. [134] reported the postsynthesis treatment for the formation of molecular-scale periodicity within the pore walls of mesoporous organosilica derived using the triblock copolymer surfactant pluronic P123. Since the discovery of this first benzene-bridged mesoporous hybrid material with crystal-like pore walls, the number of mesoporous hybrid materials exhibiting molecular-scale periodicity have increased. Kapoor et al. [135]. have also demonstrated a similar arrangement using 1,3-bis(triethoxysilyl)benzene, a nonlinear symmetrically bridged organosilica precursor, in a synthesis route similar to that described by Inagaki et al. [130]. This material also showed similar atomic-scale periodicity at 7.6 Å, as well as its higher-order diffractions (Fig. 7.7).

The series of PMOs with crystal-like pore walls were researched using organosilica precursors by Mokaya et al. [136], Sayari and Wang [137], and Fröba et al. [138]. Moyaka et al. [136] synthesized an ethylene-containing PMO with crystal-like pore walls using 1,2-bis(triethoxysilyl)ethylene and cetyltrimethylammonium bromide as template under basic conditions. This material exhibited a molecular-level periodicity because of the lamellar ordering of the ethylene groups with a basal spacing of 5.6 Å, and a high surface area $(1,300\,m^2\,g^{-1})$ with a pore size of 4 nm. Sayari and Wang [137] used 1,4-bis(triethoxysilylethane-2-vinyl)benzene as precursor and alkyltrimethylammonium chloride as surfactant under basic conditions to obtain 2D hexagonal 1,4-vinylbenzene PMO materials. Fröba et al. [138] synthesized the same vinylbenzene-bridged PMO with a pore size of 2.7 nm and specific surface area of $800\,m^2\,g^{-1}$. Another mesoporous hybrid system that showed

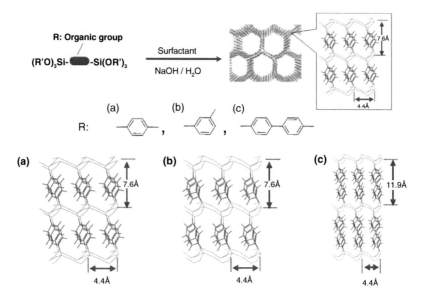

Fig 7.7. Simulated structure models of pore walls for mesoporous (**a**) 1,4-phenylene-silica, (**b**) 1,3-phenylene-silica, and (**c**) 4,4-biphenylylene-silica

periodically ordered mesopores as well as molecular-scale periodicity in the entire pore wall region was synthesized by Kapoor et al. [139] using 4,4′-bis(triethoxysilyl)biphenyl as organosilica precursor, in which hydrophilic silicate layers alternated with hydrophobic biphenylene layers (Fig. 7.7). These results clearly demonstrated that such a unique arrangement in the pore walls can also be obtained in other hierarchically ordered mesoporous solids by changing the nature of the organic linkers. The material showed a molecular-scale pore surface periodicity along the channel direction with a basal spacing of 11.6 Å. The pore diameter and BET surface area were determined to be 3.5 nm and 869 m^2 g^{-1}, respectively. Furthermore, Morell et al. [140] reported the in situ synchrotron SAXS/XRD measurement of the formation of an ordered mesoscopic hybrid biphenyl-bridged PMO with crystal-like pore walls. They demonstrated the formation of the mesostructure and the periodicity within pore walls.

In another attempt, Inagaki et al. [141] reported the synthesis of lamellar mesophases of phenylene- and biphenylylene-silica composites with periodicity within the silicate layers from mixtures of bridged organosilane precursors and surfactants at room temperature. Figure 7.8 shows a representative structural model of lamellar mesophase phenylene-silica hybrid materials. Interestingly, the phenylene-silica layers of the lamellar mesophase include a 4.2 Å periodicity that is different from the 7.6 Å periodicity observed in the pore walls of crystal-like mesoporous phenylene silicas. A periodicity of 4.2 Å, which is very close to the phenylene–phenylene distance (4.4 Å), was also observed in

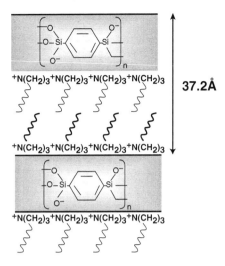

Fig 7.8. Representative structural models of the lamellar mesophase phenylene-silica hybrid material

the simulated model of the 2D hexagonal mesophase of phenylene-silica. It was shown that the periodic lamellar mesophases within the layers can also be obtained by refluxing the initial mixture at 55 and 75°C, indicating that stable lamellar mesophases are likely to be formed under a wide range of condensation temperatures. In addition, the intercalation of a biphenylylene-silica mesophase with toluene also demonstrated the uniquely flexible features of these expandable interlayered materials (Fig. 7.8). The diffraction peaks at a d-spacing of 30.0 and 14.9 Å confirm the lamellar lattice in the material. The 4.2 Å periodicity observed in the lamellar biphenylylene-silica mesophase was relatively lower compared to the lamellar phenylene-silica mesophase. The XRD peaks at a d-spacing of 46.7 and 23.4 Å indicate the swelling of the interlayer region of the lamellar hybrid material. This study is the first report on a lamellar mesophase organosilica hybrid with a crystal-like sheet structure. Its potential applications are eagerly anticipated.

7.5.2 Synthesis of PMO with Crystal-Like Pore Walls Obtained from Allyl Precursors

A wide range of synthetic routes to PMOs with different mesophases and morphologies have been reported using organosilica precursors, which consist of trialkoxy groups such as bridged organosilicon molecules $(R'O)_3Si–R–Si(OR')_3$. However, the range of suitable alkoxysilane precursors is limited because alkoxysilane precursors are difficult to obtain in high purity owing to the limitations in distillation and chromatographic separation, namely that alkoxysilane precursors are highly reactive toward hydrolysis, rendering the

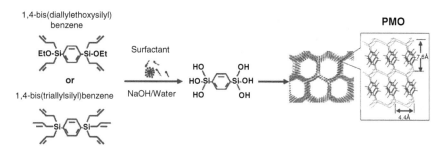

Fig 7.9. Alternate route for the synthesis of mesoporous phenylene silicas from allylorganosilane precursors

silicon compounds difficult to handle under hydrolytic conditions and during purification by silica gel chromatography. For example, the precursor 1,4-bis(diallylethoxysilyl)benzene is preferable to 1,4-bis(triethoxysilyl)benzene in terms of their handling and purification during synthesis. Shimada et al. [142, 143] have demonstrated a new method for functionalizing the surface of mesoporous silica using allylorganosilane, which is stable under regular hydrolytic conditions. Recently, Inagaki's group [144] has presented the preparation of a new family of bridged allylorganosilane precursors that, upon surfactant-assisted assembly under basic conditions, afford ordered mesoporous organosilica having crystal-like pore walls with molecular-scale periodicity (Fig. 7.9).

The XRD patterns of the as-synthesized benzene-silica hybrid mesoporous material derived from 1,4-bis(diallylethoxysilyl)benzene show a d_{100} reflection peak at 41.6 Å and sharp peaks at d = 7.6, 3.8, and 2.5 Å at intermediate scattering angles (2θ = 10–40°), indicating that the pore walls are composed of crystal-like domains with a spacing of 7.6 Å in the channel direction. The surfactant-free material also exhibited a well-defined pattern with diffraction peaks in the low-angle region (d_{100} spacing of 45.7 Å). The molecular-scale periodicity was fully retained upon surfactant removal, indicating substantial framework ordering with crystalline pore walls. The final material was identical to the materials synthesized from the alkoxy derivative of the benzene-bridged precursor [1,4-bis(triethoxysilyl)benzene], which exhibited weaker peaks related to molecular-scale periodicity [130]. The BJH pore diameter, BET surface area, and mesopore volume were 23.5 Å, 744 m^2 g^{-1}, and 0.53 cm^3 g^{-1}, respectively. Furthermore, Inagaki's group has synthesized a stable 1,4-bis(triallylsilyl)benzene precursor without any alkoxy terminal group, and attempted to form mesostructures under basic conditions. The XRD pattern clearly shows the successful formation of a mesostructure with a d_{100} spacing of 35.5 Å. The other two broad reflections at d = 9.5 and 4.4 Å were due to the molecular-scale periodicity of the pore walls. This material exhibits structural ordering, with a BET surface area of 960 m^2 g^{-1} and a BJH pore diameter of 23.2 Å. Kapoor et al. [120] have reported the synthesis of three-dimensional ($Pm - 3n$) cubic phenylene-bridged hybrid mesoporous silica material using the allylorganosilane precursor 1,4-bis(triallyl)phenylene and cetyltrimethylammonium chloride (C_{16}TMACl) as structure-directing agent in acidic medium. The specific surface area and pore size of this material are $680\,m^2\,g^{-1}$ and $37.7\,Å$, respectively. Finally, the methodology of PMO formation using both the alkoxy derivative of the precursors and the allylorganosilane precursors themselves is unique. This achievement represents a new step toward the discovery of alternative organosilane precursors for the synthesis of mesoporous organosilica, to replace the existing precursors that are difficult to obtain in high purity.

7.5.3 Functionalization of PMO with Crystal-Like Pore Walls

The ordered mesophases with crystal-like pore walls could be fabricated with interactive bridging organic spacers inside the pore walls; these mesophases allowed modification of the organic functional groups via further chemical transformations for versatile applications [145]. For instance, Yang et al. [146] developed sulfuric acid-functionalized mesoporous phenylene-silica by cocondensation of 1,4-bis(triethoxysilyl)benzene and 3-mercaptopropyltrimethoxysilane using a cationic surfactant template in basic medium, followed by oxidative transformation of the thiol (–SH) group to (–SO$_3$H) using HNO$_3$. The resulting materials showed a molecular-scale periodicity ($7.6\,Å$) similar to that previously observed for benzene-bridged mesoporous materials. Characterization results revealed that the mercaptopropyl group (C_3H_6–SH) or its oxidized derivatives (C_3H_6–SO$_3$H) are linked to silicate layers of the hybrid mesoporous solids (Fig. 7.10). The degree of oxidation of –SH to –SO$_3$H was 41.7% and the sulfonic acid group concentration was $0.70\,meq\,g^{-1}$. The materials were found to be very reactive toward esterification of acetic acid with ethanol, and the catalytic results showed a higher conversion compared to commercially available Nafion-H [147].

Further, the synthesis of sulfuric acid-functionalized biphenyl-bridged mesoporous materials with similar characteristics is also interesting because these materials have an equimolar ratio of phenylene to silica, which would enhance hydrophobicity in catalytic applications. The materials were synthesized by cocondensation of 4,4'-bis(triethoxysilyl)biphenyl precursor and 3-mercaptopropyltrimethoxysilane in basic medium and a cationic surfactant followed by an oxidation treatment similar to that described for benzene-bridged functionalized materials. The presence of –SO$_3$H groups was confirmed by acid titration. The sulfonic acid group concentration was significantly higher ($0.99\,meq\,g^{-1}$) than those obtained for benzene-bridged mesoporous materials ($0.70\,meq\,g^{-1}$) [148]. This method yields highly hydrophobic biphenyl-bridged bifunctional hybrid mesoporous solids with sulfonic acid functionalities and crystalline pore walls, which may be useful for potential applications.

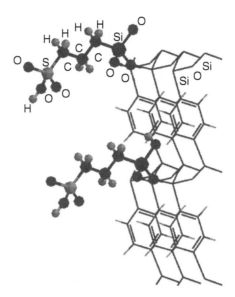

Fig 7.10. Structural image of bifunctional mesoporous benzene-silica attached with propylsulfonic acid groups on the silicate layers of ordered pore surface

Kapoor et al. [149] were able to synthesize phenylene-bridged periodic mesoporous materials (pore diameters ∼ 2.0 nm) having a spherical morphology with diameters between 0.6 and 1.0 µm under very mild basic conditions by using a dilute ethanolic ammonia solution instead of an aqueous NaOH solution with surfactants. However, the pore walls consisted of a rather poor molecular-scale periodicity compared to conventionally produced phenylene-bridged mesoporous organosilicas.

More recently, Kim et al. [150] reported hydrogen storage properties for nickel-atom-dispersed benzene-silica hybrid materials. It was demonstrated that this material provided an alternative to reversible hydrogen storage materials with high hydrogen capacities exceeding 6 wt% at moderate pressure conditions.

7.6 Summary

Organic–inorganic hybrid materials are increasingly taking their position in the free spaces between organic chemistry, inorganic chemistry, ceramics, polymer, biology, and so on. Especially, since the first synthesis of PMOs, these materials have attracted increasing research attention in materials science because these materials can organize a wide range of organic units inside silica matrices with almost no limitation. The high loading rate and uniform dispersion of organic groups in their frameworks endow PMO materials with unique advantages over pure mesoporous silica. One interesting characteristic of PMOs is that their polarity, hydrophobicity, and hydrophilicity can be tuned within a certain range by the choice of the organic component, and hence their ability to adsorb other materials can be controlled. Moreover, hybrid mesoporous materials have also shown advantages in the host–guest system and their ability to host–guest species that impart interesting properties is now being widely studied, and considerable success has already been achieved. The formation of PMOs with molecular periodicity in their walls advances the development of molecular nanotechnology and nanoscience. Pioneering research in controlling molecular self-organization would allow the use of nanostructured materials in diverse applications and nanodevices. Finally, various challenges and fundamental questions still remain; however, we feel it is well worth striving to contribute to the development of mesoporous materials.

References

1. C.T. Kresge, M.E. Leonowicz, W.J. Roth, J.C. Vartuli, J.C. Beck, Nature **359**, 710 (1992)
2. J.S. Beck, J.C. Vartuli, W.J. Roth, M.E. Leonowicz, C.T. Kresge, K.D. Schmitt, C.T.W. Chu, D.H. Olson, E.W. Sheppard, S.B. McCullen, J.B. Higgins, J.L. Schlenker, J. Am. Chem. Soc. **114**, 10834 (1992)
3. T. Yanagisawa, T. Shimizu, K. Kuroda, C. Kato, Bull. Chem. Soc. Jpn. **63**, 988 (1990)
4. S. Inagaki, Y. Fukushima, K. Kuroda, Chem. Commun. 680 (1993)
5. J.H. Clark, D.J. Macquarrie, Chem. Commun. 853 (1998)
6. R. Anwander, I. Nagl, M. Widenmeyer, G. Engelhardt, O. Groeger, C. Palm, J.T. Röser, J. Phys. Chem. B **104**, 3532 (2000)
7. K. Moller, T. Bein, Chem. Mater. **10**, 2950 (1998)
8. A. Stein, B.J. Melde, R.C. Schroden, Adv. Mater. **12**, 1403 (2000)
9. T. Maschmeyer, Curr. Opin. Solid State Mater. Sci. **3**, 71 (1998)
10. X.S. Zhao, G.Q. Lu, J. Phys. Chem. B **102**, 1556 (1998)
11. R. Anwander, C. Palm, J. Stelzer, O. Groeger, G. Engelhardt, Stud. Surf. Sci. Catal. **117**, 135 (1998)
12. C.P. Jaroniec, M. Kruk, M. Jaroniec, A. Sayari, J. Phys. Chem. B **102**, 5503 (1998)
13. M.H. Lim, A. Stein, Chem. Mater. **11**, 3285 (1999)
14. D.S. Shephard, W. Zhou, T. Maschmeyer, J.M. Matters, C.L. Roper, S. Parsons, B.F.G. Johnson, M.J. Duer, Angew. Chem. Int. Ed. **37**, 2719 (1998)
15. K. Yamamoto, T. Tatsumi, Chem. Lett. **6**, 624 (2000)
16. S. Jaenicke, G.K. Chuah, X.H. Lin, X.C. Hu, Micropor. Mesopor. Mater. **35**, 143 (2000)
17. X.H. Lin, G.K. Chuah, S. Jaenicke, J. Mol. Catal. A: Chem. **150**, 287 (1999)
18. B.M. Choudary, K.M. Lakshmi, P. Sreekanth, T. Bandopadhyay, F. Figueras, A. Tuel, J. Mol. Catal. A: Chem. **142**, 361 (1999)
19. H.P. Lin, L.Y. Yang, C.Y. Mou, S.B. Liu, H.K. Lee, New J. Chem. **24**, 253 (2000)

166   S. Fujita et al.

20. A. Cauvel, G. Renard, D. Brunel, J. Org. Chem. **62**, 749 (1997)
21. S.Y.V. Rao, D.E. DeVos, P.A. Jacobs, Angew. Chem. Int. Ed. Engl. **36**, 2661 (1997)
22. N.K. Mal, M. Fujiwara, Y. Tanaka, Nature **421**, 350 (2003)
23. N.K. Mal, M. Fujiwara, Y. Tanaka, T. Taguchi, M. Matsukata, Chem. Mater. **15**, 3385 (2003)
24. C. Tourné-Péteilh, D. Brunel, S. Bégu, F. Chiche, F. Fajula, D.A. Lernera, J.M. Devoisselle, New J. Chem. **27**, 1415 (2003)
25. D.L. Rodman, H. Pan, C.W. Clavier, X. Feng, Z.L. Xue, Anal. Chem. **77**, 3231 (2005)
26. Y. Zhu, M. Fujiwara, Angew. Chem. Int. Ed. **46**, 2241 (2007)
27. S.L. Burkett, S.D. Sims, S. Mann, Chem. Commun. 1367 (1996)
28. D.J. Macquarrie, Chem. Commun. 1961 (1996)
29. L. Mercier, T.J. Pinnavaia, Chem. Mater. **12**, 188 (2000)
30. L. Mercier, T.J. Pinnavaia, Micropor. Mesopor. Mater. **20**, 101 (1998)
31. J. Brown, L. Mercier, T.J. Pinnavaia, Chem. Commun. 69 (1999)
32. L. Mercierb, T.J. Pinnavaia, Environ. Sci. Technol. **32**, 2749 (1998)
33. W.M. Van Rhijn, D.E. De Vos, B.F. Sels, W.D. Bossaert, P.A. Jacobs, Chem. Commun. 317 (1998)
34. D.J. Macquarrie, D.B. Jackson, S. Tailland, K. Wilson, J.H. Clark, Stud. Surf. Sci. Catal. **129**, 275 (2000)
35. W.M. Van Rhijn, D.E. De Vos, W.D. Bossaert, J. Bullen, B. Wouters, P. Grobet, P.A. Jacobs, Stud. Surf. Sci. Catal. **117**, 183 (1998)
36. M.H. Lim, C.F. Blanford, A. Stein, Chem. Mater. **10**, 467 (1998)
37. I. Diaz, C. Marquez-Alvarez, F. Mohino, J. Perez-Pariente, E. Sastre, J. Catal. **193**, 295 (2000)
38. I. Diaz, C. Marquez-Alvarez, F. Mohino, J. Perez-Pariente, E. Sastre, J. Catal. **193**, 283 (2000)
39. D. Margolese, J.A. Melero, S.C. Christiansen, B.F. Chmelka, G.D. Stucky, Chem. Mater. **12**, 2448 (2000)
40. R. Richer, Chem. Commun. 1775 (1998)
41. J. Liu, Y. Shin, Z. Nie, J.H. Chang, L.Q. Wang, G.E. Fryxell, W.D. Samuels, G.J. Exarhos, J. Phys. Chem. **104**, 8328 (2000)
42. K.A. Koyano, T. Tatsumi, Y. Tanaka, S. Nakata, J. Phys. Chem. B **101**, 9436 (1997)
43. M.H. Lim, A. Stein, Chem. Mater. **11**, 3285 (1999)
44. M.H. Lim, C.F. Blanford, A. Stein, J. Am. Chem. Soc. **119**, 4090 (1997)
45. R.J.P. Corriu, A. Mehdi, C. Reyé, C. Thieuleux, Chem. Mater. **16**, 159 (2004)
46. M. Jia, A. Seifert, M. Berger, H. Giegengack, S. Schulze, W.R. Thiel, Chem. Mater. **16**, 877 (2004)
47. R. Huq, L. Mercier, Chem. Mater. **13**, 4512 (2001)
48. C. Liu, N. Naismith, L. Fu, J. Economy, Chem. Commun. 2472 (2003)
49. C.E. Fowler, B. Lebeau, S. Mann, Chem. Commun. 1825 (1998)
50. B. Lebeau, C.E. Fowler, S.R. Hall, S. Mann, J. Mater. Chem. **9**, 2279 (1999)
51. N. Liu, Z. Che, D.R. Dunphy, Y.B. Jiang, R.A. Assink, C.J. Brinker, Angew. Chem. Int. Ed. **42**, 1731 (2003)
52. S. Inagaki, S. Guan, Y. Fukushima, T. Ohsuna, O. Terasaki, J. Am. Chem. Soc. **121**, 9611 (1999)
53. B.J. Melde, B.T. Holland, C.F. Blanford, A. Stein, Chem. Mater. **11**, 3302 (1999)

7 Organic–Inorganic Hybrid Mesoporous Silica   167

54. T. Asefa, M.J. MacLachlan, N. Coombs, G.A. Ozin, Nature **402**, 867 (1999)
55. C. Yoshina-Ishii, T. Asefa, N. Coombs, M.J. MacLachlan, G.A. Ozin, Chem. Commun. 2539 (1999)
56. S. Guan, S. Inagaki, T. Ohsuna, O. Terasaki, J. Am. Chem. Soc. **122**, 5660 (2000)
57. A. Sayari, S. Hamoudi, Y. Yang, I.L. Moudrakovsk, J.R. Ripmeester, Chem. Mater. **12**, 3857 (2000)
58. M.P. Kapoor, S. Inagaki, Chem. Mater. **14**, 3509 (2002)
59. S. Hamoudi, Y. Yang, I.L. Moudrskoviski, S. Lang, A. Sayari, J. Phys. Chem. B **105**, 9118 (2001)
60. Y. Liang, R. Anwander, Micropor. Mesopor. Mater. **72**, 153 (2004)
61. T. Ren, X. Zhang, J. Suo, Micropor. Mesopor. Mater. **54**, 139 (2002)
62. A. Sayari, Y. Yang, Chem. Commun. 2582 (2002)
63. S. Hamoudi, S. Kaliaguine, Chem. Commun. 2118 (2002)
64. M.C. Burleigh, S. Jayasundera, C.W. Thomas, M.S. Spector, M.A. Markowitz, B.P. Gaber, Colloid Polym. Sci. **282**, 728 (2004)
65. M.C. Burleigh, M.A. Markowitz, M.S. Spector, B.P. Gaber, J. Phys. Chem. B **106**, 9712 (2002)
66. M.C. Burleigh, M.A. Markowitz, S. Jayasundera, M.S. Spector, C.W. Thomas, B.P. Gaber, J. Phys. Chem. B **107**, 12628 (2003)
67. O. Muth, C. Schellbach, M. Fröba, Chem. Commun. 2032 (2001)
68. M.P. Matos, M. Kurk, L.P. Mercuri, M. Jeroniec, T. Asfa, N. Coombs, G.A. Ozin, T. Kamiyama, O. Terasaki, Chem. Mater. **14**, 1903 (2002)
69. W. Guo, J.-Y. Park, M.-O. Oh, H.-W. Jeong, W.-J. Cho, I. Kim, C.-S. Ha, Chem. Mater. **15**, 2295 (2003)
70. E.B. Cho, K. Char, Chem. Mater. **16**, 270 (2004)
71. H. Zhu, D.J. Jones, J. Zajac, J. Rozière, R. Dutartre, Chem. Commun. 2568 (2001)
72. W. Guo, I. Kim, C.-S. Ha, Chem. Commun. 2692 (2003)
73. E.B. Cho, K.-W. Kwon, H. Char, Chem. Mater. **13**, 3837 (2001)
74. H. Djojoputro, X.F. Zhou, S.Z. Qiao, L.Z. Wang, C.Z. Yu, G.Q. Lu, J. Am. Chem. Soc. **128**, 6320 (2006)
75. T. Asefa, M.J. MacLachlan, H. Grondey, N. Coombs, G.A. Ozin, Angew. Chem. Int. Ed. **39**, 1808 (2000)
76. W. Wang, S. Xie, W. Zhou, A. Sayari, Chem. Mater. **16**, 1756 (2004)
77. K. Nakajima, D. Lu, I. Tomita, S. Inagaki, M. Hara, S. Hayashi, K. Domen, J.N. Kondo, Adv. Mater. **17**, 1839 (2005)
78. K. Nakajima, I. Tomita, M. Hara, S. Hayashi, K. Domen, J.N. Kondo, J. Mater. Chem. **15**, 2362 (2005)
79. K. Nakajima, I. Tomita, M. Hara, S. Hayashi, K. Domen, J.N. Kondo, Catal. Today **116**, 151 (2006)
80. X. Zhou, S. Qiao, N. Hao, X. Wang, C. Yu, K.L. Wang, D. Zhao, G.Q. Lu, Chem. Mater. **19**, 1870 (2007)
81. K. Landskron, G.A. Ozin, Science **306**, 1529 (2004)
82. T. Kimura, Chem. Mater. **15**, 3742 (2003)
83. T. Kimura, Chem. Mater. **17**, 5521 (2005)
84. T. Kimura, K. Kato, J. Mater. Chem. **17**, 559 (2007)
85. P.H. Mutin, G. Guerrero, A. Vioux, J. Mater. Chem. **15**, 3761 (2005)
86. G. Temtsin, T. Asefa, S. Bittner, G.A. Ozin, J. Mater. Chem. **11**, 3202 (2001)

168 S. Fujita et al.

87. K. Landskron, B.D. Hatton, D.D. Perovic, G.A. Ozin, Science **302**, 266 (2003)
88. W. Wang, W. Zhou, A. Sayari, Chem. Mater. **15**, 4886 (2003)
89. W.J. Hunks, G.A. Ozin, Chem. Commun. 2426 (2004)
90. J. Morell, G. Wolter, M. Fröba, Chem. Mater. **17**, 804 (2005)
91. J. Morell, M. Gungerich, G. Wolter, J. Jiao, M. Hunger, M.J. Klar, M. Fröba, J. Mater. Chem. **16**, 2809 (2006)
92. F. Hoffmann, M. Gungerrich, P.J. Klar, M. Fröba, J. Phys. Chem. C **111**, 5648 (2007)
93. T. Asefa, M. Kruk, J.N. MacLachlan, H. Grondey, M. Jaroniec, G.A. Ozin, J. Am. Chem. Soc. **123**, 8520 (2001)
94. M.C. Burleigh, S. Jayasundera, M.S. Spector, C.W. Thomas, M.A. Markowitz, B.P. Garer, Chem. Mater. **16**, 4 (2004)
95. J. Alauzun, A. Mehdi, C. Reye, R.J.P. Corriu, J. Am. Chem. Soc. **128**, 8718 (2006)
96. J. Alauzun, A. Mehdi, C. Reye, R.J.P. Corriu, J. Mater. Chem. **17**, 349 (2007)
97. R.J.P. Corriu, A. Mehdi, C. Reye, C. Thieuleux, Chem. Commun. 1564 (2003)
98. E. Besson, A. Mehdi, C. Reye, R.J.P. Corriu, J. Mater. Chem. **16**, 246 (2006)
99. V. Dufaud, F. Beauchesne, L. Bonneviot, Angew. Chem. Int. Ed. **44**, 3475 (2005)
100. R.J.P. Corriu, E. Lancelle-Beltran, A. Mehdi, C. Reye, S. Brandes, R. Guilard, Chem. Mater. **15**, 3152 (2003)
101. A. Corma, D. Das, H. Garcia, A. Leyva, J. Catal. **229**, 322 (2005)
102. H. Zhu, D.J. Jones, J. Zajac, R. Dutartre, M. Rhomari, J. Roziere, Chem. Mater. **12**, 4886 (2002)
103. L. Zhang, W. Zhang, J. Shi, Z. Hua, Y. Li, J. Yan, Chem. Commun. 210 (2003)
104. O. Olkhovyk, M. Jaroniec, J. Am. Chem. Soc. **127**, 60 (2005)
105. M. Álvaro, M. Benitez, D. Das, B. Ferrer, H. Gracía, A. Corma, Chem. Mater. **16**, 2222 (2004)
106. D.M. Jiang, Q.H. Yang, J. Yang, L. Zhang, G.R. Zhu, W.G. Su, C. Li, Chem. Mater. **17**, 6154 (2005)
107. D.M. Jiang, Q.H. Yang, J. Yang, G.R. Zhu, C. Li, J. Catal. **239**, 65 (2006)
108. C. Li, H. Zhang, D. Jiang, Q. Yang, Chem. Commun. 547 (2007)
109. C. Baleizão, B. Gigante, D. Das, M. Álvaro, H. Garcia, A. Corma, J. Catal. **223**, 106 (2004)
110. S. Polarz, A. Kuschel, Adv. Mater. **18**, 1206 (2006)
111. A. Ide, R. Voss, G. Scholz, G.A. Ozin, M. Antonietti, A. Thomas, Chem. Mater. **19**, 2649 (2007)
112. M. Álvaro, B. Ferrer, V. Fornés, H. García, Chem. Commun. 2546 (2001)
113. E. Besson, A. Mehdi, D.A. Lerner, C. Reye, R.J.P. Corriu, J. Mater. Chem. **15**, 803 (2005)
114. P.N. Minoofar, R. Hernandez, S. Chia, B. Dunn, J.I. Zink, A.-C. Franville, J. Am. Chem. Soc. **124**, 14388 (2002)
115. M. Álvaro, C. Aprile, M. Benitez, J.L. Bourdelande, H. García, J.R. Herance, Chem. Phys. Lett. **414**, 66 (2005)
116. H. Peng, J. Tang, L. Yang, J. Pang, H.S. Ashbaugh, C.J. Brinker, Z. Yang, Y. Lu, J. Am. Chem. Soc. **128**, 5304 (2006)
117. Q. Yang, M.P. Kapoor, N. Shirokura, M. Ohashi, S. Inagaki, J.N. Kondo, K. Domen, J. Mater. Chem. **15**, 666 (2005)
118. J. Liu, Q. Yang, M.P. Kapoor, N. Setoyama, S. Inagaki, J. Yang, L. Zhang, J. Phys. Chem. **109**, 12250 (2005)

7 Organic–Inorganic Hybrid Mesoporous Silica   169

119. P.L. Dhepe, M. Ohashi, S. Inagaki, M. Ichikawa, A. Fukuoka, Catal. Lett. **102**, 163 (2005)
120. M.P. Kapoor, M. Yanagi, Y. Kasama, T. Yokohama, S. Inagaki, T. Shimada, H. Nanbu, L.R. Juneja, J. Mater. Chem. **16**, 3305 (2006)
121. M.P. Kapoor, A. Bhaumik, S. Inagaki, K. Kuraoka, T. Yazawa, J. Mater. Chem. **12**, 3078 (2002)
122. M.P. Kapoor, A.K. Sinha, S. Seelan, S. Inagaki, S. Tsubota, H. Yoshida, M. Haruta, Chem. Commun. 2902 (2002)
123. A. Bhaumik, M.P. Kapoor, S. Inagaki, Chem. Commun. 470 (2003)
124. B.J. Hughes, J.B. Guilbaud, M. Allix, Y.Z. Khimyak, J. Mater. Chem. **15**, 4728 (2005)
125. Q. Yang, J. Yang, Z. Feng, Y. Li, J. Mater. Chem. **15**, 4268 (2005)
126. A. Fukuoka, H. Araki, Y. Sakamoto, S. Inagaki, Y. Fukushima, M. Ichikawa, Inorg. Chim. Acta **350**, 371 (2003)
127. S. Hudson, J. Cooney, B.K. Hondnett, E. Magner, Chem. Mater. **19**, 2049 (2007)
128. A. Fukuoka, Y. Sakamoto, S. Guan, S. Inagaki, N. Sugimoto, Y. Fukushima, K. Hirahara, S. Iijima, M. Ichikawa, J. Am. Chem. Soc. **123**, 3373 (2001)
129. Y. Sakamoto, A. Fukuoka, T. Higuchi, N. Shimomura, S. Inagaki, M. Ichikawa, J. Phys. Chem. B **108**, 853 (2004)
130. S. Inagaki, S. Guan, T. Ohsuna, O. Terasaki, Nature **416**, 304 (2002)
131. N. Bion, P. Ferreira, A. Valente, I.S. Goncalves, J. Rocha, J. Mater. Chem. **13**, 1910 (2003)
132. B. Onida, L. Borello, C. Busco, P. Ugliengo, Y. Goto, S. Inagaki, E. Garrone, J. Phys. Chem. B **109**, 11961 (2005)
133. B. Onida, B. Camarota, P. Ugliengo, Y. Goto, S. Inagaki, E. Garrone, J. Phys. Chem. B **109**, 21732 (2005)
134. Y. Goto, K. Okamoto, S. Inagaki, Bull. Chem. Soc. Jpn. **78**, 932 (2005)
135. M.P. Kapoor, Q. Yang, S. Inagaki, Chem. Mater. **16**, 1209 (2004)
136. Y. Xia, W. Wang, R. Moyaka, J. Am. Chem. Soc. **127**, 790 (2005)
137. A. Sayari, W. Wang, J. Am. Chem. Soc. **127**, 12194 (2005)
138. M. Cornelius, F. Hoffman, M. Fröba, Chem. Mater. **17**, 6674 (2005)
139. M.P. Kapoor, Q. Yang, S. Inagaki, J. Am. Chem. Soc. **124**, 15176 (2002)
140. J. Morell, C.T. Teixeira, M. Cornelius, V. Rebbin, M. Tiemann, H. Amenitsch, M. Fröba, M. Linden, Chem. Mater. **16**, 5564 (2004)
141. K. Okamoto, M.P. Kapoor, S. Inagaki, Chem. Commun. 1423 (2005)
142. T. Shimada, K. Aoki, Y. Shinoda, T. Nakamura, N. Tokunaga, S. Inagaki, T. Hayashi, J. Am. Chem. Soc. **125**, 4688 (2003)
143. K. Aoki, T. Shimada, T. Hayashi, Tetrahedron Asymmetry **15**, 1771 (2004)
144. M.P. Kapoor, S. Inagaki, S. Ikeda, K. Kakiuchi, M. Suda, T. Shimada, J. Am. Chem. Soc. **127**, 8174 (2005)
145. Q. Yang, M.P. Kapoor, S. Inagaki, J. Am. Chem. Soc. **124**, 9694 (2002)
146. Q. Yang, J. Liu, J. Yang, M.P. Kapoor, S. Inagaki, C. Li, J. Catal. **228**, 265 (2004)
147. Q. Yang, M.P. Kapoor, S. Inagaki, N. Shirokura, J.N. Kondo, K. Domen, J. Mol. Catal. A: Chem. **230**, 85 (2005)
148. M.P. Kapoor, Q. Yang, Y. Goto, S. Inagaki, Chem. Lett. **32**, 914 (2003)
149. M.P. Kapoor, S. Inagaki, Chem. Lett. **33**, 88 (2004)
150. S.Y. Kim, J.W. Lee, J.H. Jung, J.K. Kang, Chem. Mater. **19**, 135 (2007)

8

Organic–Inorganic Hybrid Zeolites Containing Organic Frameworks

Katsutoshi Yamamoto and Takashi Tatsumi

8.1 Introduction

Organic–inorganic hybridization of ordered porous materials has attracted a lot of interests because it can widen the range of their applications by not only giving new functions but also manipulating the surface properties. In most studies, mesoporous amorphous silica materials were employed as host materials for hybridization. At first, the pore surface of ordered mesoporous silica materials was directly [1–3] or postsynthetically [4, 5] functionalized by pendant organic groups. Through such functionalization, the structural stability and/or the catalytic performance of the materials were successfully improved. Later, organosilanes having bridging organic groups instead of terminal organic groups were employed by several researchers to synthesize surfactant-templated mesoporous silica materials [6–8]. In these hybrid materials, bridging organic groups such as ethylene and ethenylene groups were embedded in the mesostructured pore wall. Because the insertion of bridging organic groups does not form structural defects in the pore wall, such materials can incorporate large amounts of organic groups without deteriorating the structural order compared with mesoporous silica materials modified with terminal organic groups. They have proved to have well ordered structures and those framework organic groups can be further functionalized. After these reports were published, a large number of hybrid mesoporous silica materials having various bridging organic groups, structures, and characteristic properties were successfully synthesized [9–17].

On the other hand, not so many studies have been done on the hybridization of crystalline zeolitic materials. The organic–inorganic hybridization of zeolites started also from the functionalization with terminal organic groups. Jones et al. [18–22] succeeded in modifying *BEA-type zeolite by various terminal organic groups. They employed organosilanes such as phenethyltrimethoxysilane and 3-mercaptopropyltrimethoxysilane as a part of silicon source and synthesized *BEA-type zeolite materials named OFMSs (organic-functionalized molecular sieves). Although it seems that the synthesis of

OFMSs with other zeolitic structures such as FAU and MFI was attempted [23], detailed results have not yet published. Through this hybridization, a new function can be added to a microporous zeolite material. The researchers prepared a *BEA-type OFMS having sulfonic groups inside the pore and employed it as a shape-selective acid catalyst [18,22]. Thus synthesized OFMS material catalyzed the acetalization of small reactant molecules but was not active for a bulky reactant molecule whose size is larger than the zeolite pore opening. In a similar manner, Yan et al. employed phenylphosphonic acid as a part of phosphorous source to synthesize an ALPO$_4$ material [24] having the same crystal structure as VPI-5 [25].

Before these reports, Maeda et al. [26–29] succeeded in synthesizing a series of aluminum methylphosphate materials (AlMepO). Two isomeric AlMepO materials, AlMepO-α [27] and AlMepO-β [26, 28], were synthesized by using methylphosphonic acid as an only phosphorous source. Both of them have one-dimensional pore channels whose openings consist of 18 metal (phosphorous or aluminum) atoms. Methyl groups are located on the channels to almost completely line the inner surface. These organically lined channels were proved to have unique gas adsorption properties [30–32]. Based on the definition by the International Zeolite Association [33], AlMepO could not be called zeolites because of the presence of octahedral aluminum atoms in the framework. However, these materials seem important because, as far as we know, they are the first organically functionalized microporous crystalline materials.

In this way, the functionalization with terminal organic groups provided microporous crystals with new functions and/or characteristic surface properties. However, the introduction of terminal organic groups inevitably forms structural defects in the silicate framework, and bulky organic groups incorporated in the pore of zeolites may impair the microporosity. To overcome such drawbacks, we have synthesized a new type of organic–inorganic hybrid zeolites ZOL (zeolite with organic group as lattice) [34–38]. ZOL materials are the first hybrid zeolites, in which an organic group is incorporated as a lattice (Fig. 8.1). In this chapter, the synthesis method and the unique characteristics of ZOL materials will be described.

8.2 Synthesis and Physical Properties of ZOL

In ZOL materials, a divalent methylene group is introduced to be substituted for an oxygen atom in the zeolite framework. To introduce a methylene group, bis(triethoxysilyl)methane (BTESM), in which a methylene group bridges two silicon atoms, is employed as a silicon source (Fig. 8.2). Because Si–C bond is usually longer than Si–O bond, the insertion of methylene moieties into the zeolite framework looks difficult. However, Si–C–Si angle ($\sim$109$°$) is smaller than Si–O–Si angles in zeolite framework; for example, those in MFI-type zeolite range from 140$°$ to 170$°$ [39]. This smaller bond angle would compensate for the distance of two silicon atoms to enable the insertion of methylene

8 Organic–Inorganic Hybrid Zeolites Containing Organic Frameworks   173

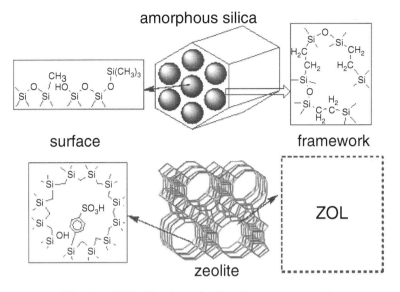

Fig. 8.1. Hybridization of ordered porous materials

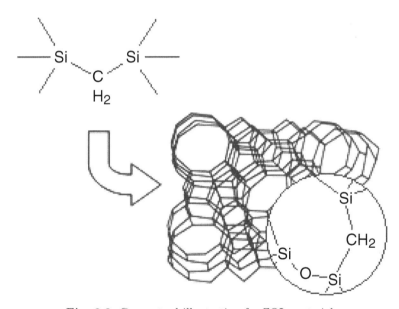

Fig. 8.2. Conceptual illustration for ZOL materials

species into the zeolite framework. Astala and Auerbach [40] theoretically demonstrated that methylene fragments can stably substitute for oxygen atoms in LTA- and SOD-type zeolite frameworks through the calculation based on the density functional theory.

Table 8.1. Typical synthesis conditions for ZOL materials

Material	Mother gel composition(molar ratio)	Temp. (K)	Time (days)	Topology
ZOL-1	$0.5BTESM:0.47TPAOH:21H_2O$	443	5	MFI
ZOL-2	$0.5BTESM:0.25TEMBr:0.13Na_2O:20H_2O$	413	20	MFI
ZOL-5	$0.5BTESM:0.018Al_2O_3:0.042Na_2O:58H_2O$	463	7	MFI
ZOL-A	$0.5BTESM:0.52Al_2O_3:1.64Na_2O:66.5H_2O$	373	14	LTA
ZOL-B(F)	$0.1BTESM:0.8TEOS:0.54TEAF:7.63H_2O$	413	14	*BEA

TPAOH tetrapropylammonium hydroxide, *TEMABr* triethylmethylammonium bromide, *TEOS* tetraethyl orthosilicate, *TEAF* tetraethylammonium fluoride

Table 8.1 summarizes typical synthesis conditions for ZOL materials. ZOL materials are synthesized in the conditions similar to those of corresponding conventional zeolites. As shown in the table, ZOL materials can be crystallized both in the presence and in the absence of structure-directing agent (SDA). The powder XRD patterns of these ZOL materials are exhibited in Fig. 8.3. ZOL materials show diffraction patterns similar to those of corresponding conventional zeolites, although the presence of amorphous products as concomitant phases is also implied in some samples. In addition to these materials, the syntheses of ITQ-21- and MOR-type ZOL materials have been reported [41].

The SEM images of ZOL materials are shown in Fig. 8.4. MFI-type ZOL materials, ZOL-2 and ZOL-5, exhibit a coffin-like crystal shape, which is a morphology typical of conventional MFI-type zeolites. ZOL-A with the LTA structure has a cubic crystal shape, also typical of LTA-type materials, although ZOL-A has a somewhat rounded crystal shape.

The ^{13}C CP/MAS and the ^{29}Si DD/MAS NMR spectra of ZOL-A and ZOL-2 are displayed in Fig. 8.5. In the ^{13}C NMR spectra of both samples, a resonance peak is observed at around 0 ppm. This peak is assigned to carbon species directly bonded to a silicon atom, indicating the incorporation of organically functionalized silicon species in the products. The presence of organically modified silicon species was also confirmed by the resonance peaks of T-type silicon species at -60 to -70 ppm in their ^{29}Si NMR spectra. On the other hand, the ^{29}Si NMR spectra also demonstrate conspicuous resonance peaks attributable to inorganic silicon species at around -90 to -110 ppm. Because only BTESM is employed as a silicon source for the synthesis of these ZOL materials, the formation of purely inorganic silicon species through the decomposition of this organosilane is implied. Thus formed inorganic silicon species would be cocrystallized with organosilanes into ZOL materials. The crystallization scheme of ZOL materials will be discussed in detail below.

In the IR spectrum of ZOL-A (Fig. 8.6a), peaks appear at 2,938 and $2,862\,cm^{-1}$, which are not observed in the spectrum of zeolite A (Fig. 8.6b). These peaks are assigned to asymmetric and symmetric C-H stretching vibrations of the CH_2 group, respectively. At the same time, the IR spectrum

8 Organic–Inorganic Hybrid Zeolites Containing Organic Frameworks   175

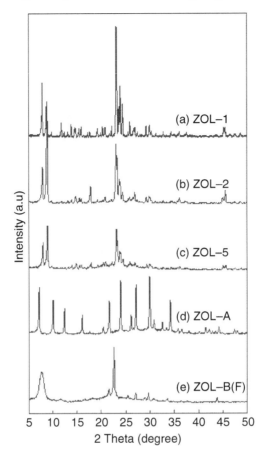

Fig. 8.3. Powder XRD patterns of as-synthesized ZOL materials. (**a**) ZOL-1, (**b**) ZOL-2, (**c**) ZOL-5, (**d**) ZOL-A, and (**e**) ZOL-B(F)

exhibits two other peaks at 2,970 and 2,880 cm^{-1}. These peaks should be attributed to the CH$_3$ group, also suggesting the hydrolysis of Si–CH$_2$–Si linkage into Si–CH$_3$ and HO–Si. These peaks attributable to CH$_2$ and CH$_3$ groups are also observed in other SDA-free ZOL materials, ZOL-2 and ZOL-5.

The SDA-free ZOL materials show microporosity like conventional zeolites. Their N$_2$ adsorption/desorption isotherms (Fig. 8.7) exhibit a noticeable adsorption step at $P/P_0 = 0$, indicating the presence of micropores, although the micropore volumes based on the t-plots (Table 8.2) are smaller than those of conventional zeolites. The presence of an amorphous impurity phase and/or a methylene group larger than a bridging oxygen atom would be responsible for this smaller micropore volume.

Because of the organic functionalization, ZOL materials are hydrophobic compared with their inorganic counterparts. Table 8.3 exhibits the adsorption

Fig. 8.4. SEM images of (**a**) ZOL-2, (**b**) ZOL-5, (**c**) ZOL-B(F), and (**d**) ZOL-A

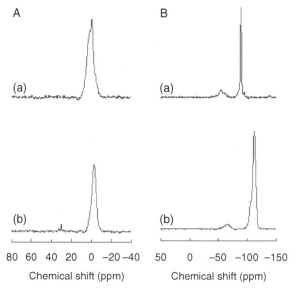

Fig. 8.5. (**A**) ^{13}C CP/MAS and (**B**) ^{29}Si DD/MAS NMR spectra of (*a*) ZOL-A and (*b*) ZOL-2. Chemical shifts are given in ppm from TMS

properties of ZOL-A and zeolite A for organic molecules. Before the adsorption measurement, samples were ion-exchanged to Ca^{2+}-form, exposed to water vapor at room temperature for 48 h, and pretreated at 323 or 473 K under vacuum for 10 h. When pretreated at 473 K, both ZOL-A and zeolite A desorb

8 Organic–Inorganic Hybrid Zeolites Containing Organic Frameworks   177

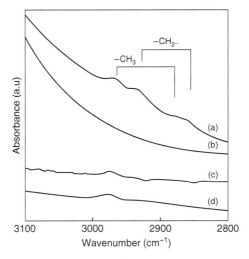

Fig. 8.6. IR spectra of (**a**) ZOL-A, (**b**) zeolite A, (**c**) ZOL-2, and (**d**) ZOL-5

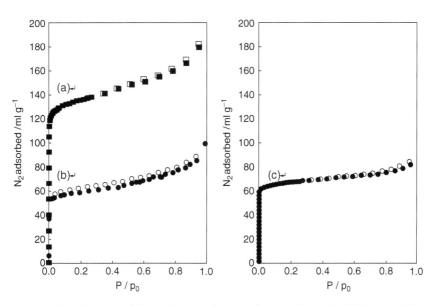

Fig. 8.7. N_2 adsorption/desorption isotherms of as-synthesized ZOL materials at 77 K: (**a**) ZOL-A, (**b**) ZOL-5, and (**c**) ZOL-2

most of the adsorbed water molecules so that they can show large adsorption capacities for n-hexane in proportion to their pore volumes. However, after the pretreatment at 323 K the amount of the water molecules remaining on ZOL-A would be much smaller than that on zeolite A due to the higher hydrophobicity of ZOL-A. As a result, ZOL-A adsorbs much larger amount

178   K. Yamamoto and T. Tatsumi

Table 8.2. Physical properties of SDA-free ZOL materials

Material	Phase	Al/(Si+Al)	C content[a] (wt%)	V_p [b] $(cm^3 g^{-1})$
ZOL-2	MFI	0	1.9	0.092
ZOL-5	MFI	0.034	3.2	0.091
ZOL-A	LTA	0.45	2.2	0.20

[a] Carbon content estimated by CHN analyses
[b] Micropore volume based on t-plot of N_2 adsorption isotherm at 77 K

Table 8.3. Adsorption properties of ZOL-A and zeolite A

Material	H_2O remaining in micropore[a]	Adsorption capacity (mL-liquid per g-zeolite)[b]	
	(g-H_2O per g-zeolite)	n-Hexane	Benzene
Pretreated at 473 K			
ZOL-A $(0.20)^c$	0.023	0.059	0.0043
Zeolite A $(0.25)^c$	0.042	0.078	0.0010
Pretreated at 323 K			
ZOL-A $(0.20)^c$	0.191	0.028	0.0019
Zeolite A $(0.25)^c$	0.241	0.0069	0.0016

[a] Values are estimated by the TG analyses for water adsorbed materials treated in saturated water vapor at room
[b] Adsorption capacity for n-hexane or benzene at 298 K at 10 Torr
[c] Values in parentheses indicate the micropore volume estimated based on t-plot of N_2 adsorption isotherm at 77 K

of n-hexane than zeolite A. On the other hand, ZOL-A can adsorb quite a small amount of benzene because bulky benzene molecules cannot enter into the micropore of ZOL-A. This shape-selective lipophilicity/hydrophobicity is one of the unique characteristics of a ZOL material, simultaneously indicating that it is not a mixture but a genuine hybrid material.

The thermal stability of methylene fragment is improved when it is incorporated into ZOL materials. Figure 8.8 shows the DTA profiles and the XRD patterns of the products synthesized from the mixture of 20% BTESM and 80% TEOS (as Si mol). After the synthesis time of 1 day, the obtained amorphous material shows the exothermic DTA peak attributable to the combustion of organic moieties at ca. 600 K. After 3 days, however, the LTA-type material is crystallized, and the exothermic peak appears at higher temperature around 700 K, indicating the higher thermal stability of organic moieties in crystalline ZOL-A. Figure 8.9 compares the DTA profiles of ZOL materials and amorphous aggregate synthesized in fluoride media. Also in this case, ZOL materials show higher thermal stability than the amorphous aggregate. Their ^{29}Si DD/MAS NMR spectra before/after the calcination (not shown) indicate that the T-type silicon peak of amorphous aggregate completely disappears after the calcination at 813 K, whereas those of ZOL materials are

8 Organic–Inorganic Hybrid Zeolites Containing Organic Frameworks   179

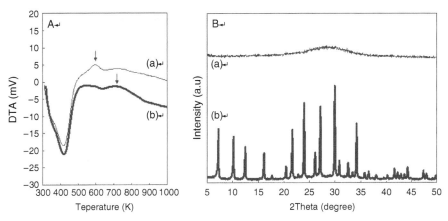

Fig. 8.8. (**A**) DTA curves of ZOL-A synthesized from the mixture of BTESM and TEOS (BTESM:TEOS = 1:4) for (*a*) 1 day and (*b*) 3 days and (**B**) XRD patterns of the corresponding materials

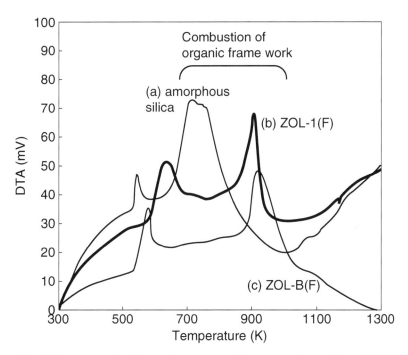

Fig. 8.9. DTA curves for (**a**) methylene-containing amorphous silica, (**b**) ZOL-1, and (**c**) ZOL-B synthesized in fluoride media

180 K. Yamamoto and T. Tatsumi

still observed after the calcination. Furthermore, the figure also demonstrates
that the organic framework of ZOL-B(F) is combusted at higher temperature
than that of ZOL-1(F). This dependence of thermal stability on the structure
would be corroborative evidence for the incorporation of organic moieties into
the framework.

8.3 Crystallization Scheme

As described above, ZOL materials contain inorganic silicon species, implying
the cleavage of Si–CH$_2$–Si linkage. Si–C bonds are usually regarded as stable
against hydrolysis in mild hydrothermal conditions. However, in the case of
Si–CH$_2$–Si, Si–CH$_2^-$ could be formed through the nucleophilic attack of
hydroxyl anion, and this carbanion is stabilized by the vacant d orbital of the
adjacent Si atom (Scheme 8.1). Consequently, Si–CH$_2$–Si linkage is cleaved
into Si–CH$_3$ and HO–Si, and the latter inorganic silicon species would be
cocrystallized with organosilanes to form ZOL materials.

Generally, the crystallization of ZOL materials takes longer time than
the crystallization of their inorganic isostructures. For example, the crystallization of ZOL-A from BTESM requires 14 days, whereas zeolite A can be
obtained from TEOS in only 1 day under the same hydrothermal conditions.
As shown in Fig. 8.10, at the beginning of the hydrothermal synthesis of ZOL-
A, inorganic silicon species are formed through the hydrolysis of Si–CH$_2$–Si
linkage. The crystallization of ZOL-A seems to start after a certain amount
of inorganic silicon species is formed. Presumably, the nucleation only from
methylene-bridged silicon species would be difficult to occur due to the ununi-
formity of the bond lengths. This hypothesis is in accordance with the fact
that ZOL-A can be obtained more shortly in 3 days from the mixture of 80%
TEOS and 20% BTESM (as Si mol%). It can also explain why the addition
of TEOS as a part of silicon sources is required for the crystallization of ZOL
in fluoride media. In nearly neutral fluoride media Si–CH$_2$–Si linkage will not
be cleaved, so the nucleation does not occur without the addition of inorganic
silicon source.

Figure 8.11 compares the ^{29}Si DD/MAS NMR spectra of zeolite materials synthesized from methylene- or methyl-functionalized silicon sources. In
the spectrum of ZOL-5, a broad peak attributable to organically functionalized silicon species is clearly observed. In contrast, Me-ZSM-5, synthesized

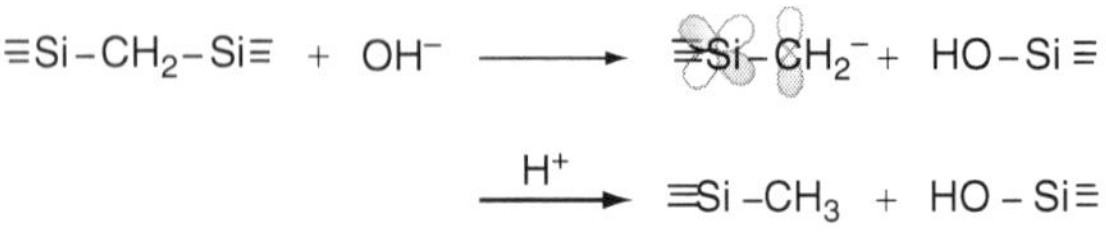

$$\equiv\text{Si}-\text{CH}_2-\text{Si}\equiv \ + \ \text{OH}^- \longrightarrow \ \equiv\text{Si}-\text{CH}_2^- + \ \text{HO}-\text{Si} \equiv$$

$$\xrightarrow{\text{H}^+} \ \equiv\text{Si}-\text{CH}_3 \ + \ \text{HO}-\text{Si}\equiv$$

Scheme 8.1. Cleavage of Si–CH$_2$–Si linkage during the hydrothermal treatment
in alkaline media

8 Organic–Inorganic Hybrid Zeolites Containing Organic Frameworks   181

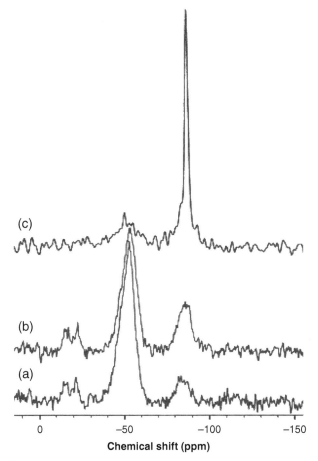

Fig. 8.10. ^{29}Si DD/MAS NMR spectra of the products synthesized from BTESM after the hydrothermal treatment for (*a*) 3 and (*b*) 7 days with an amorphous phase, and (*c*) 14 days with the LTA phase (ZOL-A). Chemical shifts are given in ppm from TMS

from the equimolar mixture of TEOS and methyltriethoxysilane (MTES), does not exhibit T-type peaks (Fig. 8.11b). This finding implies that methylated silicon species formed through the decomposition of methylene-bridged species scarcely participate in the crystallization of ZOL-5. Therefore, we suppose that methyl groups involved in ZOL-5 were formed by the hydrolysis of framework methylene group after the crystallization. Lesthaeghe et al. [42] theoretically suggested the possibility of the cleavage of a methylene linkage in zeolite frameworks to give a terminal methyl group.

On the other hand, T-type peaks are observed in the spectra of silicalite-1 and zeolite A synthesized from the mixture of TEOS and MTES, named Me-silicalite-1 and Me-A, respectively. (Fig. 8.11d,f). Therefore, methylated

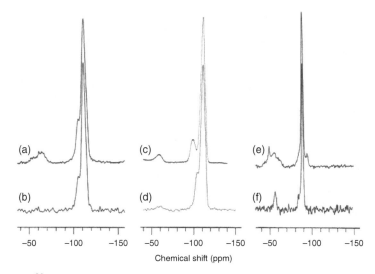

Fig. 8.11. ^{29}Si DD/MAS NMR spectra of methylene- or methyl-substituted zeolite materials: (a) ZOL-5, (b) Me-ZSM-5, (c) ZOL-1, (d) Me-silicalite-1, (e) ZOL-A, and (f) Me-A. Chemical shifts are given in ppm from TMS

silicon species in ZOL-1 and ZOL-A can be derived from methylated silicon sources formed through the decomposition of BTESM before the crystallization as well as from the cleavage of the framework methylene group. From this point of view, it is of interest that the ^{29}Si DD/MAS NMR spectrum of ZOL-A exhibits a peak at −94.5 ppm assignable to Si(3Al) (Fig. 8.11e). When the Si–CH$_2$–Si linkage is introduced into the LTA framework, Si(3Al) species is formed through the replacement of Al atoms with organically bridged Si atoms. That is, the presence of Si(3Al) peak demonstrates that methylene-bridged silicon species really participate in the crystallization of ZOL-A.

A probable crystallization scheme for ZOL materials is illustrated in Fig. 8.12. First, Si–CH$_2$–Si linkage in the silicon source is cleaved in the hydrothermal conditions in alkaline media. Thus formed inorganic silicon species are mainly used for zeolite nucleation. In the crystallization of ZOL-5, inorganic silicon species and methylene bridged silicon species are cocrystallized into zeolite. In the cases of ZOL-1 and ZOL-A, methylated silicon source would also participate in the crystallization, as described above. After the crystallization framework methylene groups are possibly cleaved into methyl and hydroxyl groups. In fluoride media, fluoride anions attack silicon species only to form stable five-coordinated silicon species without cleaving Si–CH$_2$–Si linkage. Therefore, an inorganic silicon source such as TEOS must be added for the crystallization of ZOL materials.

When seed crystals are employed, the initial step of Si–C cleavage and nucleation can be skipped. Therefore, the crystallization can start before

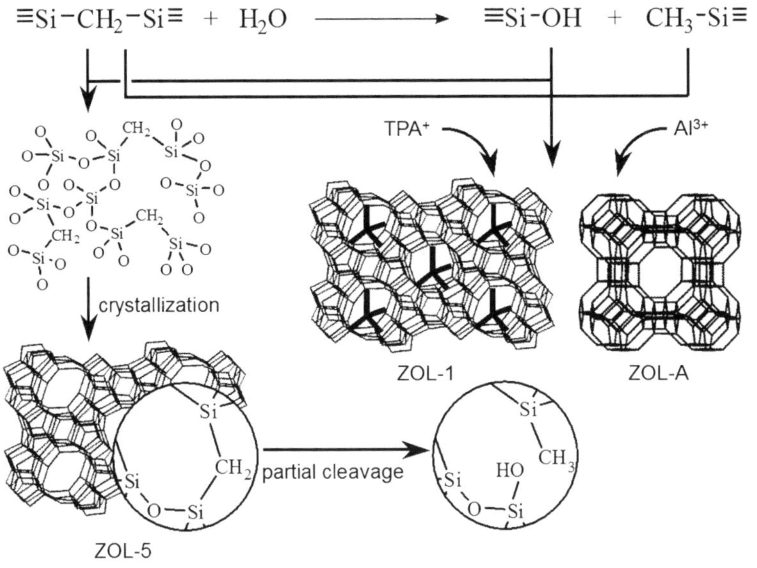

8 Organic–Inorganic Hybrid Zeolites Containing Organic Frameworks   183

Fig. 8.12. Probable crystallization scheme for ZOL materials

a large amount of $Si–CH_2–Si$ linkage is broken, possibly resulting in the formation of a ZOL material with higher organic contents. Actually, when ZOL-5 is synthesized in the presence of seed crystals, the carbon content of the resulting material is increased (Fig. 8.13). Astala and Auerbach [40] simulated the lattice parameters of the methylene-substituted SOD framework and predicted that at first incorporation of methylene group would rather shrink the unit cell due to the small Si–C–Si angle and that further incorporation would lead to the unit cell expansion owing to the long Si–C bond length. Also in our experimental results, the incorporation of a small amount of methylene (carbon content $= 1\,wt\%$) into the MFI framework makes little change of diffraction angles, although the incorporation of a larger amount of methylene (carbon content $= 3.3\,wt\%$) results in the shift to the lower angle caused by the unit cell expansion.

8.4 Proof of Organic Framework

Considering that inorganic silicon species are formed by the decomposition of the methylene-bridged silicon source, the resulting product could be a mixture of conventional zeolite and carbon-containing impurities. To prove that organic moieties are included as zeolite framework, careful ^{13}C MAS NMR measurements have been conducted.

First, the presence of methylene groups in ZOL materials is quantitatively evaluated. Figure 8.14 exhibits the semilogarithmic plots of the ^{13}C

184   K. Yamamoto and T. Tatsumi

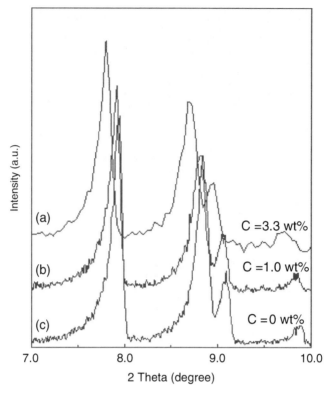

Fig. 8.13. Low-angle region of the XRD patterns of as-synthesized ZOL-1 prepared in the (*a*) presence and (*b*) absence of seed crystals and (*c*) silicalite-1

spin–lattice relaxation (T1c) decay obtained by using the Torchia's [43] pulse sequence. The T1c decay of Me-A linearly decreases, indicating the presence of only one kind of carbon species (methyl group). In contrast, the T1c decays of ZOL-A and ZOL-2 consist of two straight lines with different slopes, suggesting that these materials contain two kinds of carbon species. One with a shorter T1c decay comes from the methyl group, and the other with a longer relaxation time from the methylene group.

Based on this T1c decay, the methyl and methylene ratio ($M_{CH_3} : M_{CH_2}$) is estimated according to the following equation:

$$I = M_{CH_3} \exp(-\tau/T1_{CH_3}) + M_{CH_2} \exp(-\tau/T1_{CH_2}),$$

where I is the relative intensities of resonance peaks, M_{CH_3} is the relaxation time for carbon species in the methyl group, and M_{CH_2} is the relaxation time for carbon species in the methylene group.

Thus, estimated ^{13}C relaxation time and $M_{CH_3} : M_{CH_2}$ ratios are exhibited in Table 8.4. Through the curve fitting simulation, $M_{CH_3} : M_{CH_2}$ ratio in ZOL-A is estimated at 39:61. The percentage of methylene group of ZOL-2 (45%) is

8 Organic–Inorganic Hybrid Zeolites Containing Organic Frameworks   185

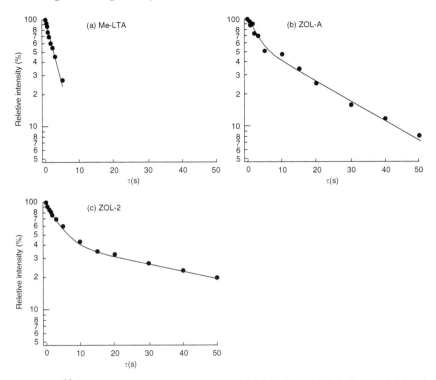

Fig. 8.14. ^{13}C T1c relaxation decay curves of (**a**) Me-A, (**b**) ZOL-A, and (**c**) ZOL-2

smaller than that of ZOL-A presumably because it is hydrothermally treated at higher temperature.

For ZOL-A having a short relaxation time, another estimation method can be used. As indicated in Fig. 8.14 and Table 8.4, the relaxation of methyl species in ZOL-A completes within 15 s ($\approx 5\times$ T1c). Therefore, the Torchia's pulse sequence with the relaxation delay time of 15 s gives a resonance peak attributable only to methylene species. On the other hand, the ^{13}C NMR spectrum using the dipolar dephasing pulse sequence with the delay time of 100 μs should exhibit a resonance peak only for the methyl group, because resonance peaks for the methylene and methine groups almost completely diminishes within 100 μs due to the dipole–dipole interaction between ^{13}C and ^{1}H. Figure 8.15 exhibits thus obtained spectra for methyl (Fig. 8.15a) and methylene (Fig. 8.15b) species in ZOL-A. The resonance peak of methylene species is observed at around 1 ppm, and that of methyl species at around -2 ppm. By the peak deconvolution of the ^{13}C DD/MAS NMR spectrum (Fig. 8.15c), the $M_{\text{CH}_3} : M_{\text{CH}_2}$ ratio is estimated at 42:58, which is similar to the value obtained above through the curve fitting simulation for the T1c relaxation decay (Table 8.4).

Table 8.4. CH$_3$:CH$_2$ ratios of ZOL materials obtained through the curve fitting simulation of ^{13}C T1c relaxation decay

Material	Silicon source	^{13}C relaxation time (s)		Molar ratio (%)	
		$T1_{CH_3}$	$T1_{CH_2}$	M_{CH_3}	M_{CH_2}
Me-A	50% MTES + 50% TEOS	3.7	–	100	–
ZOL-A	100% BTESM	2.8	23.4	39	61
ZOL-2	100% BTESM	4.0	59.3	55	45

MTES methyltriethoxysilane

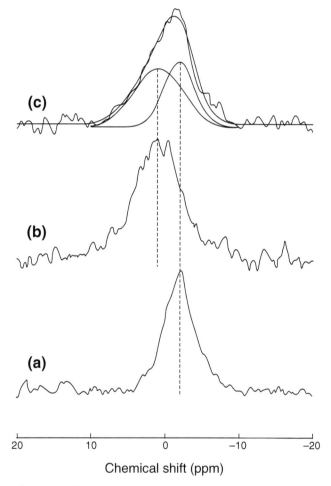

Fig. 8.15. ^{13}C MAS NMR spectra of (**a**) methyl group obtained using dipolar dephasing pulse sequence with delay time of 100 µs, (**b**) methylene group using Torchia pulse sequence with relaxation delay time of 15 s and, (**c**) all carbon species by DD/MAS mode with pulse delay = 120 s (total resonance peak and deconvoluted peaks obtained by curve fitting simulation) in ZOL-A

8 Organic–Inorganic Hybrid Zeolites Containing Organic Frameworks 187

Next, the presence of methylene group as zeolite framework is demonstrated by using a ZOL-A sample on which n-hexane molecules are adsorbed. In this NMR experiment, the Goldman–Shen pulse sequence [44] is employed. In this pulse sequence, all the ^{1}H nuclei are initially magnetized, and relaxed. After an appropriate relaxation time, mobile ^{1}H in adsorbed n-hexane molecules, which has longer relaxation time, remains magnetized selectively. Then the magnetization is diffused to ^{1}H in organic species of ZOL-A during the spin-diffusion time. The closer the distance between two ^{1}H is, the faster the magnetization diffuses. Then, the magnetization is transferred to ^{13}C via cross-polarization to be detected as the resonance.

Figure 8.16A shows the spectrum using the Goldman–Shen pulse sequence with the spin-diffusion time of 50 ms together with the usual CP/MAS spectrum. In these two spectra, the ratios of the peak area attributable to the framework organic species to that of n-hexane are almost the same. This finding means that all the organic moieties in ZOL-A are accessible to n-hexane, the magnetization of ^{1}H being spin-diffused and quickly equilibrated. In contrast, an amorphous nonporous aggregate having a similar organic content, as a control, shows a much smaller peak area ratio than that observed in the usual CP/MAS spectrum (Fig. 8.16B). In this amorphous aggregate, most

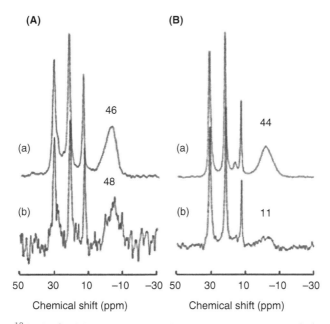

Fig. 8.16. ^{13}C CP/MAS NMR spectra for n-hexane adsorbing (**A**) ZOL-A and (**B**) amorphous aggregate prepared from BTESM: (*a*) usual pulse sequence and (*b*) the Goldman–Shen pulse sequence with the spin-diffusion time of 50 ms. *Numbers* indicated above peaks represent relative peak area, where total area for n-hexane peaks is normalized to 100

organic groups would be buried inside the aggregate as revealed by its small surface area. The spin diffusion from adsorbed n-hexane to such buried carbon species is impossible or at least very slow. As a result, the peak area cannot reach the value of the usual CP/MAS spectrum. These experimental results reasonably suggest that organic moieties in ZOL-A are exposed to the surface. That is, ZOL-A is not a physical mixture of a carbon-containing amorphous aggregate and conventional zeolite A but a genuine organic–inorganic hybrid zeolite material having framework organic groups.

8.5 Applications and Prospects

Diaz et al. [45] attempted to optimize the synthesis conditions and reported the synthesis of ZOL materials having larger amount of organic contents. As described above, one of the most important problems with ZOLs is whether organic groups are included in the crystalline framework or not. Therefore, further characterizations on the organic groups in their products will furnish the evidence for their incorporation into the framework.

Maeda et al. [46] applied the silicon source BTESM to the synthesis of $AlPO_4$ materials. The synthesis conditions of $AlPO_4$ materials are usually milder than those of aluminosilicate-based zeolites. Therefore, it might be possible to avoid the cleavage of the methylene linkage under such milder conditions, resulting in the formation of organic–inorganic hybrid aluminophosphate zeolites containing no structural defects.

Brondani et al. [47] reported that methylene-bridged 3-membered ring (Fig. 8.17, *1*) can be synthesized from BTESM. Shimojima and Kuroda [48] also succeeded in selectively preparing the D3R unit (Fig. 8.17, *2* and *3*) from the same organosilanes. If a zeolite structure can be built up from such an already-assembled building block (Landskron et al. [49] actually employed *1* for the synthesis of mesoporous silica), this approach would become a promising method for designing relevant structures of zeolites or obtaining new structures of zeolites.

It has been observed that the Si–CH_2–Si framework in ZOL materials is converted into the Si–NH–Si by the thermal treatment in the ammonia

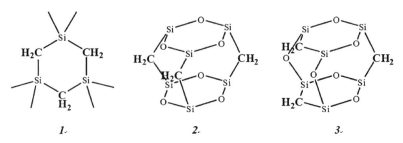

Fig. 8.17. Building blocks synthesized from BTESM

8 Organic–Inorganic Hybrid Zeolites Containing Organic Frameworks

atmosphere [50]. Elanany et al. [51] computationally calculated the acidity of the organic moieties of methylene- or amine-substituted CHA-type zeolite based on the density functional theory. Zheng et al. [52] theoretically studied the adsorption behavior of an organic reactant molecule on amine-bridged zeolite. In these researches, unique properties of ZOL materials were predicted, which would interest researchers in the catalysis field.

In this way, there have already appeared several studies on this new type of material. These experimental or theoretical studies will widen the application area of ZOL materials, to find out a new aspect of these hybrid zeolites.

References

1. S.L. Burkett, S.D. Sims, S. Mann, Chem. Commun. 1367 (1996)
2. D.J. Macquarrie, Chem. Commun. 1961 (1996)
3. A. Corma, J.L. Jorda, M.T. Navarro, F. Rey, Chem. Commun. 1899 (1998)
4. K.A. Koyano, T. Tatsumi, Y. Tanaka, S. Nakata, J. Phys. Chem. B **101**, 9436 (1997)
5. T. Tatsumi, K.A. Koyano, N. Igarashi, Chem. Commun. 325 (1998)
6. S. Inagaki, S. Guan, Y. Fukushima, T. Ohsuna, O. Terasaki, J. Am. Chem. Soc. **121**, 9611 (1999)
7. B.J. Melde, B.T. Holland, C.F. Blanford, A. Stein, Chem. Mater. **11**, 3302 (1999)
8. T. Asefa, J.M. MacLachlan, N. Coombs, G.A. Ozin, Nature **402**, 867 (1999)
9. C.Y. Ishii, T. Asefa, N. Coombs, M.J. MacLachlan, G.A. Ozin, Chem. Commun. 2539 (1999)
10. Y. Lu, H. Fan, N. Doke, D.A. Loy, R.A. Assink, D.A. LaVan, C.J. Brinker, J. Am. Chem. Soc. **122**, 5258 (2000)
11. S. Guan, S. Inagaki, T. Ohsuna, O. Terasaki, J. Am. Chem. Soc. **122**, 5660 (2000)
12. T. Asefa, M.J. MacLachlan, H. Grondey, N. Coombs, G.A. Ozin, Angew. Chem. Int. Ed. **39**, 1808 (2000)
13. T. Asefa, C.Y. Ishii, M.J. MacLachlan, G.A. Ozin, J. Mater. Chem. **10**, 1751 (2000)
14. S. Inagaki, S. Guan, T. Ohsuna, O. Terasaki, Nature **416**, 304 (2002)
15. Q.H. Yang, M.P. Kapoor, S. Inagaki, J. Am. Chem. Soc. **124**, 9694 (2002)
16. Y. Goto, S. Inagaki, Chem. Commun. 2410 (2002)
17. M. Kuroki, T. Asefa, W. Whitnal, M. Kruk, C. Yoshina-Ishii, M. Jaroniec, G.A. Ozin, J. Am. Chem. Soc. **124**, 13886 (2002)
18. C.W. Jones, K. Tsuji, M.E. Davis, Nature **393**, 52 (1998)
19. C.W. Jones, K. Tsuji, M.E. Davis, *Proceedings of the 12th International Zeolite Conference* (Materials Research Society, Warrendale, PA, 1999), p. 1479
20. K. Tsuji, C.W. Jones, M.E. Davis, Micropor. Mesopor. Mater. **29**, 339 (1999)
21. C.W. Jones, K. Tsuji, M.E. Davis, Micropor. Mesopor. Mater. **33**, 223 (1999)
22. C.W. Jones, M. Tsapatsis, T. Okubo, M.E. Davis, Micropor. Mesopor. Mater. **42**, 21 (2001)
23. K. Tsuji, Zeoraito **17**, 162 (2000)
24. W. Yan, E.W. Hagaman, S. Dai, Chem. Mater. **16**, 5182 (2004)

25. M.E. Davis, C. Saldarriaga, C. Montes, J. Garces, C. Crowder, Nature **331**, 698 (1988)
26. K. Maeda, Y. Kiyozumi, F. Mizukami, Angew. Chem. Int. Ed. Engl. **33**, 2335 (1994)
27. K. Maeda, J. Akimoto, Y. Kiyozumi, F. Mizukami, Angew. Chem. Int. Ed. Engl. **34**, 1199 (1995)
28. K. Maeda, J. Akimoto, Y. Kiyozumi, F. Mizukami, J. Chem. Soc., Chem. Commun. 1033 (1995)
29. K. Maeda, Y. Hashiguchi, Y. Kiyozumi, F. Mizukami, Bull. Chem. Soc. Jpn. **70**, 345 (1997)
30. K. Maeda, Y. Kiyozumi, F. Mizukami, J. Phys. Chem. B **101**, 4402 (1997)
31. C. Schumacher, J. Gonzalez, P.A. Write, N.A. Seaton, Phys. Chem. Chem. Phys. **7**, 2351 (2005)
32. J. Gonzalez, R.N. Devi, P.A. Write, D.P. Tunstall, P.A. Cox, J. Phys. Chem. B **109**, 21700 (2005)
33. Ch. Baerlocher, W.M. Meier, D.H. Olson, *Atlas of Zeolite Framework Types*, 5th revised edn. (Elsevier, Amsterdam, 2001)
34. K. Yamamoto, Y. Takahashi, T. Tatsumi, Stud. Surf. Sci. Catal. **135**, 299 (2001)
35. K. Yamamoto, Y. Sakata, Y. Nohara, Y. Takahashi, T. Tatsumi, Science **300**, 470 (2003)
36. K. Yamamoto, Y. Nohara, Y. Domon, K. Takahashi, Y. Sakata, J. Plèvert, T. Tatsumi, Chem. Mater. **17**, 3913 (2005)
37. K. Yamamoto, Y. Sakata, T. Tatsumi, J. Phys. Chem. B **111**, 12119 (2007)
38. K. Yamamoto, T. Tatsumi, Chem. Mater. **20**, 972 (2008)
39. I. Petrovic, A. Navrotsky, M.E. Davis, S.I. Zones, Chem. Mater. **5**, 1805 (1993)
40. R. Astala, S.M. Auerbach, J. Am. Chem. Soc. **126**, 1843 (2004)
41. T. Nakanishi, Y. Domon, K. Yamamoto, T. Tatsumi, *Proceedings of the 20th Meeting of Japan Association of Zeolite* (Japan Association of Zeolite, Tokyo, 2004), p. 34
42. D. Lesthaeghe, G. Delcour, V. van Speybroeck, G.B. Marin, M. Waroquier, Micropor. Mesopor. Mater. **96**, 350 (2006)
43. D.A. Torchia, J. Magn. Reson. **30**, 613 (1978)
44. M. Goldman, L. Shen, Phys. Rev. **144**, 321 (1966)
45. U. Diaz, J.A. Vidal-Moya, A. Corma, Micropor. Mesopor. Mater. **93**, 180 (2006)
46. K. Maeda, Y. Mito, T. Yanagase, S. Haraguchi, T. Yamazaki, T. Suzuki, Chem. Commum. 283 (2007)
47. D.J. Brondani, R.J.P. Corriu, S. El Ayoubi, J.E. Moreau, M.W.C. Man, Tetrahedron Lett. **34**, 2111 (1993)
48. A. Shimojima, K. Kuroda, Chem. Commun. 2672 (2004)
49. K. Landskron, B.D. Hatton, D.D. Perovic, G.A. Ozin, Science **302**, 266 (2003)
50. T. Nakanishi, Y. Domon, S. Inagaki, Y. Kubota, T. Tatsumi, *Proceedings of the 21st Meeting of Japan Association of Zeolite* (Japan Association of Zeolite, Tokyo, 2005), p. 102
51. M. Elanany, B.-L. Su, D.P. Vercauteren, J. Mol. Catal. A **263**, 195 (2007)
52. A. Zheng, L. Wang, L. Chen, Y. Yue, C. Ye, X. Lu, F. Deng, ChemPhysChem **8**, 231 (2007)

Part IV

Characterization and Process of Nanohybridized Materials

9
Characterization of Metal Proteins

Masaki Unno and Masao Ikeda-Saito

9.1 Introduction

Some metals are essential for life. Most of these metals are associated with biological macromolecule like DNA (deoxyribonucleic acid), RNA (ribonucleic acid), and more often with proteins: metals bind or interact with them. A number of protein molecules intrinsically contain metals in their structure. Some of these proteins catalyze unique chemical reactions and perform specific physiological functions. In this chapter, we will shed light on the several metal-containing proteins, termed *metalloproteins*, and other proteins interacting metals. We will also introduce several key techniques which have been used to characterize these proteins. Characterizing these proteins and to understand the relationships between their structures and functions shall continue to be one of the major avenues to solve the mysteries of life. At first, we introduce what are the protein structures and how these proteins interact with metals. In the next section, we discuss the physiological roles of some representative metals. Next, we show two examples of special metal cofactors those help the biological macromolecules to carry out their functions. Then we describe some functions of metalloproteins. Finally, we introduce some physical methods to characterize metalloproteins.

9.2 Proteins and Metals

A protein molecule is made from a long chain of amino acids, each linked to its neighbor through a covalent peptide bond. Each of proteins has its own particular amino acid sequence, and each of the 20 different amino acids has unique side chains with distinctive properties. Some of these side chains are nonpolar and hydrophobic, others are negatively or positively charged, some are reactive, and so on. These side chains limit greatly possible bond angles in a polypeptide chain. This constraint and other steric interactions severely restrict the variety of three-dimensional arrangement of atoms (or conformations) that are possible. The folding a protein is further constrained by many different sets of weak noncovalent bonds: hydrogen bonds, ionic bonds, hydrophobic interactions, and van der Waals attractions. As a result of all of these interactions, each type of protein has a particular three-dimensional structure, which is determined by the order of the amino acids in its chain.

The same weak noncovalent bonds that enable a protein chain to fold into a specific conformation also allow proteins to bind to each other to form larger structures in the cell. If a binding site recognizes the surface of a second protein, the tight binding of two folded polypeptide chains at this site creates a larger protein molecule with a precisely defined geometry. Hemoglobin (Hb), the protein that carries oxygen in red blood cells(see below), is a particularly well-studied example. Hb contains two identical α-globin subunits and two identical β-globin subunits, symmetrically arranged (Fig. 9.1).

Amino acids and proteins alone are not sufficient to perform all the reactions required for life. Living beings have evolved the capability to use

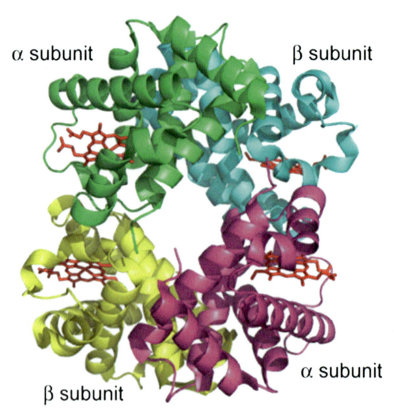

Fig. 9.1. Ribbon diagram of hemoglobin. Stick materials represent heme groups (one heme is contained in each subunit). PDB ID: 1XZ2

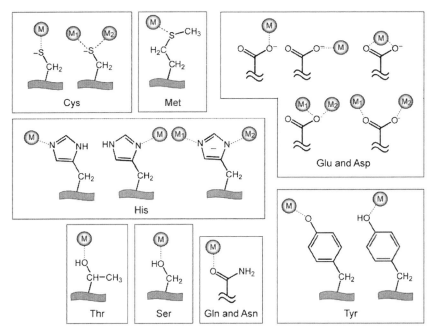

Fig. 9.2. Coordination modes of amino acid side chains. In all the examples, the metal ion is shown as a "M" in a ball, and the coordination bonds as *dashed lines*

inorganic elements for key biological processes and to defend themselves from poisoning by other elements. Metals are often found in proteins as parts of their components as seen in Hb. (The explanations of Hb are given later.) Nature has utilized the unique properties of metal ions to realize numerous special functions related to process of life.

The common metal–donor atoms in proteins are sulfur atoms of cysteines (Cys) and methionines (Met), nitrogen atoms of histidines (His), and oxygen atoms of glutamates (Glu), aspartates (Asp), and tyrosines (Tyr). Threonine (Thr), serine (Ser), and lysine (Lys) [1] side chains are also potential donors. Histidine can bind the metal ion either through its imidazole Nε or Nδ. Glutamate and aspartate can behave as monodentate, bidentate, or bridging ligands. Oxygen atoms of peptide carbonyl moieties can be donor atoms, as well. Some examples of coordination modes are summarized in Fig. 9.2.

9.3 Metals in Biology

In this section, metals of which roles in our body has been known to some extent are selected and the important knowledge for them are introduced.

9.3.1 Iron (Fe)

Fe is easily ionized to 2+ or 3+ states: it readily emits electrons. Furthermore, Fe^{2+} can be readily converted to Fe^{3+} and vice versa. Thus, Fe is an ideal metal for transferring electrons and involved in numerous physiological redox reactions. Also, Fe is contained in many proteins especially related to O_2 supply (see Sect. 9.5).

Approximately, 4–5 g of Fe are contained in our body; 60–70% of Fe exist as a part of Hb in red blood cells. Anemia is found rather in women than in men, because women lose 16–32 mg of Fe per month through menstruation. This value is larger than the amount of Fe obtained from daily diet which is less than ~1.5 mg. Daily excretion of Fe has been reported to be less than 1.0 mg.

Hb contains heme as a "prosthetic group" (see Sect. 9.4), which is the complex of Fe and porphyrin. Hb is made from four molecules (subunits), namely two α-subunits and two β-subunits (Fig. 9.1). O_2 binds Hb in lung and is transferred to myoglobin (Mb) in muscle. The heart of this system is the motion of iron toward the plane of the porphyrin ring upon conversion of deoxy to oxy Hb, which serves as a trigger for cooperative binding of O_2 by the multisubunit Hb protein (Figs. 9.1 and 9.3). The protein has two different quaternary structures designated R, for relaxed, and T, for tense. The former has high affinity for O_2, similar to that of isolated subunits, whereas the latter, tense state has a diminished O_2 affinity. These two conformational states are in equilibrium with one another. In the T state, it is prevalent when all four subunits are unligated, intersubunit interactions constrain the proximal histidine to resist movement into the porphyrin-ring plane and diminish the O_2 binding constant. After ~2 O_2 molecules have bound a Hb molecule, the quaternary structure of the protein switches to the R state, where such constraints are relaxed, and O_2 binding to the remaining two subunits is facilitated. In lung which has high O_2 pressure, Hb binds O_2, whereas, in organs such as muscle which consumes O_2, Hb immediately releases O_2.

Hb containing Fe^{3+} heme cannot bind O_2. Blood in which Hb contains the Fe^{2+}–O_2 heme is in lively red, whereas the blood containing a lot of Fe^{3+}–Hb is blown. The reason why old meat changes to be brownish is that the meat contains many oxidized Hb.

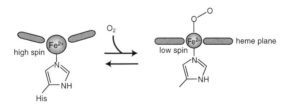

Fig. 9.3. Structural change in O_2 binding to iron–porphyrin complex

9.3.2 Copper (Cu)

Approximately 80 mg of Cu is contained in adult human body. It is proper to obtain 1–2 mg of Cu, everyday. Concentration of Cu is higher in blood plasma of women than that of man. 90–95% of Cu in blood plasma is contained in a protein called ceruloplasmin (CP). Despite extensive studies, the precise functions of CP remain unclear. One of the proposed functions of this enzyme is to catalyze the reaction from Fe^{2+} to Fe^{3+} of the iron absorbed in duodenum $(Cu^{2+} + Fe^{2+} \rightarrow Cu^+ + Fe^{3+})$ [2]. The Fe absorbed in duodenum is bound to a protein termed transferrin in blood and is transferred to bone marrow in which red blood cells are produced. If the iron was Fe^{2+} not Fe^{3+}, it could not bind to transferrin. Therefore, if one lacks Cu, one will also lack Fe and will become anemia. To prevent anemia, we can obtain sufficient Cu from nuts because 1 kg of nuts contain approximately 15 mg of Cu. The reaction is also responsible for protecting blood and membrane lipids from peroxidative damage, thus emphasizing the importance of the synergy between Cu and Fe metabolism in the body. One of the other roles of CP is to regulate activity of amine oxidase to control levels of biogenic amines in the plasma, cerebral spinal, and interstitial fluids. CP is also involved in Cu transport to deliver the metal to extrahepatic tissues.

The red color of the blood of vertebrate is owing to Hb, whereas in crustacean such as crab or prawn and in mollusk such as squids or snail, the color of blood is blue. This ascribes to a protein termed *hemocyanin* (Hc), which contains Cu in the active center and carries O_2 like Hb (see Sect. 9.5).

We have a system for detoxification of Cu in our body, whereas dogs or cats do not. Tradition says that we should not give squid or prawn to dog and cat. This is based on this scientific aspects.

9.3.3 Zinc (Zn)

Fe easily changes to ion whereas Cu is rather stable as the metal. Thus, amount of the solved Fe^{2+} was large whereas Cu was merely solved in Ancient Ocean. Consequently, living things in ancient did not utilize Cu. After appearance of O_2 in earth, Cu was oxidized and begun to be solved in the sea, and then life begun to utilize Cu. It has been thought that because abundant Zn has been solved in the sea, Zn has been given a role to deliver electrons to enzymes related to degradation of proteins or carbohydrates. Zn readily releases H^+ from its bound water to form the basic OH^- around it $[Zn^{2+}(H_2O) \rightarrow Zn^{2+}(OH^-) + H^+]$. The acid (H^+) and the base (OH^-) attack and cleave the peptide bond in proteins; hydrolysis occurs. This is likely to be one of the reasons why Zn is found in many "degradation" proteins.

The amount of Zn in adult human body is 1.4–2.3 g and ~50% of them is contained in blood and ~20% of them localizes in skin. The remaining Zn exists in hair, nail, born, etc. Especially, in male reproductive organs, the concentration of Zn is very high; the amount of Zn in prostate is twice compared to those in other organs such as liver and kidney. This is attributed to the fact that Zn is involved in DNA replication (see Sect. 9.5). Therefore, Zn localizes in the area where cell division actively occurs. For normal adult human body, the required nutrition of Zn per 1 day is 10–15 mg. Under stress condition, the concentration of Zn in blood diminishes. A lack of Zn occasionally causes disorders of taste. Zn also has roles to suppress the hazardous influence of cadmium (Cd) or mercury (Hg) those has entered the body.

9.3.4 Calcium (Ca)

Ca is very reactive: rapidly reacts with water and air, and thus always exists as the chemical compounds. Ca is the most abundant metal contained in human body. Most of them are contained in teeth and bones, naturally. Calcium ion (Ca^{2+}) is related to most physiological functions, for instance, cell division, brain function, regulation of muscle contraction, and blood coagulation.

Ca^{2+} related to muscle contraction exists in sarcoplasmic reticulum that surrounds each myofibril like a net stocking. The signal from the nerve triggers an action potential in the muscle cell plasma membrane, and this electrical excitation spreads rapidly into a series of membraneous folds, the transverse tubules that extend inward from the plasma membrane around each myofibril. When voltage-sensitive proteins in the T-tubule membrane are activated by the incoming action potential, they trigger the opening of Ca^{2+}-release channels in the sarcoplasmic reticulum. Ca^{2+} floods into the cytosol then initiates the contraction of each myofibril. The increase in Ca^{2+} concentration is transient because the Ca^{2+} is rapidly pumped back into sarcoplasmic reticulum by an abundant, ATP-dependent Ca^{2+}-pump. The Ca^{2+} dependence of vertebrate skeletal muscle contraction is due to several proteins (tropomyosin and some troponins) and actin thin filament.

Ca^{2+} is also involved in neuronal signal transmission. Neuronal signals are transmitted from cell to cell at specialized sites of contact known as synapses (Fig. 9.4). The cells are electrically isolated from one another, the presynaptic cell being separated from the postsynaptic cell by a narrow synaptic cleft. A change of electrical potential in the presynaptic cell triggers it to release small signal molecules. The neurotransmitter diffuses rapidly across the synaptic cleft and provokes an electrical change in the postsynaptic cell by binding to transmitter-gated ion channels. For example, influx of Ca^{2+} into the glomus cells causes the release of dopamine. Glomus cells are equipped with a full complement of ion channels, any of which could mediate rapid depolarization in response to hypoxia. In glomus cell, large-conductance Ca^{2+}- and voltage-gated potassium (BK) channel are allosterically activated by intracellular Ca^{2+} and membrane depolarization.

Because Ca^{2+} is also involved in glycogen breakdown, an intake of too much sugar makes one to be lack of Ca^{2+}. It is required for adult people to take Ca^{2+} of 0.6 g per 1 day. Vitamin D has a role to facilitate to absorb Ca^{2+}.

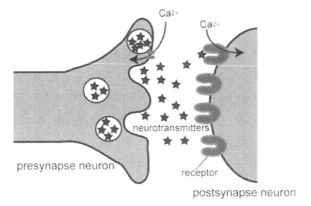

Fig. 9.4. Drawing of a neurotransmission. Neurotransmitters released by activated presynaptic nerve terminal open the receptor channels, allowing Ca^{2+} influx that depolarizes the postsynaptic membrane

9.4 Special Metal Cofactors

The small molecule cofactor is sometimes called a "prosthetic group" indicating its ability to help the macromolecule carry out its function. Typically, metal ions are cofactors that are used to catalyze reactions or to help provide the required geometry of a protein. To realize the biological demand, nature utilizes the inorganic or organic ligands for the metal center. The two common constructs of nonprotein origin are iron–sulfur clusters and tetrapyrroles (including hemes, chlorophylls, and corrins). Nature tries to use them for as many purposes essential to life as possible.

9.4.1 Iron–Sulfur Cofactor

Proteins containing Fe–S clusters are found throughout nature. The most common types of iron–sulfur clusters are illustrated in Fig. 9.5. The [1Fe–0S] does not really represent an iron–sulfur cluster. However, in this type, Fe is coordinated to four Cys residues from a protein and this class is treated together with iron–sulfur cluster. The [2Fe–2S] is the simplest type of iron–sulfur cluster. It usually reveals the occurrence of an $Fe_2S_2(S–Cys)_4^{2-}$ cluster in which two sulfide ligands bridge the two tetrahedrally coordinated iron atoms and four Cys residues from the protein coordinate the [2Fe–2S] cluster. The [3Fe–4S] has the geometry of nonplanar structure that can be derived from [4Fe–4S] core by removing one of the iron ion and the corresponding Cys. In the protein containing [4Fe-4S] clusters, each iron is coordinated by one protein donor (most often a Cys) and by three sulfide ions that in turn bridge three metals. The most ubiquitous iron–sulfur clusters in biology fall into this class. The basic structure is a distorted cube with alternating Fe and S atoms at the corners. The result is two interpenetrating concentric

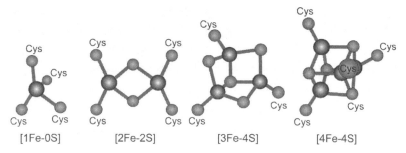

Fig. 9.5. Structures of [Fe–S] clusters. Iron is shown as *bigger ball* and sulfur is shown as *smaller ball*. Cys is the coordinated cysteine residue which is from the protein environment

tetrahedral of four iron and four sulfide atoms. The cube is anchored to the protein by four cysteinyl sulfur atoms at the four Fe corners, resulting in a distorted tetrahedral coordination geometry.

Starting with purified apoprotein, it is sometimes possible to form a functional Fe–S cluster by direct addition of high concentrations of Fe^{2+} and S^{2-} [3] However, in cell, formation of intracellular [Fe–S] clusters does not occur spontaneously but requires complicated biosynthetic machinery, because high concentrations of free metal and sulfide ions both are toxic. Synthesis of Fe–S clusters in bacterial proteins is thought to initiate by the generation of sulfane sulfur. Then, the sulfur is donated to acceptor scaffold proteins, and Fe is also incorporated into the scaffold proteins (perhaps requiring ferredoxin or a ferredoxin-like domain for an electron-transfer step). Finally, the intact cluster is incorporated into target apoproteins by using the molecular chaperons or other cellular components [4–10]. Eukaryotes possess a related, but more complex pathway for Fe–S cluster assembly [11]. Most significantly, the reducing environment of the mitochondrion may be necessary for formation of clusters required in both the mitochondria and cytoplasm. In addition to scaffold proteins or chaperon proteins that have close bacterial counterparts, the mitochondrial protein frataxin has been essential for cluster biosynthesis in eukaryotes [12, 13]. Frataxin acts as an iron chaperone protein to modulate mitochondrial aconitase activity [14]. Furthermore, a mitochondrial membrane transporter is required for Fe–S proteins in cytoplasm [15]. Maturation of cytoplasmic Fe–S proteins also requires glutathione, perhaps functioning to deliver clusters to the apoproteins or assisting in the transfer or insertion of the cofactor [16].

Owing to their remarkable structural plasticity and versatile chemical/elecstronic features [Fe–S] clusters participate in electron transfer, substrate binding/activation, iron/sulfur storage, regulation of gene expression, and enzyme activity [17]. The ability to delocalize electron density over both Fe and S atoms makes Fe–S clusters ideally suited for their primary role in

9 Characterization of Metal Proteins 201

mediating biological electron transport. For example, Fe–S cluster predominates in the early part of the respiratory chain.

9.4.2 Hemes

Of the various redox active prosthetic groups found in nature, iron–porphyrins are most diverse. They are categorized as heme a, heme b, heme c, heme d, heme d_1, heme o, heme P460, and siroheme. Their structures show a common skeleton constituted by the tetrapyrrole ring, but differ in their side chain structures which extend form the edge of the porphyrin. Although extensive electronic delocalization of the porphyrin favors planarity, heme nevertheless possesses a relatively high degree of plasticity. Significant deviations from planarity are often encountered in proteins.

Heme b (or protoporphyrin IX) is the simplest representative (Fig. 9.6). This heme has methyl groups at positions 1, 3, 5, and 8, two propionate groups at positions 6 and 7, and two vinyl groups at positions 2 and 4. Protons are present at the so-called α-, β-, γ-, and δ-mesopositions. In the free heme, the coplanar conformation of the vinyl groups with respect to the heme is energetically favored by the interaction between π orbitals of the vinyl group and the aromatic heme plane. However, once incorporated in proteins, vinyl group orientation is governed by the steric requirements of the protein site and hydrophobic interactions with protein amino acids. Heme b is noncovalently bound to the protein; this is the prosthetic group found in Hb, Mb, cytochrome P450, and cytochrome b5. C-type cytochromes contain heme c which differs from heme b in that the vinyl side chains are covalently linked to two Cys residues by thioether bonds. Other heme derivatives include heme a, present in cytochrome c oxidase and heme d_1 found in nitrite reductase. Heme P460 in the N-terminal domain of hydroxylamine oxidoreductase is attached to the protein matrix through two thioether bonds analogous to those of

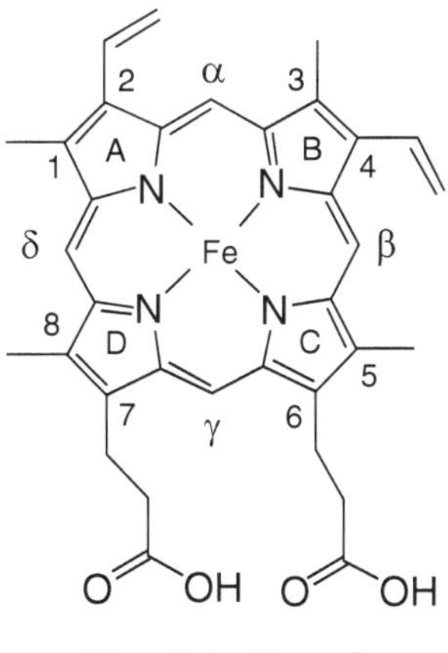

Fig. 9.6. Heme b

c-type cytochromes; here an additional covalent bond exists between heme α-mesocarbon and the Cϵ of Tyr on an adjacent subunit.

Heme biosynthesis is a multistep process that involves many enzymes. Depending organism, enzymes convert either succinyl-CoA or glutamate to 5-aminolevulinic acid (ALA). Two ALA combine to form porphobilinogen (PBG), and four PBG are condensed to uroporphyrinogen III. A series of decarboxylations and oxidations yield protoporphyrin. Ferrochelatase catalyzes the final step in heme synthesis, the insertion of Fe to generate protoheme IX [18].

The biological heme degradation and turnover is initiated by a family of enzymes termed *heme oxygenases* (HO) that catalyze oxidative degradation of Fe^{3+} protoporphyrin IX to biliverdin IX, Fe^{2+} and CO in the presence of reducing equivalents [19]. In mammals, the biliverdin is further reduced by biliverdin reductase to bilirubin which is subsequently conjugated with glucuronic acid and excreted.

Until fairly recently heme was considered critical in three classes of proteins: heme enzymes like peroxidases and P450, the O_2 storage/transport globins, and electron-transfer cytochromes (see Sect. 9.5). To this list, regulatory proteins that bind heme resulting in downstream signaling processes and various gas sensing proteins must now be added.

9.5 Functions of Metalloproteins

9.5.1 O_2 Transport

One of the special functions of metalloproteins is respiration. Three kinds of proteins those transfer oxygen (O_2) have been known: hemoglobin (Hb)-myoglobin (Mb) type, hemocyanin (Hc), and hemerythrin (Hr). In these, O_2 molecule binds to the iron center or the copper center. In Hb and Mb, the O_2 binding site is at the iron–porphyrin complex (Fig. 9.6), which changes the structure upon binding of O_2 as discussed above (Fig. 9.3). In other two respiration protein families, one pair of metal ions is utilized in O_2 binding reaction. In Hc, O_2 binds between two Cu atoms; whereas in Hr, O_2 molecule coordinates at terminal of Fe_2 unit. In these two cases, the ferrous or deoxy metals are oxidized by O_2 and then, the peroxo or hydroperoxo forms are generated. Although these proteins have different structures in their reaction centers from that found in Hb and Mb, O_2 binds to their metals reversibly as found in Hb and Mb. Chemical strategies of reactions in the three kinds of proteins are similar. Nature provided different metals to different living species for the same function.

9.5.2 Electron Transfer

Net electron transfer occurs in protein molecules and causes reduction–oxidation (redox) reaction without overall chemical change in a substrate molecule. O_2 as Lewis base binds reversibly to the iron in the porphyrin as Lewis acid. These proteins donate or accept of the electrons required for redox reactions to realize the specific functions as, for instance, N_2 fixation.

There are two electron-transfer centers widely found in bioinorganic chemistry; iron–sulfur clusters and cytochromes. The iron–porphyrin groups in cytochromes resemble that found in Hb, which is used for O_2 transport.

9.5.3 Structural Roles for Metal Ions

Several families of proteins that regulate the expression of genes have recently been found to contain Zn^{2+} ion. In these proteins, the metal ions play a structural role, forming the central core of small nucleic-acid-binding domains referred to as "zinc fingers" (Fig. 9.7). They were first identified in a transcription factor, TFIIIA [20]. The DNA-binding properties of zinc fingers have been explored in exquisite detail. In the class of proteins, the metal-binding domains appear to be directly involved in interactions with DNA. Other classes of Zn^{2+}-containing proteins involved in gene regulation have been discovered in which metal-stabilized domains may play other roles. For example, Zn^{2+} ions are present in most DNA and RNA polymerases, and the functions of the metal ions in these important enzymes remain to be elucidated in detail.

9.5.4 Function of Metalloenzymes

Metalloenzyme which catalyzes the specific reaction is one of the subclasses of metalloproteins. One may systematize the field in terms of particular elements that define classes of enzymes (those containing Fe, Cu, Zn, etc.), or, one can organize the subject in terms of reaction types (oxygenase, ligases, proteases, etc.). A net chemical reaction occurs in the substrate by the actions of metalloenzymes. They contain interesting reactions that are not faked by synthetic chemistry of small molecules. For example, they are catalytic reduction from N_2 to NH_3, oxidation of water to O_2, and reduction of gem-diols to monoalcohols, etc. Metalloenzymes are classified according to their reactions.

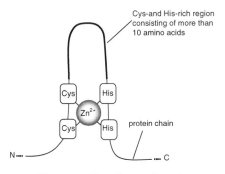

Fig. 9.7. Zinc finger domain

Within each category there are usually several kinds of metal centers that can catalyze the required chemical transformation, a situation analogous to that already encountered for respiratory proteins. The reasons for this diversity are shrouded in evolutionary history, but most likely include bioavailability of a given element at the geosphere/biosphere interface during the initial development of a metalloenzyme, as well as pressure to evolve multiple biochemical pathways to secure the viability of critical cellular functions.

9.6 Physical Methods in Characterizing Metalloproteins

9.6.1 X-Ray Crystallography

To understand cellular processes, knowledge of the three-dimensional structure of enzymes and other macromolecules is vital. It is easier understood how something works if one knows what it looks like, and it is true also for proteins. X-ray crystallography is one of the most powerful methods to achieve it.

The pioneering work by Perutz and Kendrew [21] on the structure of Hb and Mb in the 1950s led to increase in the number of proteins whose structure was determined using X-ray diffraction. The introduction of sophisticated computer hardware and software dramatically reduced the time required to determine a structure while increasing the accuracy of the results. Recombinant DNA technology has further stimulated interest in protein structure determination. A protein that was difficult to isolate in sufficient quantities from its natural source can often be produced in arbitrarily large amounts using expression of its cloned gene in a microorganism. Also, a protein modified by site-directed mutagenesis of its gene can be created for scientific investigation and industrial application. Here, X-ray diffraction plays a key role in guiding the molecular biologist to the best amino acid positions for modification. Moreover, it is often important to learn what effect a change in a protein's sequence will have on its three-dimensional structure. Chemical and pharmaceutical companies have become very active in the field of protein structure determination because of their interest in protein-based drug design.

Metalloprotein Crystals

The first essential step in determining the X-ray structure of a protein is to grow crystals of sufficient size and quality. A crystal of organic material is a three-dimensional periodic arrangement of molecules. When the material precipitates from a solution, its molecules attempt to reach the lowest free energy state. It is surprising to observe that even large protein molecules follow this principle, although occasionally they unfortunately do not crystallize.

The search for a minimum free energy and as a consequence the regular packing of molecules in a crystal lattice often leads to a symmetric relationship between the molecules. However, protein crystals do not have all the 230 space groups those are permitted in crystals of small molecules, because, in protein crystals, the application of mirror planes and inversion centers would change the asymmetry of the amino acids: L-amino acid would become a D-amino acid, but these are never found in proteins. Protein crystals always consist of not only protein molecules but also a large number of water molecules, quite different from small molecule crystals.

The high-energy photons of X-rays have a harmful effect on living tissue. This also applies to protein crystals, which undergo radiation damage upon X-ray exposure. Radiation causes the formation of radicals and photoelectron in the crystal, leading to subsequent chemical reactions, which gradually destroy the crystalline order. Consequently, the X-ray pattern from the crystal dies away. This problem can be usually overcome by cooling the crystals down to liquid N_2 temperature (but see [22]). However, in the metalloprotein crystallography, another problem remains even under such a low temperature condition. Electron librated within the protein crystal by X-rays during crystallographic data collection can alter the redox state of the active center [23]. Especially, redox enzymes have evolved to channel electrons efficiently into an oxidized active site. Sometimes, metals in proteins are reduced by X-rays at dose smaller than those required for the collection of a complete diffraction data set. To monitor the redox state of the active center of the metalloprotein, in some cases microspectrophotometer has been used in the recent crystallographic studies for metalloproteins [24]. Details of microspectrophotometer techniques will be discussed later.

Crystals of metalloproteins also have some advantages. First, they are commonly colored, thus visual inspection tells us whether these crystals are made from proteins or salts (Fig. 9.8). Further, we can easily pick up and treat the crystals for the following experiments. Second, phasing problems could be solved by using anomalous diffraction methods without the attachment of heavy-atom-containing reagents, because the metalloproteins themselves contain the metal(s) that have the unique X-ray absorption edges. The phase problem will be discussed in the next paragraph.

Fig. 9.8. Crystals of a metalloprotein. The crystals of the complexes of heme and HmuO, a heme oxygenase from *Corynebacterium diphtheriae*

Phase Problems

The scattering is an interaction between X-rays as electromagnetic waves and the electrons. The wave scattered by the crystal may be described as a summation of the enormous number of waves, each scattered by one electron in the crystal. The scattered wave has zero amplitude unless the Laue conditions (each of indexes hkl is integer) and Bragg's law ($2d \sin \theta = \lambda$, here d, θ, λ represent lattice plane distance in the crystal, an angle between the incident beam and the reflected beam, and wavelength, respectively) are fulfilled.

From the direction of the diffracted beams the dimensions of the unit cell can be derived. But we are, of course, more interested in the content of the unit cell, that is, in the structure of the protein molecules. The molecular structure and the arrangement of the molecules in the unit cell determine the intensities of the diffracted beams. Therefore, a relationship must be found between the intensities of the diffracted beams and the crystal structure. The result of an X-ray structure determination is the electron density in the crystal, because X-rays are scattered almost exclusively by the electrons in the atoms and not by the nuclei.

The fundamental equation for its calculation is

$$\rho(xyz) = \frac{1}{V} \sum_h \sum_k \sum_l |F(hkl)| \exp[-2\pi i(hx + ky + lz) + i\alpha(hkl)].$$

In this equation $\rho(xyz)$ is expressed as a Fourier transformation of the structure factors $\mathbf{F}(hkl)$. The amplitude of these structure factors is obtained from the intensity of the diffracted beam after application of certain correction factors: $I(hkl) = |F(hkl)|^2$. The phase angles $\alpha(h\,k\,l)$ cannot be derived in a straightforward manner, but can be found in an indirect way. In principle, four techniques exist for solving the phase problem in protein X-ray crystallography:

1. The isomorphous replacement method, which requires the attachment of heavy atoms to the protein molecules in the crystal.
2. The multiple wavelength anomalous diffraction method. It depends on the presence of sufficiently strong anomalously scattering atoms in the protein structure itself. Anomalous scattering occurs if the electrons in an atom cannot be regarded as free electrons.
3. The molecular replacement method for which the similarity of the unknown structure to an already known structure is a prerequisite.
4. Direct methods, the methods of the future, still in a stage of development toward practical application for proteins.

In the conventional X-ray structure determinations, the electrons in the atom have been regarded as free electrons. Under this condition, the intensities of a reflection hkl and its Bijvoet mate $-h-k-l$ are equal. However, this is no longer true if the X-ray wavelength approaches an absorption edge wavelength.

9 Characterization of Metal Proteins   207

Whereas free electrons have phase difference of 180° with respect to the incident beam, the diffracted beam of the inner shell electrons of the heavy atoms does not differ 180° in phase from the incident beam. The atomic scattering factor can be written as

$$f_{anom} = f + \Delta f + \mathrm{i}f'' = f' + \mathrm{i}f''.$$

Δf is the change in the electron scattering factor along real axis and $\mathrm{i}f''$ along the imaginary axis. Therefore, the intensities of a reflection hkl and its Bijvoet mate are no longer equal. The effects (anomalous effect and wavelength dependence) greatly contribute to determine the phase angles of the reflections, although intensities must be measured with extreme accuracy. In principle the anomalous scattering by heavy atoms contributes to the determination of the protein phase angles as much as the isomorphous replacement does. The principle of this method is not explained here (see general reference 4).

9.6.2 Optical Absorption Spectroscopy

X-ray crystallography does not delineate the electronic structure of the metal center(s). Light absorption spectroscopy is one of the most convenient and useful tools for probing electronic structure of the metal centers. Molecules possess not one but several accessible electronic energy levels. Transitions between various electronic states lead to the absorption of energy in the ultraviolet, the visible, and, for many transition-metal complexes, the near infrared regions of the electromagnetic spectrum. There are three major sources of electronic spectra in metal complexes, namely, internal ligand bands such as are found in porphyrins, transitions associated purely with metal orbitals such as d–d transitions, and charge transfer bands between metal and ligand. A large number of spectral features of metalloproteins have been documented. These spectral properties have been used to assign metal oxidation states, follow reactions, and identify chemical species in newly discovered systems.

Mb is rich with spectroscopic opportunities. Its red color is derived from the absorption of blue light by the heme group, and the oxidation state can be discerned by the naked eye. The strong Soret band at approximately 400 nm is accompanied by a weaker absorption in the 500–600 nm range whose number and location of maxima depend on the ligation species. These optical spectra can be used to identify the ligation and the spin state of the iron. An example is shown in Fig. 9.9, an absorption spectrum of the O_2 bound form of Mb at 100 K [25].

Absorption spectroscopy is also a very versatile method in that measurements can be made using liquid, gaseous, or solid, samples. As mentioned in Sect. 9.6.1, in recent crystallographic studies for metal-containing macromolecules, the microphotometer is sometimes used to identify the electronic state of the metals in centers of the protein in the crystal. Single-crystal microspectrophotometry in parallel with X-ray data collection allows

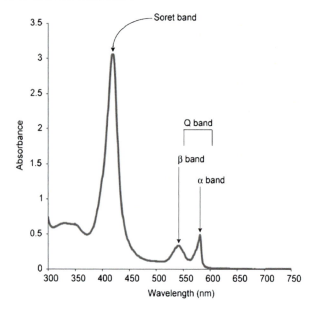

Fig. 9.9. Absorption spectrum of frozen oxy-Mb solution measured at 100 K. Courtesy of Dr. G.S. Sligar

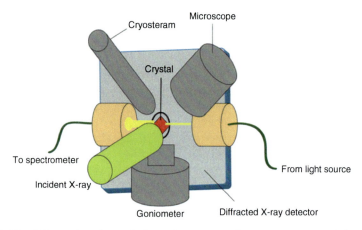

Fig. 9.10. Schematic view of the microspectrophotometer setup. The sample mounted in a nylon loop is cooled by a cryostream system delivering cryogenic gaseous nitrogen. The sample is positioned at the intersecting focal points of the mirror objectives with a goniometer and can be visually inspected with a microscope

structural enzymologist to be confident that structural data relate to the particular reaction state in which they are interested (Fig. 9.10).

Horse radish peroxidase (HRP) catalyzes oxidation of various organic compounds through catalytically active HRP Fe(IV) species which are generated by the reaction with hydrogen peroxide. Crystal structure determination of the Fe(IV) peroxidase intermediates had been proven difficult, because photoelectrons generated during X-ray exposure reduce chemically active high-valence Fe(IV) intermediates generated by the reaction with hydrogen peroxide. Hajdu and coworkers [26] have shown that, in the case of HRP, X-ray derived photoelectrons are used to cleave the bound O_2 molecule. To investigate this photoelectron-driven reduction of O_2, X-ray dose was spread over multiple crystals and composite data sets were generated with equivalent radiation dose. A single-crystal microspectrophotometer has been used to monitor the changes in redox state during X-ray exposure, correlating the spectrally observed species to the crystal structures observed with increasing radiation dose. This strategy allows, for the first time, determination of the structures of the X-ray-sensitive high-valence redox intermediate of HRP.

9.6.3 X-Ray Absorption Spectroscopy

Because X-rays have wavelengths on the order of atomic dimensions, these highly energetic photons can be used to study the molecular structure of materials. X-ray absorption spectroscopy (XAS) can yield limited molecular structural information on noncrystalline samples. The absorption of X-rays can excite the $1s$ (K edge) or $2s$, $2p$ (L edge) electrons of an element to empty localized orbitals or, for higher energy X-ray photons, the continuum. This phenomenon, which has been known for decades, became of great utility following the availability of high-intensity, tunable X-ray beams from synchrotron radiation source.

A typical X-ray absorption spectrum is shown in Fig. 9.11. At a well-defined X-ray photon energy, a sharp rise in absorption coefficient is observed. This rise is called an X-ray absorption *edge* (e.g., in the case of Cu, the K-absorption edge is 1.380 Å) and is due to electron dissociation from a core level of one type of atom (the absorbing atom) in the sample. Spectral features in the edge region are sensitive to the electronic structure of the absorbing atom and can often be used to identify the geometric arrangement of atoms around the absorbing atom. Modulation of the X-ray absorption energy spectrum by back-scattering from neighboring atoms produces an extended X-ray absorption fine structure (EXAFS) from which details about metal-coordination geometries could be extracted. The method has been widely applied in metalloprotein structural studies and has the great advantage, because noncrystalline solid and even solution samples could be studied.

XAS has also been used in the crystal structure analysis of metalloproteins to determine the wavelength for the intensity data collection. The anomalous scattering by heavy atoms contributes to determination of the protein phase angles as mentioned above. Condition for application of the method is that the wavelengths are carefully chosen to optimize the difference in intensity between Bijovet pairs and between the diffraction at the selected wavelength. To determine the absorption edge of the protein containing the

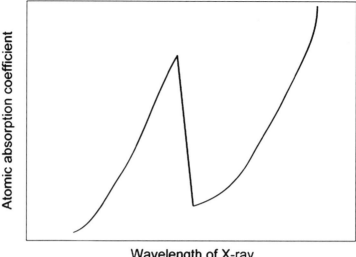

Fig. 9.11. A typical X-ray absorption spectrum. The atomic absorption coefficient or "atomic cross section" for absorption μ_a (cm^2) is defined by $\mu_a = (\mu/\rho) \times (A/N)$, where μ/ρ is the mass absorption coefficient (cm^2 g^{-1}), ρ is the density of the absorber, A is its atomic weight, and N is the Avogadro's number. μ (cm^{-1}) is the total linear absorption coefficient defined by $I = I_0 \exp[-\mu t]$ with t the thickness of the material (cm), I_0 the intensity of the incident, and I the intensity of the transmitted beam

anomalous scatter, X-ray absorption fine structure (XAFS) is commonly used in crystallographic data collection at synchrotron facilities.

9.6.4 Vibrational Spectroscopy

As might be imagined, the vibrational spectra of proteins are extremely complex. However, metalloproteins in which the metal center has associated electronic transitions offer a special advantage producing resonance Raman (RR) spectra in which the ligand vibrations are markedly enhanced. Therefore, RR spectroscopy is capable of providing useful structure probe of metal complexes and of metal centers in biological systems. Metal centers are frequently involved in allowed electronic transitions, due to ligand–metal charge-transfer (LMCT) transitions, or due to the influence of the metal on π–π* transitions of a ligand (e.g., porphyrin), and consequently they give wide scope to the application of RR spectroscopy. The d–d electronic transitions that often give simple coordination compounds their characteristic colors are electric-dipole forbidden, and are ineffective in resonance enhancement.

The porphyrin vibrational modes have been well characterized via extensive studies of heme proteins and model heme complexes, and certain functionally critical modes such as iron-histidine (Fe-His) stretching mode as well as iron-exogenous ligand modes have been identified. Advances in instrumental techniques allow for sensitive detection of differences in heme vibrational modes induced by changes in the protein environment. RR spectroscopic investigations of those chromophores can be used to identify cysteine, tyrosine, or exogenous azide (N_3^-) coordination, respectively. If the vibrational frequency of a coordinating ligand falls in a spectral region having little or no overlap with protein bands, or where the band intensity is very strong, both infrared and RR methods can be employed. Solvent and protein matrix background signals can often be averaged out by difference spectroscopy, where spectra of the metalloprotein in two forms are subtracted from one another. The two forms might be with and without added substrate or inhibitor molecules, or isotopically substituted ligand derivatives such as ^{16}O–^{16}O and ^{18}O–^{18}O, or H_2O and D_2O.

9.6.5 Electron Paramagnetic Resonance Spectroscopy

Electron paramagnetic resonance (EPR) is a technique that probes certain properties and environment of a paramagnetic center by characterizing the interaction of that center with an applied magnetic field. This method requires samples with unpaired electrons and is ideally suited for studying many metalloproteins, including those containing Cu^{2+}, Co^{2+}, Fe^{2+}, Fe^{3+}, Mn^{2+}, Mn^{3+}, Mo^{5+}, and metal clusters such as Fe_2O^{3+} and $Fe_4S_4^{+,3+}$. The protein part is diamagnetic and therefore "EPR silent." The method is very sensitive, being able to detect high-spin ferric ions at concentration as low as the μM range. The sensitivity of an EPR measurement increases with decreasing temperature. Thus, EPR spectra are most commonly recorded at low temperatures, 4–77 K. Interestingly, EPR spectroscopy was also used in the isolation and purification of iron-containing proteins, such as ferredoxins. Since these electron-transfer proteins have no measurable catalytic activity, their purity could not be established by the classic biochemical approach of enzymatic activity measurements. Instead, the appearance and integrated signal intensity of their EPR spectra were monitored following the usual column-chromatographic purification steps.

EPR signals arise from the interaction of the molecular magnetic dipole moment $(-\beta(L + g_eS))$, where L is the angular momentum, S is the total spin, β is the Bohr's magneton, and g is approximately 2.002319 and is the free electron g-value (but in practice, electrons in metal system typically exhibit g-values quite different from 2.0) with an externally applied magnetic field.

We see that there are several important structural interactions that dramatically affect the appearance of the EPR spectrum. The major one of these is the spin–orbit interaction. For example, the crystals of ferricytochrome c, which is a low-spin heme protein and thus contains a single unpaired electron, clearly show that the g-value is orientation dependent. The implication, therefore, is that the magnetic moment associated with the single unpaired electron of the heme iron is not a simple number, it exhibits the property of being anisotropic.

There are processes those impose additional structure on EPR envelope. The more common are the consequences of the paramagnetic electron being associated with nucleus that possesses a nuclear spin; these are the nuclear hyperfine and superhyperfine interactions. Less common are the magnetic effects due to the paramagnetic electron being situated close to a second paramagnetic center such as is found when two or more iron–sulfur clusters are present in the same protein.

An important variant of the EPR method is electron-nuclear double resonance spectroscopy (ENDOR), in which the EPR transition is saturated while a second irradiation signal is scanned across the nuclear magnetic resonance (NMR) frequency range of an isotope in the complex that might cause hyperfine splitting of the EPR lines. Usually the latter cannot be directly resolved. Irradiation of the sample at specific NMR frequencies causes the EPR signal to become unsaturated, and an ENDOR signal results. A typical application is study of ^{17}O-labeled substrates, such as H_2O, O_2, or H_2O_2, with EPR active metal centers, such as iron porphyrins.

9.6.6 Nuclear Magnetic Resonance Spectroscopy

NMR spectroscopic methods have been highly useful for the study of molecular structure and dynamics. From ^1H and, to a lesser extent, ^{13}C, ^{15}N, and ^{31}P NMR spectra distance (from nuclear Overhauser studies) and torsion angular (from coupling constants) information about the three-dimensional structure of biopolymers, including bound metal ions, can be obtained. Structure determination through NMR spectroscopy has several advantages with respect to X-ray, even if in general it is more time consuming and has severe limitations in the size of the proteins and in resolution that can be afforded [27]:

- NMR is able to provide both dynamic and structural information on proteins in a variety of different states in solution at atomic resolution.
- NMR is unique in characterizing internal mobility at atomic level over a large time range, as well as hydration properties, solvent accessibility, and exchange processes.
- NMR has the power to study transient complexes and weak protein–protein and protein–ligand interactions.
- NMR has the power to characterize proteins with some level of internal disorder or in noncompletely folded states, to probe residual structure, and to study low molecular weight aggregates. This aspect is particularly relevant as it is becoming more and more evident that a relatively large number of proteins exist in vivo in partially unfolded states. This property allows the proteins a facile interaction with their partners, and also makes them more prone to aggregation.

9 Characterization of Metal Proteins

Problems and challenges need to be overcome when determining structures of metalloproteins. Indeed, metal ion, which in most cases has no NMR accessible nucleus, constitutes a point of discontinuity in the network of inter-nuclear dipolar and scalar couplings. Even more severe effects need to be taken into account when the metal ion is paramagnetic, i.e., it contains one or more unpaired electrons.

A large number of metalloproteins contain paramagnetic metal ions which posse unpaired electrons, or metal ions which cycle between different oxidation states, one of which is paramagnetic. The presence of a paramagnetic center not only affects the NMR parameters and might create problems in signal detection, but it can also dramatically reduce the intensity of nuclear Overhauser effects (NOEs) and the efficiency in the transfer of scalar couplings in homonuclear and heteronuclear experiments, thus making spectral assignment and determination of structural restraints are not possible through standard approaches. These problems can sometimes be overcome or reduced with suitable, tailored experiments and the effects induced by a paramagnetic center on the NMR parameters can even be turned into precious sources of information on its surroundings, thus compensating for the lack or reduction of the classical structural restraints. Therefore, NMR is particularly suited for structural determination of systems containing metal ions which may undergo redox reactions under irradiation with X-ray, and it can also be efficiently used to characterize oxidation state-dependent structural and dynamical changes.

9.6.7 Mössbauer Spectroscopy

This method, in which γ-rays emitted from the nuclei of source element in an excited nuclear state are absorbed by the same element in a sample, is especially valuable for studies of ^{57}Fe in bioinorganic chemistry. Iron-57 is a stable isotope with 2.2% natural abundance. It is important to know that the Mössbauer phenomenon rests on the fact that γ-radiation (high-energy photons) can be emitted or absorbed without imparting recoil energy ($E_R = E_\gamma^2/2M = 1.95 \times 10^{-3}$ eV, where $E_\gamma = 14.4$ KeV and M is the mass of the ^{57}Fe nucleus). In a solid, most of the recoil energy is converted into lattice vibration energy. Mössbauer has shown, however, that there is a certain probability, described by the recoil-free fraction f, that γ-emission and absorption take place without recoil. To observe the Mössbauer effect, ^{57}Fe nucleus must be placed in a solid or frozen solution matrix.

In ^{57}Fe Mössbauer spectroscopy, transitions between the nuclear ground state of ^{57}Fe (nuclear spin $I_g = 1/2$; nuclear g-factor, $g_g = 0.181$) and a nuclear excited state at 14.4 KeV ($I_e = 3/2$, $g_e = -0.161$, nuclear quadrupole moment Q) are observed (Fig. 9.12). The energy of this transition depends on the immediate environment of the absorber and the shift relative to a given standard (i.e., iron foil) is called "isomer shift" (IS). The $M_I = \pm 1/2, \pm 3/2$ components of the $I = 3/2$ excited state of ^{57}Fe are split in energy by the electric quadruple interaction (quadrupole splitting, QS). The physical origin

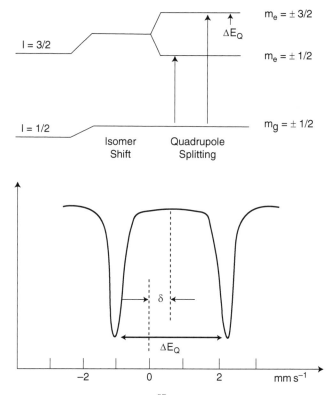

Fig. 9.12. Quadrupole splitting of the ^{57}Fe excited state and the isomer shift. Also shown is a typical Mössbauer spectrum for a sample containing randomly oriented molecules. In the absence of heterogeneities or relaxation effects the absorption lines have Lorentzian shape. The velocity scale is drawn relative to the centroid of Fe metal

of this effect is the interaction of the nuclear quadrupole moment (existent for all nuclei with $I > 1/2$) with the electric field gradient at the iron nucleus. From the IS some information about the metal oxidation state as well as the types of ligands coordinated to iron can be obtained (Table 9.1). More powerful from a structural point of view, however, is the QS, which reveals the asymmetry of the electric field surrounding the metal center. Both the IS and QS are sensitive reporters of the spin state, valence state and covalency of the iron sites.

Mössbauer spectroscopy can be used to study magnetic coupling phenomena and, for mixed valence species, to provide an estimate of the rate of internal electron-transfer reactions. The time scale of Mössbauer method, 10^{-7} s, is well suited for this task. ^{57}Fe spectroscopy has been especially valuable in sorting out the various kinds of iron–sulfur and Fe–Mo–S clusters in nitrogenase and the iron formal oxidation states. These studies are facilitated

9 Characterization of Metal Proteins 215

Table 9.1. Values for ΔE_Q and δ for some compounds of biological interest[a] (from general reference 3)

Oxidation state	Spin State	Ligand Set	ΔE_Q (mm s^{-1})	δ (mm s^{-1})
Fe(IV)	$S = 2$	Fe–(O,N)	0.5–1.0	0.0–0.1
	$S = 1$	Hemes	1.0–2.0	0.0–0.1
		Fe–(O,N)	0.5–4.3	−0.20–0.10
Fe(III)	$S = 5/2$	Hemes	0.5–1.5	0.35–0.45
		Fe–S	< 1.0	0.20–0.35
		Fe–(O,N)	0.5–1.5	0.40–0.60
	$S = 3/2$	Hemes	3.0- -3.6	0.30–0.40
	$S = 1/2$	Hemes	1.5–2.5	0.15–0.25
		Fe–(O,N)	2.0–3.0	0.10–0.25
Fe(II)	$S = 2$	Hemes	1.5–3.0	0.85–1.0
		Fe–S	2.0–3.0	0.60–0.70
		Fe–(O,N)	2.0–3.2	1.1–1.3
	$S = 0$	Hemes	< 1.5	0.30–0.45

[a] The entries give typical numbers at 4.2 K. The isomer shifts are quoted relative to the centroid of Fe metal at 298 K, the standard most commonly used. [Fe–(O,N)] represents hexacoordinate and pentacoordinate Fe sites with oxygen and/or nitrogen ligands

by temperature-dependent spectral measurements, down to 4.2 K, in the presence of applied external magnetic fields, and by computer simulation of the spectrum to provide IS, QS, and nuclear-electron magnetic hyperfine coupling parameters. Although ^{57}Fe Mössbauer spectroscopy has dominated bioinorganic applications of the technique, other possible nuclei are ^{40}K, ^{61}Ni, and ^{67}Zn.

General References

1 S.J. Lippard, J.M. Berg, *Principles of Bioinorganic Chemistry* (University Science Books, Mill Valley, CA, 1994)

2 I. Bertini, H.B. Gray, E.I. Stiefel, J.S. Valentine, *Biological Inorganic Chemistry – Structure & Reactivity* (University Science Books, Mill Valley, CA, 1994)

3 L. Que Jr., *Physical Methods in Bioinorganic Chemistry – SPECTROSCOPY and MAGNETISM* (University Science Books, Sausalito, CA, 2000)

4 J. Drenth, *Principles of Protein X-ray Crystallography* (Springer, New York, 1999)

5 B. Alberts, A. Johnson, J. Lewis, M. Raff, K. Roberts, P. Water, *The Cell* (Garland Science, New York, 2002)

6 J. Kuhar, R.P. Hausinger, Chem. Rev. **104**, 509 (2004)
7 H. Sakurai, *Why Are Metals Needed for Human Body?* (Japanese) (Kodansha, Tokyo, 2000)
8 WALK Co. Ltd., supervised by T. Masumoto, *Small Dictionary for Everything About Metals* (Japanese) (Kodansha, Tokyo)

References

1. C.G. Mowat, E. Rothery, C.S. Miles, L. McIver, M.K. Doherty, K. Drewette, P. Taylor, M.D. Walkinshaw, S.K. Chapman, G.A. Reid, Nat. Struct. Mol. Biol. **11**(10), 1023 (2004)
2. N.E. Hellman, S. Kono, G.M. Mancini, A.J. Hoogeboom, G.J. De Jong, J.D. Gitlin, J. Biol. Chem. **277**(48), 46632 (2002)
3. R. Malkin, J.C. Rabinowitz, Biochem. Biophys. Res. Commun. **23**(6), 822 (1966)
4. P. Yuvaniyama, J.N. Agar, V.L. Cash, M.K. Johnson, D.R. Dean, Proc. Natl. Acad. Sci. USA **97**(2), 599 (2000)
5. C. Krebs, J.N. Agar, A.D. Smith, J. Frazzon, D.R. Dean, B.H. Huynh, M.K. Johnson, Biochemistry **40**(46), 14069 (2001)
6. S. Ollagnier-de-Choudens, T. Mattioli, Y. Takahashi, M. Fontecave, J. Biol. Chem. **276**(25), 22604 (2001)
7. H.D. Urbina, J.J. Silberg, K.G. Hoff, L.E. Vickery, J. Biol. Chem. **276**(48), 44521 (2001)
8. A.D. Smith, J.N. Agar, K.A. Johnson, J. Frazzon, I.J. Amster, D.R. Dean, M.K. Johnson, J. Am. Chem. Soc. **123**(44), 11103 (2001)
9. H. Mihara, S. Kato, G.M. Lacourciere, T.C. Stadtman, R.A. Kennedy, T. Kurihara, U. Tokumoto, Y. Takahashi, N. Esaki, Proc. Natl. Acad. Sci. USA **99**(10), 6679 (2002)
10. M. Nuth, T. Yoon, J.A. Cowan, J. Am. Chem. Soc. **124**(30), 8774 (2002)
11. R. Lill, G. Kispal, Trends Biochem. Sci. **25**(8), 352 (2000)
12. T. Lutz, B. Westermann, W. Neupert, J.M. Herrmann, J. Mol. Biol. **307**(3), 815 (2001)
13. A. Rotig, D. Sidi, A. Munnich, P. Rustin, Trends Mol. Med. **8**(5), 221 (2002)
14. A.L. Bulteau, O'H.A. Neill, M.C. Kennedy, M. Ikeda-Saito, G. Isaya, L.I. Szweda, Science **305**(5681), 242 (2004)
15. G. Kispal, P. Csere, C. Prohl, R. Lill, EMBO J. **18**(14), 3981 (1999)
16. K. Sipos, H. Lange, Z. Fekete, P. Ullmann, R. Lill, G. Kispal, J. Biol. Chem. **277**(30), 26944 (2002)
17. H. Beinert, R.H. Holm, E. Munck, Science **277**(5326), 653 (1997)
18. H.A. Dailey, T.A. Dailey, C.K. Wu, A.E. Medlock, K.F. Wang, J.P. Rose, B.C. Wang, Cell Mol. Life Sci. **57**(13–14), 1909 (2000)
19. M. Unno, T. Matsui, M. Ikeda-Saito, Nat. Prod. Rep. **24**(3), 553 (2007)
20. J. Miller, A.D. McLachlan, A. Klug, EMBO J. **4**(6), 1609 (1985)
21. J.C. Kendrew, M.F. Perutz, Annu. Rev. Biochem. **26**, 327 (1957)
22. E. Garman, J.W. Murray, Acta Crystallogr. D Biol. Crystallogr. **59**(Pt. 11), 1903 (2003)
23. B. Chance, P. Angiolillo, E.K. Yang, L. Powers, FEBS Lett. **112**(2), 178 (1980)

24. A.R. Pearson, A. Mozzarelli, G.L. Rossi, Curr. Opin. Struct. Biol. **14**(6), 656 (2004)
25. M. Ibrahim, I.G. Denisov, T.M. Makris, J.R. Kincaid, S.G. Sligar, J. Am. Chem. Soc. **125**(45), 13714 (2003)
26. G.I. Berglund, G.H. Carlsson, A.T. Smith, H. Szoke, A. Henriksen, J. Hajdu, Nature **417**(6887), 463 (2002)
27. F. Arnesano, L. Banci, M. Piccioli, Q. Rev. Biophys. **38**(2), 167 (2005)

10

Electron Microscopy Characterization of Hybrid Metallic Nanomaterials

Daisuke Shindo and Zentaro Akase

10.1 Introduction

In order to understand the excellent properties of nanoscale hybridized materials, it is very important to investigate the microstructures and interfaces of these materials at the nanometer scale. In this chapter, we present the basic principles of transmission electron microscopy and its applications to these materials. In addition to high-resolution transmission electron microscopy (HREM) and high-angle annular dark-field (HAADF) scanning transmission electron microscopy (STEM), analytical electron microscopy, including energy dispersive X-ray spectroscopy (EDS) and electron energy-loss spectroscopy (EELS) as well as elemental mapping methods using these spectroscopy techniques will be presented. Also, the electron holographic technique for characterization of magnetic fields of nanohybridized materials will be explained. In addition to electron microscopic observation techniques, recently developed specimen preparation techniques, which are indispensable for obtaining homogeneous and thin films of nanohybridized materials, will be presented. In particular, a focused ion beam (FIB) method will be emphasized. The nanohybridized materials discussed in this chapter include carbon-based core–shell structure, nanocrystalline soft magnetic materials, nanocomposite magnets, and high-T_c superconducting oxides. Application data will be provided in order to explain the usefulness of these analytical techniques for characterization of nanohybridized materials.

10.2 Principles of Electron Microscopy

10.2.1 Transmission Electron Microscopy

We first describe the main principles of transmission electron microscopes. The formation of images in a transmission electron microscope can be understood in simple terms using an optical ray diagram with an optical objective lens,

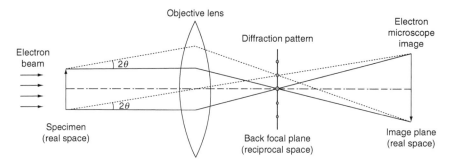

Fig. 10.1. Optical ray diagram with an optical objective lens showing the principle of the imaging process in a transmission electron microscope

as shown in Fig. 10.1. When a crystal with lattice spacing d is irradiated with electrons of wavelength λ, diffracted waves will be produced at specific angles 2θ, satisfying the Bragg condition, i.e.,

$$2d \sin \theta = \lambda. \tag{10.1}$$

The diffracted waves form diffraction spots on the back focal plane. In electron microscopes, with the help of electron lenses, the regular arrangement of the diffraction spots are projected on the screen and the so-called electron diffraction pattern can be observed. If the transmitted beam and the diffracted beams are interfered onto the image plane, a magnified image (electron microscope image) is observed. The space where the diffraction pattern forms is called the reciprocal space while the latter space at the image plane or at a specimen is called the real space. The scattering from the specimen to the back focal plane, in other words, the transformation from the real space to the reciprocal space is mathematically given by the Fourier transform.

In a transmission electron microscope, by adjusting the electron lenses, i.e., by changing the focal lengths of the electron lenses, both the electron microscope image (information in real space) and diffraction pattern (information in reciprocal space) can be observed. Thus, in the analysis of microstructures in materials, both observation modes can be combined successfully. For example, in the investigation of electron diffraction patterns, we first observe an electron microscope image. Then, by inserting an aperture over a specific area (selected area aperture) and by adjusting the electron lenses, we obtain a diffraction pattern of the area. This observation mode is called selected area diffraction. With a selected area diffraction pattern obtained from an interface region of hybrid materials, crystal structures, and mutual crystal orientation relationships between adjacent materials can easily be clarified. The selection of an area is usually limited to about 0.1 μm in diameter, but recently the microdiffraction method wherein the incident electrons are converged on a specimen can be utilized to obtain a diffraction pattern from an area smaller than several nm in diameter.

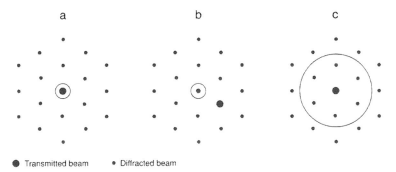

Fig. 10.2. Three observation modes in electron microscopy using an objective aperture. The center of the objective aperture is assumed to be set to the optical axis. (**a**) Bright-field method; (**b**) dark-field method; and (**c**) high-resolution electron microscopy (axial illumination)

On the other hand, in order to investigate an electron microscope image, we first observe an electron diffraction pattern. Subsequently, by selecting the transmitted beam or one of the diffracted beams with an aperture (objective aperture) and changing to the imaging mode, we can observe an image with enhanced contrast where impurities and lattice defects can easily be identified. As shown in Fig. 10.2a, this observation mode of electron microscope images, by selecting the transmitted beam, is called the bright-field method and the image observed is a bright-field image. When one diffracted beam is selected (Fig. 10.2b), this is called the dark-field method and the observed image is a dark-field image.

It is also possible to form electron microscope images using a large objective aperture and selecting more than two beams on the back focal plane as shown in Fig. 10.2c. This observation mode is called high-resolution electron microscopy and the image observed is a high-resolution electron microscope image [1]. In high-resolution electron microscopy, one of the most important parameters which determines the capability of electron microscopes is the resolution limit. The theoretical resolution limit of an electron microscope d is given by

$$d = 0.65 C_s^{1/4} \lambda^{3/4}, \qquad (10.2)$$

where C_s and λ are the spherical aberration constant of the objective lens and the wavelength of the incident electrons, respectively. Thus, much effort has been devoted in attempting to obtain small values of C_s and λ to achieve good resolution of electron microscopes. In addition to improvements of objective lenses, using a small spherical aberration constant, high-voltage electron microscopes with short wavelengths have been constructed. Recently, an electron-optical device, the "C_s corrector," which reduces the C_s value dramatically, has been made practicable [2–4]. In the C_s corrector, a function of a concave lens, which enables correction of the spherical aberration, is achieved

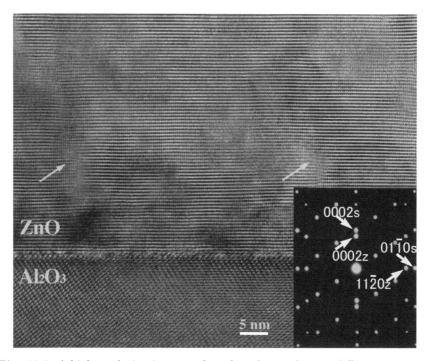

Fig. 10.3. A high-resolution image and a selected area electron diffraction pattern from the interface of ZnO and Al$_2$O$_3$. In the diffraction pattern, z and s correspond to ZnO and sapphire (Al$_2$O$_3$)

with multipoles and transfer lenses. It should be noted that the effective resolution limit is determined not only by C_s but also by C_c (chromatic aberration coefficient). Thus, in the introduction of the C_s correction, C_c should be kept to be a small value. The effective resolution limit can be estimated by investigating the Fourier diffractogram of high-resolution electron microscope images of thin amorphous films. Currently, the best resolution obtained is around 0.1 nm.

Figure 10.3 shows an example of an electron diffraction pattern and a cross sectional high-resolution image of the interface between ZnO and Al$_2$O$_3$ (sapphire). The electron diffraction pattern clearly shows the mismatch between the ZnO and Al$_2$O$_3$ lattices. On the other hand, in a high-resolution electron microscope image, while the lattice fringes of Al$_2$O$_3$ are sharp, there are some defect contrasts in ZnO as indicated by the arrows in the figure. These defect contrasts are considered to result from threading dislocations [5]. As demonstrated in this example, electron diffraction is useful to confirm the lattice spacing, while high-resolution electron microscopy can be efficiently used to identify lattice defects, especially at the interface.

10.2.2 High-Angle Annular Dark-Field STEM

Elastically scattered electrons get distributed at large scattering angles. On the other hand, inelastically scattered electrons are distributed at small scattering angles. Therefore, the elastically scattered electrons can be selected by detecting the scattered electrons at large scattering angles. In this method, the transmitted electrons, which are at the center of the diffraction pattern, are not detected. Thus, the signal obtained by the beam scanning method with a STEM forms a dark-field STEM-image. The distribution of the scattered electrons, except for the Bragg reflection, has rotational symmetry, and thus an annular shaped detector is used to make high detective efficiency. This detecting mode is called the high-angle annular dark-field method. Figure 10.4 shows the principle of HAADF.

It is to be noted that the signal intensity of HAADF is proportional to the square of the atomic number Z [6], and thus, a HAADF image is sometimes called a Z-contrast image or a Z^2-contrast image. The interpretation of the HAADF image is straightforward, and bright image contrast directly indicates heavy elements if the specimen thickness is uniform. HAADF imaging has thus attracted wide attention because of the ease of image interpretation and the possibility of electronic amplification of the image contrast. On the other hand, attention should be paid to any specimen thickness changes and diffraction contrast when the image contrast is interpreted quantitatively. Figure 10.5 shows an atomic-resolution HAADF image of a grain boundary in a semiconducting $SrTiO_3$ ceramic condenser [7]. The image was obtained with a 200 kV FE-STEM which provides a beam diameter of less than 0.2 nm.

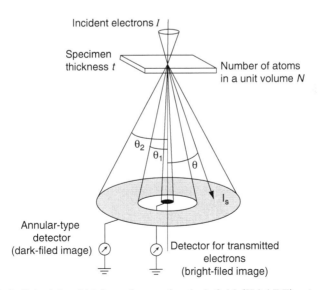

Fig. 10.4. Principle of high-angle annular dark-field (HAADF) microscopy

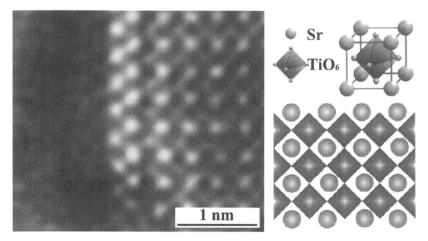

Fig. 10.5. Atomic-resolution HAADF image of a grain boundary in a semiconducting SrTiO$_3$ ceramic condenser. A 200 kV FE-STEM (JEM-2010F) was used to obtain the image. The structure model is shown on the *right* [8]. Copyright 2002, Springer-Verlag

Also, the attached HAADF detector corrected the electrons at the scattering angle of 50–110 mrad. The columns of Sr with large Z appear as bright dots, and the columns consisting of Ti and O are also clearly seen. The structure model is shown on the right in the figure.

10.2.3 Analytical Electron Microscopy

Outline of Electron Energy-Loss Spectroscopy and Energy Dispersive X-Ray Spectroscopy

Inside a transmission electron microscope column, a specimen is illuminated with high-energy electrons, and there may be various interactions between the specimen and the incident electrons. Electron scattering caused by the specimen can be classified into two groups: elastic scattering and inelastic scattering. In elastic scattering, the direction of the scattered electron changes, but the velocity (or energy) does not. Diffracted electrons and back-scattered electrons belong to the elastic scattering category. On the other hand, electrons suffering a change in their velocity (or energy) belong to the inelastic scattering category. While imaging modes such as the bright-field method, dark-field method, and high-resolution electron microscopy mainly utilize elastically scattered electrons, analytical electron microscopy utilizes inelastically scattered electrons.

The spectroscopy of inelastically scattered electrons, taking into account these scattering processes, is called electron energy-loss spectroscopy. The spectroscopy of characteristic X-rays resulting from inner-shell excitation is termed energy dispersive X-ray spectroscopy (EDS or EDX, EDXS). EELS and EDS are the two methods most popularly used in analytical electron microscopy [8].

The principles of EELS and EDS can be explained using one of the inelastic electron scattering processes, i.e., the excitation of an inner-shell electron. Figure 10.6 schematically shows the change in the electronic structure due to the excitation of an electron in the K-shell. The resulting energy-loss spectrum and X-ray spectrum observed are shown at the bottom of Fig. 10.6. Here, we consider the case where an incident electron transfers energy to the specimen and the electron in the K-shell (1s orbital) is excited. Since the energy levels below the Fermi level are all occupied by electrons in the ground state, one of the electrons in the K-shell can only transit to the unoccupied state above the Fermi energy. Thus, when the electron loses an energy larger than ΔE, which corresponds to the energy difference between the energy at the K-shell and the Fermi energy, the probability of transition from the K-shell to the unoccupied density of state increases drastically, and eventually a sharp peak appears at the energy ΔE in an energy-loss spectrum. In this excitation process of the inner-shell electron, the peak tends to accompany the tail in a higher energy region. From this shape, the peak appearing in the energy-loss spectrum is generally called an edge. The threshold energy of the edge is specific to each element, and thus the specimen can be identified with the energy value ΔE. In

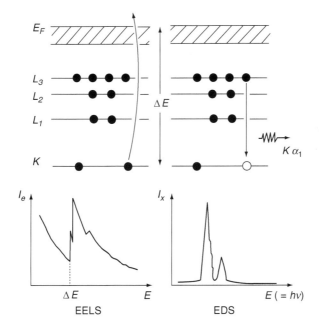

Fig. 10.6. Inner-shell electron excitation, a resultant electron energy-loss spectrum, and an energy dispersive X-ray spectrum

addition, information about the content of the element can be obtained from the integrated intensity of the edge. Furthermore, from the accurate value of the threshold energy and the shape of the edge, information about chemical bonding can be obtained.

When an atom changes from an excited state to the ground state, surplus energy is emitted as a characteristic X-ray or an Auger electron. In both cases, one of the electrons in a higher energy level transits to the hole in the lower energy level in a manner satisfying the selection rule. Similar to EELS, the energy of the characteristic X-ray can be used for specifying a constituent element since the energy at the X-ray peak position is specific to each element. Also, from the integrated intensity, the composition of the material can be evaluated. In Fig. 10.6, the characteristic X-ray emission resulting from the transition of the electron from the L_3 shell to the K-shell is illustrated, and the resultant X-ray emitted is called a characteristic $K_{\alpha 1}$ X-ray. There are also several other characteristic X-rays which are frequently used for compositional analysis, such as $K_{\beta 1}$ and $L_{\alpha 1}$ corresponding to transitions from the M_3 shell to the K-shell and from the M_5 shell to the L_3 shell, respectively.

Elemental Mapping with EDS and EELS

The analysis method wherein the incident electron beam is stopped at a point in a specimen, and the energy-loss electrons and X-rays from the area are detected is called point analysis. The electron beam can be scanned on the specimen with a beam-scanning system, and the energy-loss electrons or specific X-ray intensity is measured. When the brightness signal corresponding to the energy-loss electrons or the characteristic X-ray intensity measured is displayed on the CRT by synchronizing it with the position signal, two-dimensional intensities of electrons or X-rays can be obtained. This observation mode is called the elemental mapping method and is effective for analyzing the distribution of the constituent element in two dimensions.

On the other hand, various kinds of energy filter systems on the basis of electron energy-loss spectrometers have been installed in electron microscopes. These energy filters have been categorized as being of two types (Fig. 10.7). The first is the so-called in-column type, in which the energy filter is inserted in the microscope column. The omega type [9] and the Castaing–Henry type [10] filters are examples of the in-column type. The other is the so-called postcolumn-type filter, where the filter is installed at the bottom of an electron microscope (i.e., under the camera chamber). Typically for the postcolumn type, a sector-type energy filter is popular [11,12]. In both cases, energy-loss spectra can be observed on the screen or the monitor, and electrons with a specific energy can be selected by the energy-selecting slit to record energy-filtered images and diffraction patterns.

Figure 10.8 shows an example of a HAADF–STEM and elemental mapping methods for a $Sm_2(Fe_{0.95}, Mn_{0.05})_{17}N_{4.2}$ powder sample consisting of crystalline and amorphous phases [13]. Since the intensity of the HAADF–STEM

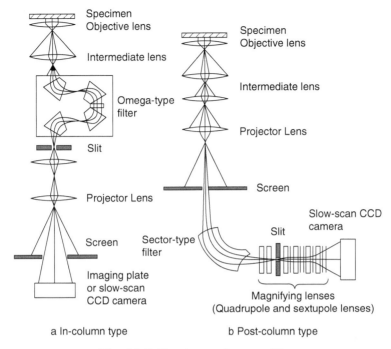

Fig. 10.7. Two types of energy filters

image is proportional to the square of the atomic number Z, i.e., $I_s \propto Z^2$, the dark region corresponds to the area consisting of light elements. The amorphous layers appear as dark bands, and thus they are considered to consist of light elements. Figure 10.8b,c indicates the intensity profiles of the one-dimensional elemental mapping images obtained by EDS and EELS, respectively, at the line A–B in Fig. 10.8a. In the intensity profiles of the EDS elemental mapping image in Fig. 10.8b, the amorphous regions are rich in Mn, as indicated by the black arrows, and are poor in Fe. Based on the intensity profiles of the EELS elemental mapping image in Fig. 10.8c, the amorphous regions are found to be rich in N as well as Mn. In the lower part of Fig. 10.8, these features are clearly visualized in the two-dimensional elemental mapping images with EELS, where the amorphous phases enriched with N and Mn appear as brighter regions.

10.2.4 Electron Holography

Among the various electron microscopy techniques, electron holography provides a unique method for detecting the phase shift of the electron wave due to a magnetic field and an electric field. Here, the phase shift indicates the phase change relative to the electron plane wave traveling in vacuum, $\exp[i(kz - \omega t)]$,

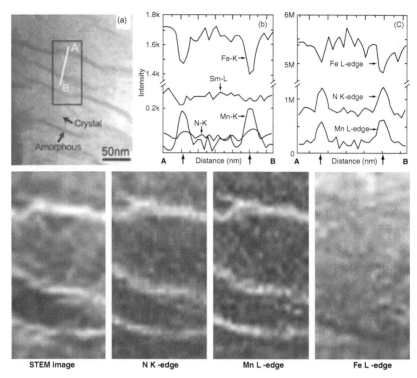

Fig. 10.8. (a) HAADF–STEM image of a Sm$_2$(Fe$_{0.95}$,Mn$_{0.05}$)$_{17}$N$_{4.2}$ powder sample. One-dimensional intensity profiles of line A–B (see the elemental mapping images observed in (**a**)) obtained by EDS (**b**) and EELS (**c**) are shown. The lower panels provide a STEM image and two-dimensional elemental mapping images with EELS showing elemental distribution (N, Mn, Fe) obtained from the rectangular area in (**a**) [13]. Copyright 2005, Elsevier

where

$$k = \frac{2\pi}{\lambda}, \quad \omega = 2\pi\nu, \tag{10.3}$$

where ν is the frequency.

Electron holography is performed through a two-step imaging process. In the first step, a hologram is formed by interfering an object wave and a reference wave using a biprism. In the second step, the phase shift is extracted from the hologram by using the Fourier transform. Figure 10.9 shows the geometrical configuration for forming a hologram in an electron microscope. An electron beam emitted from a field-emission tip is accelerated and then collimated to illuminate an object through a condenser lens system. An object is located in one half of the object plane which is illuminated with a collimated electron beam.

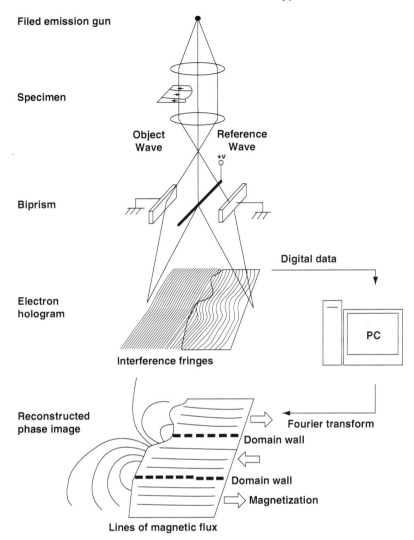

Fig. 10.9. Geometric configuration for forming a hologram in an electron microscope and schematic illustration of the two-step imaging process in electron holography

Assuming that the object is illuminated by a plane wave of unit amplitude having a wave vector parallel to the optical axis, the change in the scattering amplitude of the plane wave due to the object is, in general, described as

$$q(\mathbf{r}) = A(\mathbf{r})\exp[i\phi(\mathbf{r})], \qquad (10.4)$$

where $A(\mathbf{r})$ and $\phi(\mathbf{r})$ are real functions and describe the amplitude change and the phase shift due to the object, respectively. In conventional electron microscopy, the image intensity, i.e., the squared absolute value

$$|q(x, y)|^2 = A^2(x, y) \qquad (10.5)$$

is observed. Thus, the resultant contrast shows the amplitude change due to the absorption and/or diffraction.

On the other hand, in electron holography, the interference between the wave outside the specimen is utilized, and through the Fourier transform operation on the interference image (the so-called hologram), not only the amplitude $(A(\mathbf{r}))$ but also phase information $(\exp[i\phi(\mathbf{r})])$ is reconstructed.

In the following section, the intensity of the reconstructed phase image $I_{\mathrm{ph}}(\mathbf{r})$ is represented by a cosine function, i.e.,

$$I_{\mathrm{ph}}(\mathbf{r}) = \cos[\phi(\mathbf{r})]. \qquad (10.6)$$

As noted in [14], we could obtain information on the magnetic flux distribution through $I_{\mathrm{ph}}(\mathbf{r})$.

10.3 Specimen Preparation

In order to observe fine electron microscope images of hybrid nanomaterials, it is necessary to prepare thin films without causing them to break at their interfaces. For this purpose, it is important to select appropriate specimen preparation methods suppressing surface contamination. In this section, three typical specimen preparation methods useful for hybrid nanomaterials, i.e., ultramicrotomy, ion milling, and focused ion beam methods, are considered.

10.3.1 Ultramicrotomy

This method has been used for preparing thin sections of biological specimens and thin films of inorganic materials, which are not as hard to cut. One of the advantages of ultramicrotomy is the small contamination on the sections. Specimens of thin films or powders are usually fixed in a resin and trimmed with a glass knife, before being sliced by a diamond knife. This process is necessary so that the specimens in the resin can be sliced easily by a diamond knife. Acrylic or epoxy is used as a resin for fixing specimens. When using an acrylic resin, a gelatin capsule is used as a vessel. The acrylic is easily sliced. Epoxy takes less time to solidify than acrylic. Also, epoxy is considerably resistant to electron irradiation. In general, skill is needed to appropriately set the geometrical configurations of a diamond knife and a specimen after trimming, and to slice a specimen into homogeneous thin sections.

Figure 10.10 illustrates the principle of ultramicrotomy. Each time the arm holding the specimen comes down and back, it steps forward and the specimen is sliced with the diamond knife at the head of the boat filled with water. The sliced sections are floated on the water and are handled by a thin wood stick

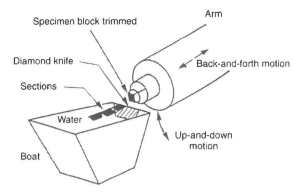

Fig. 10.10. Principle of ultramicrotomy. The *hatched region* corresponds to a diamond knife

with an eyelash to place the thin sections onto a special grid covered with a collodion or carbon thin film. While practicing analysis, care should be taken for the background in EELS and EDS from the resin and supporting film. The acrylic can be removed with chloroform after slicing. If the trimming and slicing are not carried out correctly, the expensive diamond knife will be damaged. Also, during slicing, lattice strain is frequently introduced into the sections.

10.3.2 Ion Milling

This method is widely used to obtain thin regions of materials, especially ceramics, semiconductors, and multilayer films. The principle of the sputtering phenomena, wherein atoms are ejected from the surface by irradiating it with accelerated ions, is used. First, a thin plate less than 0.1 mm is prepared from a bulk specimen using a diamond cutter and by mechanical thinning. Then, a disk 3 mm in diameter is made from the plate using a diamond cutter or an ultrasonic cutter, and a dimple is formed around the center of the surface with a dimple grinder (Fig. 10.11a). If it is possible to thin the disk directly to 0.03 mm thickness by mechanical thinning without using a dimple grinder, the disk needs to be strengthened by covering the edge with a metal ring such as a Mo ring (Fig. 10.11b). Ar ions are typically used for the sputtering, and the incidence angle with respect to the disk specimen and the accelerating voltage are set to 10°–20° and several kilovolts, respectively. In a conventional ion milling system, the ion milling is automatically stopped when a hole is formed in the specimen by detecting the laser beam passing through (Fig. 10.11c). While ion milling is continued for some time, compositional changes sometimes occur at the surface due to the differences in the sputtering efficiencies for the constituent atoms, and amorphous layers tend to form on the surface due to ion irradiation damage. In order to avoid these effects, the ion milling conditions need to be optimized, i.e., by using different ions, lowering the

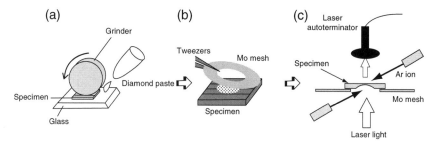

Fig. 10.11. Specimen preparation procedure using the ion milling method

accelerating voltage, and adjusting the incident beam angle. To minimize any increase in the specimen temperature during sputtering, the use of a cooling stage with liquid nitrogen is effective. If the incidence angle becomes too small, a metal ring used for strengthening the disk is irradiated with the ions and the specimen plate is coated with the metal. The existence of such impurity phases can be detected directly from energy-loss spectra and energy dispersive X-ray spectra.

10.3.3 Focused Ion Beam

This method is currently attracting much attention. It may be especially useful for hybrid materials containing a boundary between different materials, wherein it may be difficult to thin the boundary region homogeneously by other methods such as ion milling. This method was originally developed for the purpose of fixing semiconductor devices. In principle, ion beams are sharply focused on a small area and the specimen is thinned very rapidly by sputtering. Usually Ga ions are used with an accelerating voltage of about 30 kV and a current of about $10\,\mathrm{A\,cm^{-2}}$. The probe size is several tens of nanometers. By detecting the secondary electrons emitted from the specimen while irradiating it with ion beams, a secondary electron image of the surface can be displayed similar to a scanning electron microscope (SEM) image. Thus, by observing the secondary electron image, one can accurately select the appropriate region for thinning.

Currently, an FIB which is equipped with both an electron gun and an ion gun is becoming popular. Figure 10.12 shows schematic illustrations of the two-guns type FIB. This instrument not only has a Ga ion gun, but also has an electron gun, so that SEM image observations during the FIB process is possible. One of the advantages of this instrument is that the SEM creates less damage in the specimen compared to SIM. Figure 10.12c shows a thin specimen of silicon prepared with the FIB method [15]. A TEM specimen fabricated by the FIB method typically has a damaged layer on the surface. In order to remove the damaged layer, a low-energy Ar ion milling method is an effective solution and can be used.

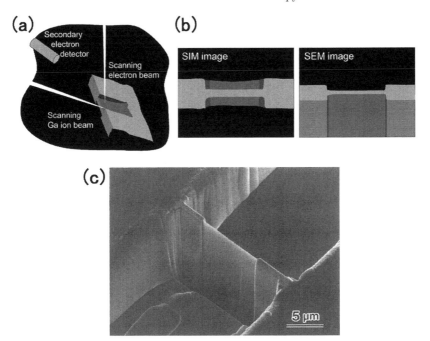

Fig. 10.12. (a) Schematic figure of an FIB, which is equipped with an electron gun and an ion gun. (b) Schematic figures of SIM image and SEM image obtained by the two-guns type FIB. (c) The thin part of the silicon prepared with the FIB method and observed with secondary electrons [8]. Copyright 2002, Springer-Verlag

10.4 Applications

In the following, we apply electron microscopic characterization techniques to five hybrid metallic nanomaterials, including the carbon–metal system, an oxide–oxide system, a metal-based compound system, a metal–oxide system and a polymer–metal system. In Table 10.1, specimen preparation for these material systems and the information obtained by TEM analysis are summarized.

10.4.1 Carbon–Metal System (Carbon-Encapsulated Metal Nanoparticles)

The first example of electron microscopy applications is the characterization of carbon-encapsulated metal nanoparticles. Currently, magnetic nanoparticles are of considerable interest due to their potential applications in magnetic fluids [16], magnetic recording materials [17], biomedicine [18], etc. The three ferromagnetic transition metals, Fe, Ni, and Co, and their alloys, especially when formed as nanometer-sized particles, readily oxidize upon exposure to

Table 10.1. List of TEM techniques and the information obtained

System (thinning process)	Electron microscopy techniques	Information obtained
Carbon–metal (no process)	HREM (+FFT) EDS	Crystal structure Clarification of core–shell structure
Oxide–oxide (ion milling)	Bright- and dark-field image Electron holography	Microstructure Magnetic flux distribution
Metal-based compound (FIB)	Bright-field image Electron diffraction Electron holography EDS	Microstructure Identification of constituent phases Magnetic flux distribution Specification of constituent phases
Metal–oxide (ultramicrotomy, FIB)	Bright-field image Electron diffraction Elemental mapping (EELS) Electron holography	Microstructure Preferred orientation onfirmation of constituent particles Magnetic flux distribution
Polymer–metal (FIB)	Piezodriving holder	Conductivity

air, resulting in deterioration of their magnetic properties. Hence, the coating of ferromagnetic nanoparticles with an oxygen-impermeable shell on the nanometer scale is becoming an important issue for potential applications in the field of nanomagnetism.

Here, cobalt nanoparticles encapsulated in graphitic shells which were synthesized using an advanced and cost-effective synthesis method are characterized by high-resolution electron microscopy [19]. These nanoparticles were synthesized using an electric plasma discharge generated in an ultrasonic cavitation field of liquid ethanol, followed by separation, drying, and annealing of the particles. Since the specimens are nanosized particles, they can be directly observed in a TEM without the need for a specific specimen thinning process.

Figure 10.13a shows as-prepared carbon nanocapsules; the average diameter of the nanocapsules was measured to be 8.5 nm. The cores of the carbon nanocapsules were isolated from each other by graphite-like carbon layers about 1–2 nm thick; lattice images of the carbon with 0.36 nm spacing are also clearly seen. This value falls within the range reported for the interplanar spacing of graphite (d_{002} = 0.34–0.39 nm) in carbon nanotubes of varying diameter [20]. HREM analysis of the as-prepared sample indicates weak particle crystallinity; however, there are some well-crystallized particles and thus FFT (fast Fourier transform operation) was carried out to clarify the structure of some of the encapsulated nanoparticles. Figure 10.13b shows an

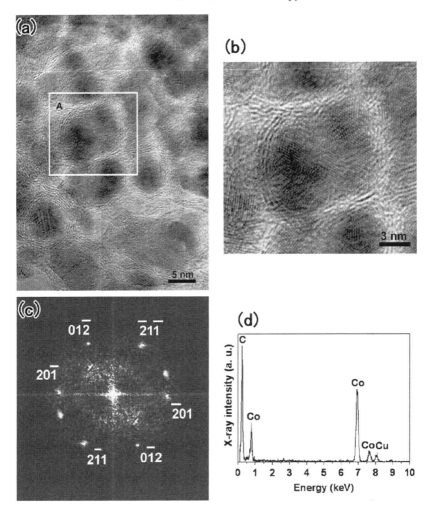

Fig. 10.13. (a) HREM image of carbon nanocapsules before annealing; (b) enlarged image of the carbon nanocapsule marked by rectangle A in (a); (c) digital diffractogram of the carbon nanocapsule presented in (b), showing that the encapsulated nanoparticles are orthorhombic Co$_3$C, and (d) EDS spectrum taken from the area shown in (b). The copper peak is due to the copper grid supporting the specimen [19]. Copyright 2007, Elsevier

enlarged image of the carbon nanocapsule in the area marked by rectangle A in Fig. 10.13a. A Fourier transform (Fig. 10.13c) of the crossed lattice image of the carbon nanocapsule core presented in Fig. 10.13b gives the arrangement of spots at spacings of 0.22, 0.21, and 0.205 nm, consistent with the (1 2 0), (0 1 2), and (1 2 1) planes, respectively, for the [1 4 2] zone-axis pattern of orthorhombic Co$_3$C (ASTM card # 26-450). None of the labeled spots at the interlayer spacing of 0.21 nm in the digital diffractogram (Fig. 10.13c) belong to the [1 4 2] zone-axis pattern. Elemental analyses by EDS (Fig. 10.13d) of the area that includes the carbon nanocapsule (Fig. 10.13b) shows the presence of only cobalt and carbon. Random EDS scans of different parts of the as-prepared powder sample after etching on the sample microgrid revealed the presence of cobalt, carbon, and a small amount of oxygen (less than 1 atm%). The minor amount of oxygen must be a result of the incomplete etching of the cobalt oxides. We rarely find lattice fringes, indicative of cobalt oxides, for the exposed particles visible in the HREM images. Thus, the elemental analysis implies that a carbon shell prevents the oxidation of the cobalt nanoparticles.

HREM observations of the annealed samples reveal that the cobalt nanoparticles exist both in the fcc (β-Co) and hcp (α-Co) forms. Figure 10.14a shows an HREM image of cobalt nanoparticles encapsulated in graphite-like carbon annealed at 873 K. The average diameter of the uniformly dispersed cobalt cores was estimated to be about 5 nm, with a fairly narrow size distribution (having a standard deviation of 1.1 nm). In size and form, the particles resemble those in magnetic films synthesized by codeposition of Co and carbon using ion-beam sputtering [21]. An FFT (Fig. 10.14c) of a digital image of the crystalline core annealed at 873 K (Fig. 10.14b) yields spots with a spacing of 0.205 nm in accordance with the (1 1 1) planes in fcc β-Co. The cobalt cores are completely isolated from each other by graphite-like carbon layers; lattice images of carbon with about 0.36 nm spacing are clearly seen. The thickness of the graphite-like carbon is estimated to be about 1–2 nm.

10.4.2 Oxide–Oxide System ($YBa_2Cu_3O_y$ High-T_c Superconductor)

As an example of the application of TEM to an oxide–oxide system, the characterization of bulk $YBa_2Cu_3O_y$ high-T_c superconductor made by the QMG (Quench and Melt Growth) method is described. This material consists of two phases. One is the so-called Y123 ($YBa_2Cu_3O_y$) and the other is the so-called Y211 (Y_2BaCuO_5). The former shows superconductivity but the latter does not. The Y211 particles works as pinning centers for the magnetic vortex, and therefore this material is expected to keep large critical current density. Figure 10.15a shows a bright-field image of Y123 containing Y211 particles. The incident beam is parallel to the [1 0 0] direction. In the figure, the lamella contrasts correspond to twin boundaries. Y123 has a phase transformation from a tetragonal structure to an orthorhombic structure around 600°C. The twin lamella results from the phase transformation. The twin plane is {1 1 0}. Figure 10.15b shows a dark-field image of Y123 containing Y211 particles using a 200 reflection of Y123. The bright contrast indicated by the white arrow in the figure corresponds to a stacking fault. The contrast of Y123 around the Y211 particles is brighter than other regions. This contrast corresponds to a strain field. In addition, dislocations are observed as well.

10 Electron Microscopy Characterization 237

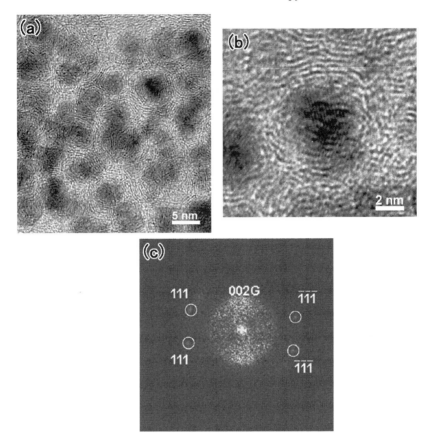

Fig. 10.14. (a) HREM image of cobalt nanoparticles encapsulated in graphite shells annealed at 873 K; (b) magnified image of a typical carbon nanocapsule; and (c) digital diffractogram of the carbon nanocapsule shown in (b) indicating fcc cobalt spacings [19]. Copyright 2007, Elsevier

In order to observe the superconducting state, the specimen needs to be cooled under its critical temperature [22]. A TEM specimen can be cooled down to around 20 K by using a liquid Helium cooling stage. Figure 10.16a shows an electron hologram near the surface of the specimen, and (b) and (c) show reconstructed phase images indicating the magnetic flux distribution at room temperature and under the critical temperature, respectively. In Fig. 10.16c, the magnetic fluxes are trapped at extremely small areas on the specimen surface and spread out into the free space. It is likely that the Y211 particle may exist near the small regions as a pinning center, and the magnetic fluxes which are present in Y123 at room temperature are pushed out into the Y211 particle under the critical temperature.

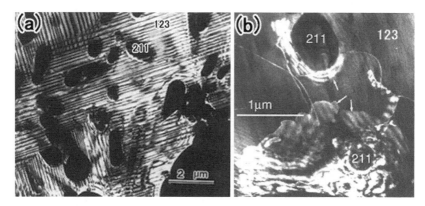

Fig. 10.15. (a) A bright-field image of QMG YBa$_2$Cu$_3$O$_y$ bulk material. The incident beam direction is [0 0 1]. (b) A 200 dark-field image around a Y211 particle. The *white arrows* indicate stacking faults

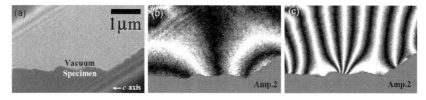

Fig. 10.16. (a) An electron hologram observed near the surface of Y123. Reconstructed phase images at room temperature (b), and 20 K

10.4.3 Metal-Based Compound System (Nd–Fe–B Nanocomposite Magnet)

Recently, nanocomposite magnets have attracted significant attention due to their superior hard-magnetic properties and low content of rare earth metals such as Nd. The nanocomposite magnet consists of two phases, i.e., one is a hard magnetic phase and the other is a soft magnetic phase with high saturation magnetization. Clarification of the microstructure and magnetic flux distribution in these nanocomposite magnets are important to understand their magnetic properties. In this section, TEM analysis of the nanocomposite magnet Nd$_{4.5}$Fe$_{74}$B$_{18.5}$ is presented. A Nd$_{4.5}$Fe$_{74}$B$_{18.5}$ nanocomposite magnet was produced via a crystallization process from amorphous alloys [23].

Figure 10.17 shows a bright-field image and a selected area diffraction pattern. From the bright-field image, it is seen that the composite magnet consists of nanosized grains of about 30 nm size. From the Debye–Scherrer rings of the electron diffraction pattern, the nanosized grains are identified to be mainly Nd$_2$Fe$_{14}$B and Fe$_3$B.

Figure 10.18a shows a reconstructed phase image of Nd$_{4.5}$Fe$_{75}$B$_{18.5}$. It can be noticed that the lines of the magnetic flux are almost parallel, and this suggests strong exchange coupling between the soft and hard magnetic

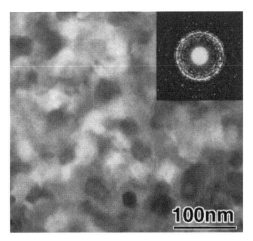

Fig. 10.17. Bright-field image of $Nd_{4.5}Fe_{77}B_{18.5}$ with the diffraction pattern inset [24]. Copyright 2003, Elsevier

phases. In the enlarged image of the closed area (Fig. 10.18b), the soft (S) and hard (no symbol) phases, as identified by energy dispersive X-ray spectroscopy, are specified wherein the Nd-L lines are the signal from the hard phase (Fig. 10.18c). The result demonstrates that the lines of magnetic flux pass through many grains of both the soft and hard phases, and the lines orient almost along the same direction due to the strong exchange coupling.

10.4.4 Metal–Oxide System (Co-CoO Recording Tape)

One of the most popular recording media currently is a Co-CoO obliquely evaporated tape deposited on a base of polyethylene terephthalate (PET). This tape has been widely used in several recording systems, for example, consumer digital video cassettes (DVC), 8-mm videocassettes (Hi-8), and computer data recording systems (e.g., AIT). It is known that in the Co-CoO film layer, Co crystallites (hexagonal close packed, hcp) are in a ferromagnetic state, whereas CoO crystallites (face-centered cubic, fcc) are in a paramagnetic state at room temperature. Furthermore, the magnetic properties are sensitive to the microstructure consisting of Co and CoO crystallites. It should be noted that for understanding the magnetic recording characteristics, investigation of the magnetic domain structure in the recorded tape is important.

The specimen investigated was a magnetic film of a Co-CoO magnetic tape available in the market (thickness 180 nm, composition $Co_{75}O_{25}$ (atm%) as determined by Auger electron spectroscopy, Bs = 701 mT, Hc = 114 kA m^{-1}). It is expected that the thin films sliced using ultramicrotomy have a clean surface, that is, there is no damaged layer as introduced by the FIB method. Thus, these specimens are suitable for analytical electron microscopy and HREM studies. However, ultramicrotomy tends to introduce some strain in

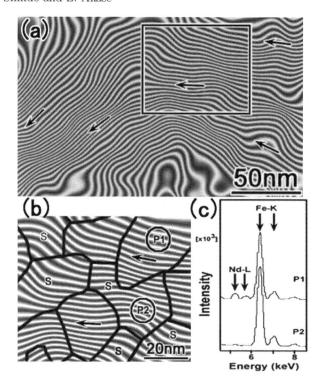

Fig. 10.18. (a) Reconstructed phase image of Nd$_{4.5}$Fe$_{77}$B$_{18.5}$ annealed at 973 K. The *arrows* indicate the direction of the lines of magnetic flux and the phase amplification is 20. (b) Enlarged reconstructed phase image of the rectangular region in (a), and (c) EDS obtained from positions P1 (hard phase) and P2 (soft phase) [24]. Copyright 2003, Elsevier

the film, which modifies the magnetic domain structure drastically. Hence, an FIB method that does not introduce strain in the film is utilized for magnetic domain structure analysis by electron holography.

Figure 10.19 shows an example of a TEM image of a sliced film prepared by ultramicrotomy. It is clearly seen that the film exhibits a columnar structure, where the growth direction is tilted about 40° off the normal to the film. It is to be noted that the growth direction changes gradually when approaching the top of the film. The gradual change in the growth direction is considered to result from the Co evaporation process, where the distance and the direction between the Co crucible and the base film rotating on a cooling drum changed gradually during the Co deposition [26]. Figure 10.20 shows electron diffraction patterns obtained from the sliced film. Figure 10.20a corresponds to a conventional electron diffraction pattern, and Fig. 10.20b is the energy-filtered electron diffraction pattern that was obtained with a slit of energy width of 20 eV to select the zero-loss peak. It is seen that due to energy filtering, the background of the electron diffraction is effectively reduced, and the peak of

10 Electron Microscopy Characterization 241

Fig. 10.19. TEM image of a cross-sectional film of a Co-CoO evaporated tape [25]. Copyright 2004, Cambridge University Press

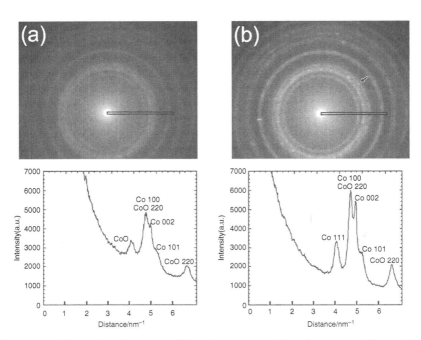

Fig. 10.20. Conventional electron diffraction pattern (**a**) and an energy-filtered electron diffraction pattern (**b**) obtained from a sliced film. The intensity profile along the marked area is given at the bottom of each diffraction pattern [25]. Copyright 2004, Cambridge University Press

Fig. 10.21. Remanent state of the tape: (**a**) lines of magnetic flux outside the tape and (**b**) inside the tape, obtained by electron holography [27]. Copyright 2006, IEEE

each Debye–Scherrer ring for Co or CoO crystallites can be clarified and thus easily indexed. The energy-filtered diffraction pattern shows a trend of preferred orientation; the *c*-axis of the Co crystallites tends to align along the column axis as evidenced by the distribution of the 0 0 2 reflections (arrow), although the crystallites oriented to different directions are also present.

Figure 10.21a,b shows reconstructed phase images of the outside and inside of the magnetic tape. The arrows indicate the direction of the magnetic flux. The periodic closed magnetic flux distribution with alternate chiralities in the direction the magnetic flux indicates the recorded signals 0 1 0 1 0 ... The detailed magnetic flux distribution is discussed with the simulation of the magnetic flux taking into the account the microstructure, clarified as above [27].

10.4.5 Polymer–Metal System (Silver–Epoxy Paste)

The last specimen presented in this chapter is solidified conductive paste consisting of Ag and an epoxy resin, which has recently been widely used in various electronic devices. In addition to clarifying the microstructure, local electrical conductivity measurements are especially important to understand the electrical properties of this conductive paste. For the local electrical conductivity measurements, we utilized a newly developed double probe piezodriving holder [28].

Figure 10.22a shows the schematic diagram of the "double-probe piezodriving holder" developed in this study. Figure 10.22b shows a micrograph of the tip portion of this specimen holder. There are two arms for mounting the probes. The two probes can be manipulated independently in three dimensions by the motions of "arm 1" and "arm 2." These arms are driven by micrometers and/or piezoelectric elements, which are located in the tail portion of the holder. Coarse movement of the probes toward the specimen is achieved by using manual micrometers. Fine movement for bringing the probe tip in contact with the specimen is achieved by using these piezoelectric elements. Commercial Pt–Ir needles, which are supplied for scanning tunneling

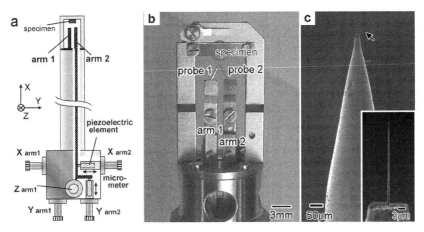

Fig. 10.22. (a) Schematic representation of the double-probe piezodriving holder. The incident electrons are parallel to the Z direction. (b) Optical micrograph showing the tip portion of the developed specimen holder. (c) Scanning ion microscope image of a Pt–Ir needle, the tip of which is sharpened by a focused ion beam. The *inset* shows an enlarged view of the needle tip [28]. Copyright 2006, American Institute of Physics

microscopy (STM) studies, were used in the transmission electron microscopes for the local conductivity measurements. The tips of these needles were sharpened into the shape shown in Fig. 10.22c by using a FIB. The tip diameter was approximately 50 nm or less in order to measure the conductivity in a nanoscale area – see the inset in Fig. 10.22c.

Figure 10.23 shows the effectiveness of the developed system for local resistivity measurements. The specimen is a solidified conductive paste of fine Ag particles and epoxy resin. The thin film is prepared using FIB as shown in Fig. 10.23a. It is a promising alternative to lead-bearing solders. It should be noted that a conductive paste is primarily required for fine pitch patterning in electronic circuits. Moreover, the probes can be easily brought in contact with the fine Ag particles (see Fig. 10.23b). As shown in the electron micrograph, it is likely that the adjacent particles A and B are connected; refer to the circle in Fig. 10.23b. However, the electric resistance between these particles is measured to be greater than $10^7\,\Omega$ (above the limit of the resistometer). The result reveals that the resistance is exceedingly larger than the resistivity of each Ag particle, which is measured to be less than $3\,\Omega$. Apparently, particle A is electrically disconnected from particle B. It is emphasized that this type of evaluation can be performed by using two probes that can be moved independently in the transmission electron microscope. In the present chapter, three typical electron microscopy characterization techniques, i.e., structure analysis, magnetic field analysis, and electric conductive analysis were presented. It was demonstrated that these three techniques combined together are very

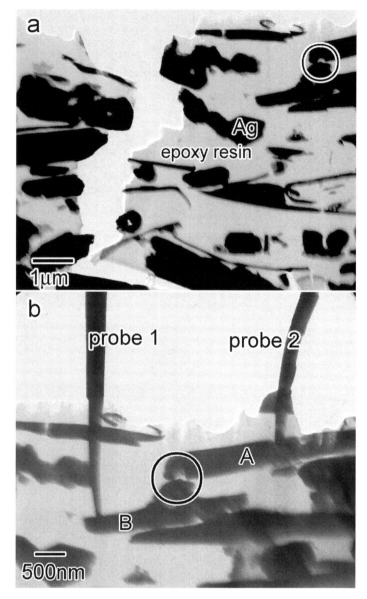

Fig. 10.23. (a) Transmission electron microscope image showing the dispersion of fine Ag particles in a solidified conductive paste. (b) Resistivity measurements for adjacent Ag particles A and B in the transmission electron microscope. The *circled portion* in (b) corresponds to that in (a) [28]. Copyright 2006, American Institute of Physics

promising to clarify the relation between the microstructure and properties of advanced nanomaterials.

Acknowledgments

The authors owe the study of the carbon-encapsulated metal nanoparticles to Prof. T. Nakamura, Dr. E. Shibata, and Dr. R. Sergiienko at Institute of Multidisciplinary Research for Advanced Materials (IMRAM), Tohoku University. They also wish to thank Dr. Morita at Nippon Steel Co., Ltd for providing high-T_c superconductor specimens, and Dr. S. Hirosawa at Neomax Co., Ltd for Nd–Fe–B nanocomposite magnet specimens. They are grateful to Dr. W. Xia at IMRAM, Dr. T. Ito, Dr. Y. Iwasaki, and Dr. J. Tachibana at Sony Co., Ltd for the cooperation in the study of recording tape, and Dr. Y. Murakami, Mr. N. Kawamoto at IMRAM for the study of silver–epoxy paste.

References

1. D. Shindo, K. Hiraga, *High-Resolution Electron Microscopy for Materials Science* (Springer, Tokyo, 1998)
2. H. Rose, Optik **85**, 19 (1990)
3. M. Haider, G. Braunshausen, E. Schwan, Optik **99**, 167 (1995)
4. G. Benner, M. Matijevic, A. Orchowski, P. Schlossmacher, A. Thesen, M. Haider, P. Hartel, Microsc. Microanal. **10**(Suppl. 3), 6 (2004)
5. S.-H. Lim, D. Shindo, J. Vac. Sci. Technol. B **19**, 506 (2001)
6. S.J. Pennycook, S.D. Benerger, R.J. Culbertson, J. Microsc. **144**, 229 (1986)
7. M. Kawasaki, T. Yamazaki, S. Sato, K. Watanabe, M. Shiojiri, Philos. Mag. A **81**, 245 (2001)
8. D. Shindo, T. Oikawa, *Analytical Electron Microscopy for Materials Science* (Springer, Tokyo, 2002)
9. G. Zanchi, J.P. Perez, J. Sevely, Optik **43**, 495 (1975)
10. R. Castaing, J.F. Hennequin, L. Henry, G. Slodgian, in *Focusing of Charged Particle*, ed. by A. Septier (Academic, New York, 1967), p. 265
11. O.L. Krivanek, A.J. Gubbens, N. Dellby, Microsc. Microanal. Microstrct. **2**, 315 (1991)
12. H. Hashimoto, Y. Makita, N. Nagaoka, *Proceedings of the 50th Annual EMSA Meeting*, Boston, 1992, p. 1994
13. A. Yasuhara, H.S. Park, D. Shindo, T. Iseki, N. Oshimura, T. Ishikawa, K. Ohmori, J. Magn. Magn. Mater. **295**, 1 (2005)
14. D. Shindo, Mater. Trans. **44**, 2025 (2003)
15. Y. Ikematsu, T. Mizutani, K. Nakai, M. Fujinami, M. Hasebe, W. Ohashi, Jpn. J. Appl. Phys. **37**, L196 (1998)
16. S. Odenbach, J. Phys. Condens. Matter **16**, R1135 (2004)
17. M. Hanayama, T. Ideno, Jpn. Patent JP 06-152793
18. P. Tartaj, M.P. Morales, S. Veintemillas-Verdaguer, T. Gonzalez-Carren-o, C.J. Serna, J. Phys. D: Appl. Phys. **36**, R182 (2003)
19. R. Sergiienko, E. Shibata, Z. Akase, D. Shindo, T. Nakamura, G. Qin, Acta Mater. **55**, 3671 (2007)

246   D. Shindo and Z. Akase

20. C.H. Kiang, M. Endo, P.M. Ajayan, G. Dresselhause, M.S. Dresselhause, Phys. Rev. Lett. **108**, 841 (1998)
21. T. Hayashi, S. Hirono, M Tomita, S. Umemura, Nature **381**, 772 (1996)
22. M. Nakata, Z. Akase, D. Shindo, M. Morita, T. Oikawa, Microsc. Microanal. **13**(Suppl. 2), 1004 CD (2007) DOI: 10.1017/S1431927607073655.
23. S. Hirosawa, H. Kanekiyo, Mater. Sci. Eng. A **217/218** 367 (1996)
24. D. Shindo, Y.G. Park, Y. Murakami, Y. Gao, H. Kanekiyo, S. Hirosawa, Scripta Mater. **48**, 851 (2003)
25. D. Shindo, M. Hosokawa, Z. Liu, Y. Murakami, T. Ito, Y. Iwasaki, J. Tachibana, Microsc. Microanal. **10**, 116 (2004)
26. T. Ito, Y. Iwasaki, H. Tachikawa, Y. Murakami, D. Shindo, J. Appl. Phys. **91**, 4468 (2002)
27. W.X. Xia, K Tohara, Y. Murakami, D. Shindo, T. Ito, Y. Iwasaki, J. Tachibana, IEEE Trans. Magn. **42**, 3252 (2006)
28. Y. Murakami, N. Kawamoto, D. Shindo, I. Ishikawa, S. Deguchi, K. Yamazaki, M. Inoue, Y. Kondo, K. Suganuma, Appl. Phys. Lett. **88**, 223103 (2006)

11

Supercritical Hydrothermal Synthesis of Organic–Inorganic Hybrid Nanoparticles

T. Adschiri and K. Byrappa

11.1 Introduction

Nanoparticles of gold are known to man since Roman times. In nature nanomaterials exist ever since the earth came into existence. These materials, whether organic, inorganic, or biological are created naturally under vivid environmental conditions. However, Michael Faraday was the first one to seriously experiment with gold nanoparticles starting in the 1850s. That is the beginning of the scientific approach to nanomaterials investigations. The term nanotechnology was first used by the Japanese researcher Taniguchi in 1974 when he referred to the ability to engineer materials at the nanometer scale [1]. The main idea behind the usage of this terminology was the miniaturization in the electronics industry. Today the term nanotechnology is popularly used to define technology where dimensions and tolerances are in the range 1–100 nm, larger than size of atom until about wavelength of light. Of course, it could be argued that nanomaterials are not new, given that Raney nickel catalysts, nanoporous aluminosilicate zeolites, and activated carbons have been used for many decades. What is certainly new, however, is the availability of an extraordinarily wide range of nanosized materials, many of which have unique properties and exciting applications [2]. The nanomaterials are known for their unique mechanical, chemical, physical, thermal, electrical, optical, electronic, magnetic, and specific surface area properties, which in turn define them as nanostructures, nanoelectronics, nanophotonics, nanobiotechnology, nanoanalytics, etc. During 1980s, the concept of size quantization was formulated for semiconductors and accordingly nanoparticles possess unique optical and electronic properties not observed for corresponding bulk samples [3, 4]. The field of functionalization is almost endless and represents an enormous space to explore. Over the past 1.5 decades considerable progress has been achieved in the area of synthesis as a tool-set for new materials and nanotechnology. A great variety of inorganic and organic compounds have been fabricated in nanoscales as tubes, ribbons, spheres, wires, belts, dots, fibers, dendrites, mushrooms like, etc. Accordingly several processing methods have been employed using both bottom-up and top-down principles. Most commonly used methods are physical vapor deposition, colloidal chemistry approach, mechanical alloying, sol-gel, mechanical grinding, hydrothermal method, laser ablation, CVD, electrodeposition, plasma synthesis, microwave techniques, etc. A large variety of nanomaterials and devices with new capabilities have been generated employing nanoparticles based on metals, metal oxides, ceramics (both oxide and nonoxide), silicates, organics, polymers, etc. In the last few years, a trend has been set in the processing of a new class of nanomaterials including coordination polymers, in which isolated metal ions or clusters are linked into arrays by organic ligands, and extended inorganic hybrids, such as hybrid metal oxides in which there is M–O–M connectivity in one, two, or three dimensions. Some of these three-dimensional networks with open structures that are the lightest crystalline materials ever made at $0.1 \, \mathrm{gm \, cc^{-1}}$ [5]. Such hybrid frameworks represent an enormous class of new materials that can harness the advantages and versatility of both organics and inorganics with a scope that is vastly greater than the combined fields of coordination chemistry and organometallics. These new class of organic–inorganic hybrid materials have a tremendous application potential as sensors, biological tags, catalysts, bioelectronics, solar energy conversion, mechanics, membranes, protective coatings, transparent pigments, electromagnets, cosmetics, etc. [6,7]. In fact, the possibility of combining properties of organic and inorganic materials is an old challenge which began from the time of industrial era. Some of the earliest and most well-known inorganic–organic representatives are where inorganic pigments or fillers are dispersed in the organic components such as solvents, surfactants, plastics, polymers, etc., to improve optical or mechanical properties of materials. However, the concept of "hybrid inorganic–organic" materials exploded only very recently with the birth of soft inorganic chemical routes, where mild synthetic conditions allow access to chemically designed hybrid inorganic–organic materials [8,9]. The small size and high surface to volume ratio of the individual nanoparticles imparts distinct size tunable physical and electronic properties that have prompted some to refer to them as "artificial atoms." As the number of nanoparticle systems under strict synthetic control has expanded, the parallel to the development of a "new periodic table" is also under consideration [10]. A highly controlled self-assembly of these hybrid nanoparticles when dispersed in organic solvents into two-dimensional and/or three-dimensional ordered structures or superlattice structures remains a relatively unexplored area. In recent years, there are lots of publications on such self-assembled nanoparticle structures in the literature. However, it is to be noted that a major challenge in the nanomaterial fabrication whether organic, or inorganic or composite nanoparticles, is the preparation of highly dispersible unagglomerated nanoparticles with an accurate control over the size and shape, which in turn is directly linked with the nanomaterial processing method. Owing to the higher surface energy of the particles in nanoscale, there is a greater tendency for nanoparticles agglomeration and aggregation. At the same time it is difficult to break down the aggregation of nanoparticles into individual particles size. Therefore, a new strategy for an effective solution to this problem has to be worked out. In this context, the hydrothermal technique can be very effective to combat these issues in nanomaterials fabrication. On the whole, the hydrothermal technique is gaining popularity in recent years for processing a wide range of materials in the polyscale, i.e., from bulk to nanosize [11–13]. In this chapter, the authors discuss the hydrothermal technique in general and the supercritical hydrothermal technique in particular for the fabrication of organic–inorganic nanohybrid particles with an emphasis on the basic principles, methodology, surface modification, size and shape control, self-assembly, and the properties of the nanoparticles obtained with appropriate examples.

11.2 Supercritical Hydrothermal Technique

The term *hydrothermal* can be defined as any homogeneous or heterogeneous chemical reaction in the presence of a solvent (whether aqueous or non-aqueous) above the room temperature and at pressure greater than 1 atm in a closed system [11]. There are several other related terms commonly used in hydrothermal like solvothermal, supercritical hydrothermal, glycothermal, alcothermal, ammonothermal, carbonothermal, lyothermal, and so on. These different terms are basically used depending upon the type of solvents employed in the reactions. Each one of these techniques has some specific advantages and suitable for processing of a particular type of materials. The conventional hydrothermal refers to the reactions in the presence of an aqueous solvents, and solvothermal refers to the nonaqueous solvents. The ammonothermal refers to the ammonium-based solvents, which are probably the best for the processing of nitrides including GaN. Similarly, the supercritical hydrothermal refers to the reactions near or above critical temperature. In nanotechnology, hydrothermal processing has an edge over the other conventional processes because it facilitates the issues like simplicity, cost effectiveness, energy saving, use of large volume of equipment (scale-up process), better nucleation control, pollution free (since the reaction is carried out in a closed system), higher dispersion, higher rate of reaction, better shape control, and lower temperature operation in the presence of an appropriate solvent, etc. The hydrothermal technique has a lot of other advantages as it accelerates interactions between solid and fluid species, phase pure and homogeneous materials can be achieved, reaction kinetics can be enhanced, the hydrothermal fluids offer higher diffusivity, low viscosity, facilitate mass transport, and higher dissolving power. Most important is that the chemical environment can be suitably tailored. In recent years, much attention has been paid on the hydrothermal solution chemistry through thermodynamic calculations, which facilitate the selection of a proper solvent and appropriate pressure–temperature range that not only helps in synthesizing the products, but also in controlling the size and shape with a significant reduction in the experimental duration. In the last couple of years, researchers are using the concept of instant hydrothermal reactions to obtain the desired nanoparticles in a shortest possible time and some even imagining a system like vending machine to produce the desired nanoparticles with definite physical properties [14,15]. A great variety of materials have been obtained using the hydrothermal method. Amongst them the native elements, metal oxides, hydroxides, silicates, carbonates, phosphates, sulfides, tellurides, nitrides, selenides, etc., both as particles and nanostructures like nanotubes, nanowires, nanorods, etc., are the popular ones. The method has been successfully used for the synthesis of a variety of carbon forms like sp^2, sp^3, and intermediate types.

The basic principle used in the hydrothermal synthesis of nanoparticles is that the solute raw material or the precursor material along with the appropriate solvent are taken in an autoclave which is maintained at a high temperature either below or above the critical temperature of the solvent over a period of time, which allows the solute to dissolve and recrystallize to the desired product. The solvent becomes highly corrosive under elevated temperature and pressure conditions and majority of the substances show higher solubility under hydrothermal conditions. Both experimental and theoretical calculations of the phase equilibria data have yielded enormous amount of literature data to select the proper processing conditions for a variety of inorganic materials [16–18]. The thermochemical computation data help in the intelligent engineering of the hydrothermal processing of advanced nanoparticles. Such a modeling tool can be successfully applied to very complex aqueous electrolyte systems over wide ranges of temperature and concentration and is widely used in both industry and academy.

Here, it is appropriate to discuss the supercritical hydrothermal in greater detail, since the present chapter is devoted to this. The ability of the supercritical fluids to dissolve nonvolatile solids has been known for more than a century [19] and it was extensively used to understand the natural mineralization [20]. But, the commercial exploitation of this quality of supercritical hydrothermal technique (also known as supercritical fluid technique) has begun only recently. It has been successfully employed as an effective energy saving process in the field of solvent extraction. A fluid is supercritical when its temperature and pressure are higher than their critical point values (T_c, P_c). Most of the interesting applications of supercritical fluids occur at $1 < T/T_c < 1.1$ and $1 < P/P_c < 2$ [21]. Under these conditions the fluid exists as a single phase having some of the advantageous properties of both liquid and gas phases. Further, supercritical hydrothermal is also known for making new materials under mild conditions, since it increases the chemical reactions, improves the mass transfer, and greatly assists in the stability of selected or desired products [22–24].

The supercritical hydrothermal method is an extension of hydrothermal technology. The distinct difference between the conventional hydrothermal and supercritical hydrothermal technology in the present-day context is that hydrothermal is treated as more closer to the soft solution processing or mild technique [25, 26], whereas the supercritical hydrothermal technology deals with the reactions at temperatures just near or above the critical temperature. The supercritical hydrothermal has witnessed a seminal growth in its application to the processing of a variety of materials owing to the availability of a wide range of solvents, and the most commonly used ones are water and carbon dioxide. The use of carbon dioxide has lots of advantages and replacing organic solvents in a number of chemical processes, including nanoparticle fabrication, food processing (such as decaffeination of coffee beans), chemical manufacturing, extraction, dry cleaning, semiconductor wafer cleaning, polymer processing, recycling, waste treatment, organic decomposition, particles coating, drug manufacturing, etc. [27,28]. The supercritical hydrothermal method with water as a solvent is highly suitable for the nanoparticle fabrication of a wide range of high-melting inorganic compounds and also for hybrid organic–inorganic hybrid nanoparticles. There are several variants in the supercritical fluid technology like static supercritical fluid process (SSF), rapid expansion of supercritical solutions (RESS), particles from gas-saturated solutions (PGSS), gas antisolvent process (GAS), precipitation from compressed antisolvent (PCA), aerosol solvent extraction system (ASES), supercritical antisolvent process (SAS), solution enhanced dispersion by supercritical fluids (SEDS), supercritical antisolvent process with enhanced mass transfer (SAS-EM), depressurization of an expanded liquid organic solution (DELOS), supercritical-assisted atomization (SAA), hydrothermal synthesis under supercritical conditions via flow reactor (HTSSF), hydrothermal synthesis under supercritical conditions via batch reactor (HTSSB), supercritical fluids drying (SCFD), supercritical fluid extraction emulsions (SFEE), etc. [13]. Since the subject is so vast, for the benefit of the reader, this chapter has been restricted to a systematic discussion of the supercritical hydrothermal processing of organic–inorganic hybrid nanoparticles using the supercritical water medium as a solvent.

11.3 Apparatus

There is a wide range of apparatus used in the supercritical hydrothermal research to suit the specific conditions and the product. However, the most commonly used apparatus are batch reactors and flow reactors. The batch reactors are the cone-cold sealed type autoclaves usually made of hastelloy or stainless steel (SS316) with an ease to handle because of simple reactor design. A typical inner volume of these batch reactors is usually 5 mL. Sometimes the platinum lining or platinum capsules are provided to these reactors to prevent the corrosion. Depending upon the pressure and temperature of the reaction and density of water in those conditions, a required amount of precursors are loaded into the reactor. These reactors can be stirred or shaken

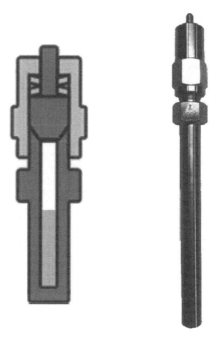

Fig. 11.1. Batch reactor

during the experiments. A series of such batch reactors can be mounted on a single frame and experiments can be run conveniently by varying several experimental parameters. The experimental duration is normally varied from 5 min to several hours. Figure 11.1 shows the schematic diagram and a typical batch reactor. The product recovery process from the batch reactors is very important while dealing with nanoparticles fabrication.

Figure 11.2 shows the typical experimental setup used in the flow type of hydrothermal reactor. An aqueous metal salt solution is first prepared and fed into the apparatus in one stream. In another stream, distilled water is pressurized and then heated to a temperature above the desired reaction temperature. The pressurized metal salt solution stream and the pure scH_2O stream are then combined at a mixing point, which leads to rapid heating and subsequent reactions in the reactor. After the solution leaves the reactor, it is rapidly quenched. In-line filters are used to remove larger particles. Pressure is controlled with a back-pressure regulator. Fine particles are collected in the effluent. By this rapid heating method, the effect of the heating period on the hydrothermal synthesis is eliminated: thus specific features of supercritical hydrothermal synthesis can be elucidated. There are several advantages in supercritical hydrothermal flow reactors, which provide nanoparticles with desired shape, size, and composition in the shortest possible residence time. Hence, this technique has an edge over the conventional methods. Figure 11.3 shows the semipilot plant flow reactor used in the fabrication of advanced hybrid organic–inorganic nanoparticles in T. Adschiri's laboratory.

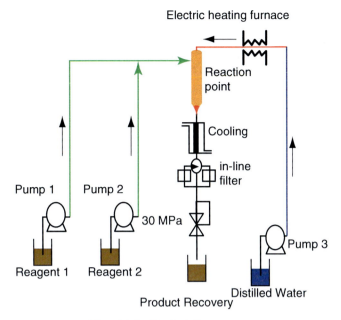

Fig. 11.2. Typical flow reactor

11.4 Basic Principles of Supercritical Hydrothermal Synthesis

It is well established that water is in a supercritical state above the critical temperature (374°C) and pressure (22.1 MPa). Figure 11.4 shows the PVT relation of water [29]. Above the critical point, the density of water varies greatly with a little change in temperature and pressure. Because of the drastic change in density, all the fluid properties change greatly around the critical point, including the dielectric constant that is a controlling factor of reaction rate, equilibrium, and solubility of metal oxides. Therefore, it is necessary to understand the basic principles, which insist upon the understanding of the properties of water, including density, dielectric constant (Fig. 11.5) [30], and ion product, varying greatly around the critical point of water and result in a specific reaction atmosphere. The dielectric constant of water at room temperature is 78 and with raise in temperature at a constant pressure, this value decreases greatly and around critical point of water it is nearly equal to the dielectric constant of polar organic solvents.

Distribution of soluble chemical species changes greatly at the critical point range. Various models have been proposed to describe the variation of reaction rate or equilibrium over a range of supercritical state [31, 32]. Due to the variation in the properties of water, phase behavior changes greatly around the critical point. Since supercritical water is of high density steam, light gases like oxygen or hydrogen form a homogeneous phase with supercritical water.

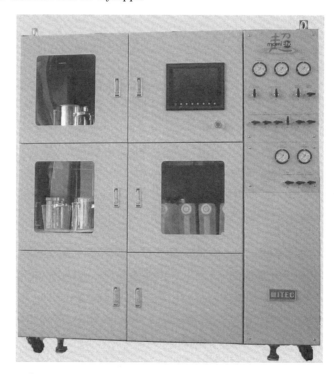

Fig. 11.3. Semipilot plant flow reactor used in the laboratory of T. Adschiri

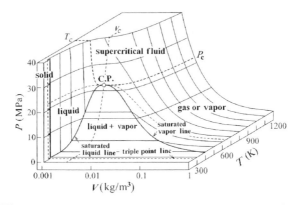

Fig. 11.4. PVT surface of water: *C.P.* critical point [29]

Figure 11.6 shows the critical loci for binary systems of organic compound and water [33]. In the right-hand side of these curves, a homogeneous phase is formed at any composition. The organic compounds and water are not miscible at a low-temperature range, but forms a homogeneous phase at higher temperatures, which is because of the reduced dielectric constant of water. It should be noted that when the pressure is very high (i.e., the density of fluid is as high as of liquid), even if both water and organic compound are in supercritical state, phase separation occurs, just like liquid water–liquid oil phase separation.

Adschiri and Arai [23,24,34–38] have pioneered the supercritical hydrothermal synthesis of metal oxides in detail during 1990s. A complete mechanism of the formation of various metal oxides have been discussed extensively by these authors. This forms the foundation for the processing of advanced hybrid organic–inorganic nanoparticles. Therefore, it is highly important to discuss the basic approaches in the fabrication of metal oxides nanoparticles using supercritical hydrothermal technique briefly before going to the hybrid organic–inorganic nanoparticle synthesis.

The knowledge on the solubility of the crystalline substances formed and the identification of the ionic species are essential to understand the supercritical processes. For example, the main equilibrium reactions for AlOOH dissolution in aqueous solution are given in Table 11.1. For the estimation of metal oxides, the equilibrium constant K for each reaction is required. There are many models to estimate this. HKF model is very widely used for K at high temperature/pressure including supercritical state. The following equation is a concept for the estimation of K:

$$\ln K(T, \rho) = \ln K(T_0, \rho_0) + \frac{\Delta H}{R}\left(\frac{1}{T} - \frac{1}{T_0}\right) - \frac{\omega_j}{RT}\left(\frac{1}{\epsilon(T,\rho)} - \frac{1}{T_0, \rho_0}\right),$$

where K is the equilibrium constant, T is the absolute temperature, T_0 is the room temperature, R is the gas constant, ϵ is the dielectric constant, ω is a parameter determined by the reaction system, ρ is the density, and ρ_o is the density at the ambient conditions.

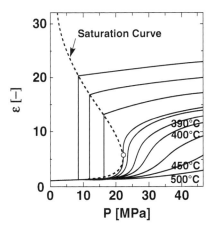

Fig. 11.5. Dielectric constant of water around the critical point [30]

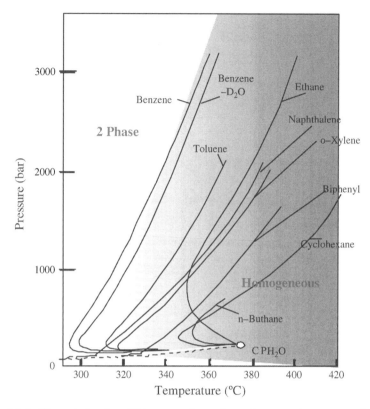

Fig. 11.6. Phase behavior of binary systems for water-organic compounds [33]

Table 11.1. Reactions related to AlOOH dissolution [23]

AlOOH + H$_2$O = Al^{3+} + 3OH$^-$	Al^{3+} + OH$^-$ = Al(OH)$^{2+}$	NaNO$_3$ = Na$^+$ + NO$_3^-$
AlOOH + H$_2$O = Al(OH)$^{2+}$ + 2OH$^-$	Al(OH)$^{2+}$ + OH$^-$ = Al(OH)$_2^+$	NaOH = Na$^+$ + OH$^-$
AlOOH + H$_2$O = Al(OH)$^+$ + 2OH$^-$	Al(OH)$_2^+$ + OH$^-$ = Al(OH)$_3$	HNO$_3$ = H$^+$ + NO$_3^-$
AlOOH + H$_2$O = Al(OH)$_4^-$ + 3OH$^+$	Al(OH)$_3$ + OH$^-$ = Al(OH)$_4^-$	H$_2$O = H$^+$ + OH$^-$
	Al(NO$_3$)$_3^-$ = Al^{3+} + 3(NO$_3$)$^-$	

The solubility and the distribution of ionic species in solution are also important for the understanding of supercritical reaction atmosphere. Solubility and distribution of ionic species can be predicted by calculating K for each reaction (Table 11.1), mass balance, and charge balance. There are several models available today to estimate the solubility of the species of interest based on the dissociation reaction in a given system. Even by using the value estimated from the ionic radius reported in the literature, the solubility behavior trend can be precisely estimated. It is well known that the experimental results could be well correlated with various models to express the increase of solubility with increasing temperature (for positive solubility materials) and increase of solubility with decrease of temperature (for negative solubility materials), and the dramatic decrease of solubility around the critical point [39]. That is the greatest advantage of supercritical hydrothermal conditions unlike the conventional hydrothermal or solvothermal conditions. Figures 11.7 and 11.8 show how significant the solubility of metal oxides and kinetics of hydrothermal synthesis change around the critical point due to the variation in the properties of water. The solubility of CuO in water over a

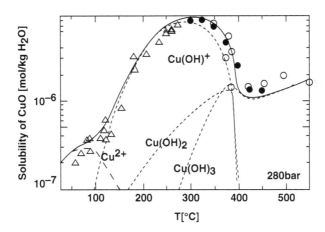

Fig. 11.7. Solubility of CuO in high-temperature water [40]

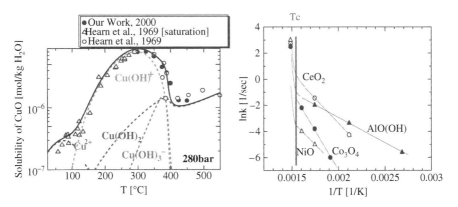

Fig. 11.8. Kinetics of supercritical hydrothermal synthesis [38]

258   T. Adschiri and K. Byrappa

wide range of temperature is shown in Fig. 11.7. The solubility first increases
with an increase in temperature and then decreases. This decrease in solubility is due to the lowering of the density and dielectric constant values.
An estimation method of metal oxide solubility is described in detail by Sue
et al. [40]. Figure 11.8 shows the Arrhenius plot of the first-order reaction rate
constant of hydrothermal synthesis, which was evaluated under supercritical
hydrothermal conditions using a flow type of apparatus [36, 38]. The Arrhenius plot of kinetic constant shows a straight line below the critical point, but
above the critical point increased two orders of magnitude. This is because of
the decrease of dielectric constant, and the details of such a mechanism has
been discussed elsewhere [38].

11.5 Supercritical Hydrothermal Synthesis of Metal Oxides

A wide range of metal oxide nanoparticles have been obtained under supercritical hydrothermal conditions for applications not only in ceramics, coatings, catalysts, sensors, semiconductors, magnetic data storage, solar energy
devices, ferrofluids, but also in medical field such as hyperthermia, bioimag-
ing, cell labeling, special drug delivery systems, and so on. However, their
application potential is dictated by their surface nature, particle size, and
also shape. Although the synthesis of metal oxides is not new, their applications were limited. With the discovery of size quantization effect in these
materials during 1980s, there is a seminal progress in the synthesis of these
metal oxides with desired properties for several new applications. Researchers
like Adschiri, Arai, Johnston, Lester, Poliakoff, etc., have contributed extensively on the synthesis of metal oxides and proposed a systematic mechanism [23, 24, 41, 42]. Metal oxides like Al_2O_3, Ga_2O_3, In_2O_3, SiO_2, GeO_2,
ZnO, V_2O_5, TiO_2, CeO_2, ZrO_2, CoO, α-Fe_2O_3, γ-Fe_2O_3, NiO, Co_3O_4, Mn_3O_4,
γ-MnO_2, Cu_2O, $CoFe_2O_4$, $ZnFe_2O_4$, $ZnAl_2O_4$, Fe_2CoO_4, $BaZrO_3$, $BaTiO_3$,
$BaFe_{12}O_{19}$, $LiMn_2O_4$, $LiCoO_2$, La_2O_3, etc., have been prepared by the above
method. Usually the particles obtained in the subcritical water conditions are
larger than those in the scH_2O, because, there is a particle growth with an
increase in the residence time, whereas under supercritical conditions such a
phenomenon has not been observed. The hydrothermal reaction rate in scH_2O
is higher, and the solubility of the metal oxides is much lower than that in subcritical water. This leads to the generation of higher degree of supersaturation.
The nucleation rate is expected by the function of degree of supersaturation
and the surface energy according to the nucleation theory. Thus, extremely
high nucleation rate can be expected at supercritical conditions, which leads
to the formation of nanosize particles. Figure 11.9 shows the mechanism for
the fine CeO_2 particles formation in scH_2O. More or less a complete list of the
materials obtained under supercritical hydrothermal conditions is available.

11 Supercritical Hydrothermal Synthesis 259

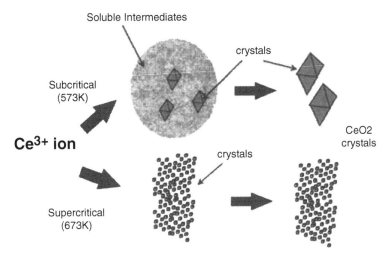

Fig. 11.9. Mechanism of CeO$_2$ nanoparticles formation under supercritical hydrothermal conditions

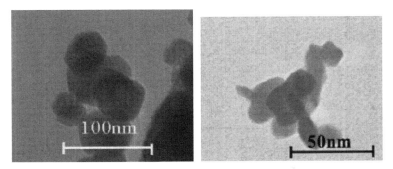

Fig. 11.10. TEM image of YAG:Tb nanoparticles synthesized at 400°C and 30 MPa [44]

The works of Reverchon and Adami [43] and Byrappa and Adschiri [12]. Basically hydrothermal synthesis method is available for the metal oxides by conventional hydrothermal synthesis method. The point to be noted here is that under supercritical hydrothermal conditions, nanometer size metal oxides could be synthesized and crystallinity of the nanoparticles would be much higher when compared to the metal oxides obtained under conventional hydrothermal conditions, wherein bulk single crystals are formed. This sometimes leads to the specific characteristics of the products. Phosphor nanoparticles show high luminescence without heat treatment which is usually necessary for the products obtained by low-temperature wet method [35, 44]. For example, Fig. 11.10 shows TEM images of YAG:Tb phosphors nanoparticles obtained using highly controlled supercritical hydrothermal conditions from

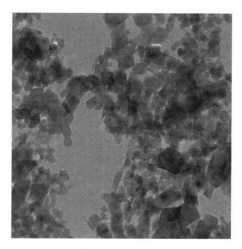

Fig. 11.11. TEM image of LiMn$_2$O$_4$ nanoparticles synthesized at 400°C and 30 MPa [44]

Fig. 11.12. Ni nanoparticles obtained from nickel acetate and formic acid at 400°C and 30 MPa

a flow reactor [44]. Similarly, Fig. 11.7 shows nanoparticles of LiMn$_2$O$_4$ with a particle size ranging from 10 to 20 nm synthesized from LiOH, Mn(NO$_3$)$_2$ and H$_2$O$_2$ at 400°C and 30 MPa. These particles do not show decay of its capacity even after the charge-discharge cycles, which has been considered as a major breakthrough point of these solid electrolyte materials [37, 45]. Kanamura et al. [46] have discussed this mechanism in detail and concluded that these particles are single crystals of LiMn$_2$O$_4$ and are totally different from those obtained by other methods. In the synthesis of LiMn$_2$O$_4$ nanoparticles, oxidizing reaction atmosphere has to be controlled by regulating oxygen gas partial pressure in the system. In this case, Mn^{2+} of Mn(NO$_3$)$_2$ should be oxidized into Mn^{3+}. To achieve this H$_2$O$_2$ was fed into the system.

H_2O_2 decomposes at supercritical conditions into oxygen gas, which forms a homogeneous phase with supercritical water to provide an excellent oxidizing atmosphere [37, 45, 46]. Figure 11.12 shows Ni nanoparticles formed from nickel acetate and formic acid at 400°C and 30 MPa. In this case, HCOOH was introduced as a reducing agent with nickel acetate solution [47, 48]. In supercritical water, HCOOH is decomposed into hydrogen and carbon dioxide. An important point is that these gases and supercritical water form a homogeneous phase, and this mixture of gas (H_2 and CO_2) shows higher reducing ability than H_2 gas, as was reported in the literature [49, 50].

Although supercritical hydrothermal technology gives highly crystalline nanoparticles with a homogeneous composition, still there is a problem of larger size, coagulation, and poor dispersibility of nanoparticles in aqueous solutions like water as seen from Figs. 11.10–11.13.

Poor dispersibility of such nanoparticles is due to the fact that there is an aggregation between nanoparticles which leads to worse dispersion and after a while, precipitated particles will appear at the bottom of the bottles as shown in Fig. 11.13. Hence, there is a trend to obtain small nanoparticles with a perfect control over the size and morphology and high dispersibility with a modern approach in materials processing like use of organic ligands, capping agents, surfactants, chelating agents, etc., which generate a new class

Fig. 11.13. Dispersibility of nanoparticles: $CoAl_2O_4$ (*blue*); ZnO (*white*); Fe_2O_3 (*brown*) in water

of nanomaterials, viz., organic–inorganic nanomaterials. This strategy is based on the miscibility of the organic ligand molecules with scH_2O due to the lower dielectric constant of the water; and the nanocrystal shape control by selective reaction of organic ligand molecules to the specific inorganic crystal surface. For this purpose even some amino acids, peptides, proteins, or DNA are used to modify the particle surfaces and such particles can be specifically combined with other proteins or DNA, which leads to the applications to new biotechnology or IT devices using directed assembly of semiconductor nanoparticles. The remaining part of this chapter has been devoted to the supercritical hydrothermal synthesis of hybrid organic–inorganic nanoparticles.

11.6 Supercritical Hydrothermal Synthesis of Hybrid Organic–Inorganic Nanoparticles

The organic–inorganic hybrid nanoparticles are considered to be most promising new class of materials that show the trade off functions between polymers/organics and inorganics (light and high mechanical strength, high thermal conductivity and electrical resistance, transparent and flexible electroconducting films, etc.). There is a huge opportunity for discovering hybrid analogues of classical oxide systems that exhibit a wide range of physical properties. This para ends with Future targets should include metallic hybrids (analogues of conducting polymers), lasers, and even superconductors.

11.6.1 Organic–Inorganic Hybrid Nanoparticles

Although the study of these hybrid nanoparticles is considered to be a recent development, the literature survey shows that the approach to process organic–inorganic materials began during 1970s itself. Several chemical routes were employed to prepare such hybrid nanoparticles and disperse them in the organic solvents. Silane coupling agents have been in common use for decades providing enhanced adhesion between a variety of inorganic and organic agents. The general formula of these organosilane coupling agents is $R_n SiX_{(4-n)}$ having dual functionality. The majority of silane coupling agents contain a hydrolyzable group (X), typically methoxy or ethoxy which readily reacts with a proton to give methanol or ethanol as byproducts of the coupling. Metal oxide has hydroxyl groups which provide the necessary proton for the coupling reaction. The "R" group is a nonhydrolyzable organic group designed to provide hydrophobicity nature for surface of metal oxide nanoparticles. In this case as shown in Fig. 11.14, by dehydration reaction between silane coupling agent and OH groups of metal oxide, stable bonds of M–O–Si–C will be formed resulting in the surface modification and changing the surface property of the metal oxides from hydrophilic to hydrophobic [51–54].

The greatest disadvantage of this route is the presence of Si shell in the structure of hybrid organic–inorganic metal oxide particles introducing significant changes in the original properties of these metal oxides and limiting their applications. For example, silane is very sensitive to UV radiation and this type of hybrid metal oxides like ZnO cannot be used for such applications as sun block. Further the silica shell around the metal oxide significantly increases the size of the modified particles. Thus, by considering such an ex situ surface modification with functional groups and silane coupling agent, the problem of surface modification of individual particles may not be possible, also there is a poor dispersibility of the modified particles and there is no control over the size and shape of the particles. Hence, researchers proposed an in situ surface modification of the nanoparticles to overcome all the above mentioned problems. Perhaps the first steps in this direction leading to a rational nanoparticle assembly strategy were taken by the groups of Mirkin [55, 56] and Alivisatos [57], who demonstrated that DNA-modified colloidal gold nanoparticles could be assembled into superstructures by hybridization of complementary base sequences in the surface-bound DNA molecules. The motive behind such studies was to obtain a perfect dispersion of inorganic nanoparticles in solvents and plastics and to achieve a change in surface properties to the nanoparticles to hydrophobic, then, it is necessary to control surface of nanoparticles by organic modifiers and make a new generation of advanced materials, viz., hybrid organic–inorganic nanoparticles. Thus, a combination of inorganic materials at the nanosize with the organic molecules could solve several problems encountered in the application of nanoparticles, and it led to the emergence of the in situ surface modification of nanoparticles with a great variety of organic surface modifiers, which bring in a perfect dispersion of the nanoparticles in solvents or in polymers. So far, tremendous efforts have been made to fabricate nanoparticle dispersed polymer, but it has been considered a difficult task to disperse the nanoparticles in organic solvents or in polymers, especially for the particles synthesized under hydrothermal conditions. This is because metal oxide particle surface is hydrophilic, and for the case of nanoparticles, it shows extremely high surface energy, which leads to the formation of aggregates.

11.6.2 Supercritical Hydrothermal Organic–Inorganic Hybrid Nanoparticles

To resolve the problem of particle aggregation, to achieve the perfect control over the size and morphology of the particles, and to obtain a desired surface property to the nanoparticles, a new processing strategy has been proposed group utilizing the supercritical hydrothermal technology [13, 58–62]. Figure 11.15 shows a schematic representation of highly effective strategy for the synthesis of metal oxide nanocrystals in the organic-ligand-assisted supercritical hydrothermal technology [58]. The method yields perfect hybrid organic–inorganic nanocrystals with very high dispersibility, and a precise

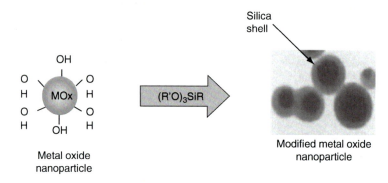

Fig. 11.14. Modified metal oxide particles with silane coupling agent

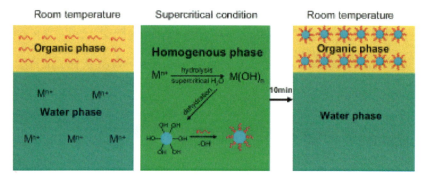

Fig. 11.15. The strategy for the synthesis of metal oxide nanocrystals in the organic-ligand-assisted supercritical hydrothermal technique [58]

control over the size and shape of the nanoparticles. The organic components are introduced into the system during the hydrothermal synthesis and in situ surface modification is obtained with a ultrathin layer of organics surrounding the inorganic unlike the case of silane coupling on the metal oxides. The organic ligands and supercritical water form a homogeneous phase and it is known that under these conditions water molecules themselves work as acid or base catalyst for various organic reactions. Depending upon the applications of nanoparticles, one can select a suitable functional groups to introduce hydrophobicity or hydrophilicity property to the surface of the modified nanoparticles.

11.6.3 Mechanism of Formation of Organic–Inorganic Hybrid Nanoparticles

A knowledge on the mechanism of formation of organic–inorganic hybrid nanoparticles is very important and it deals with the interaction of the organic ligand molecules with the inorganic metal oxide surfaces. Adschiri and group have worked out in detail the theory and mechanism of the hybrid organic–inorganic nanoparticle formation under supercritical hydrothermal conditions. To understand the mechanism of this hybrid nanoparticles formation, it is necessary to investigate the charge distribution on the surface of the metal oxide particles, which in turn reveals the chemical interaction between the modifier and the surface of the metal oxide nanoparticle. In aqueous systems, metal oxide particles are hydrated and M–OH groups cover completely their surface and the surface is neutral. The M–OH sites on the surface of particles can react with H^+ or OH^- ions from dissolved acids or bases, and positive ($M-OH^{2+}$) or negative ($M-O^-$) charges develop on the surface, and type of the reaction depends on pH of the solution. In the absence of specific adsorption of ions, amphoteric metal oxides have a characteristic pH, the pH of the point of zero charge (PZC), where the net surface charge is zero, i.e., the positive and negative sites are in equal amount (isoelectric point, iep). At pH lower than iep, the pure metal oxide surface is positively charged, while it has negative charge above it [63–65]. Normally the modifiers attach onto the surface of the nanoparticles either by physisorption or chemisorption through strong hydrogen bond. If we consider R–COOH as modifier reagents in a highly acidic pH of the reaction medium, the dissociation of modifier does not occur and the conjugation is due to the strong hydrogen bonding between hydroxylic groups on the surface of nanoparticles and functional groups of modifier [66]. If the pH of the reaction medium is in the range of modifier dissociation, another type of interaction between the modifier and the metal oxide surfaces is expected [67]. Similar chemical bonding is expected even for amines or alcohol or aldehyde as the organic ligands in the system.

The addition of surface modifiers also helps to inhibit the crystal growth that facilitates the smaller particles size with a narrow particle size distribution. Figure 11.16 shows nanoparticles of Fe_2O_3, Co_3O_4, CeO_2, and $CoAl_2O_4$ obtained under supercritical hydrothermal conditions without and with organic modifiers. As is clearly seen from Fig. 11.16, those nanoparticles without modifier are aggregated and they do not disperse in the aqueous solvents. Also, the particle size is larger compared to the modified particles. When the pH of the reaction medium is highly acidic or highly basic, very small particles along with the large particles are formed leading to a broader size distribution. It seems that because of redissolving of nanoparticles at very high and low pH, Ostwald repining occurs. Therefore, in surface modification, pH of the medium, isoelectric point (iep) and another important parameter, viz., dissociation constant (pKa) of the modifiers are very important. At pH below pKa, the modifier does not dissociate. Moreover, below iep, the surface

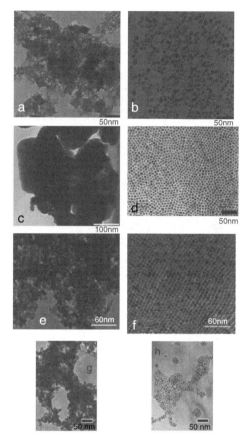

Fig. 11.16. TEM images of hybrid nanoparticles synthesized under supercritical hydrothermal conditions: (**a**) Fe_2O_3 (without modifier); (**b**) Fe_2O_3 (with modifier); (**c**) Co_3O_4 (without modifier); (**d**) Co_3O_4 (with modifier); (**e**) CeO_2 (without modifier); (**f**) CeO_2 (with modifier); (**g**) TiO_2 (without modifier); and (**h**) TiO_2 (with modifier) (from the works of Adschiri)

of metal oxide nanoparticles is surrounded by positive charges (major) and hydroxylic groups (minor). Under these conditions there is no chemical reaction occurring between the modifier and the metal oxide nanoparticle surface, but it is only through a strong hydrogen bonding the modifier can attach to the nanoparticles surface. In contrary, at higher pH than pKa, dissociation of modifier takes place and results in chemical reaction between dissociated part of modifier and OH_2^+ from particles surface. Thus, by dehydration reaction modifier attaches to the surface of the particles. The possible different types of conjugation between the surface modifier and the surface of the nanoparticles are schematically represented in Fig. 11.17. By considering the chemical

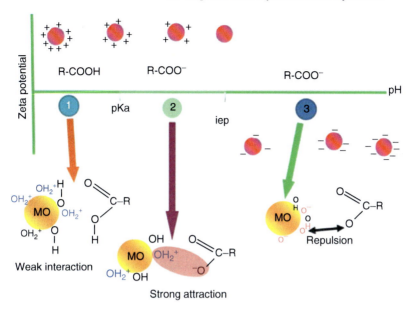

Fig. 11.17. Reaction mechanism of in situ surface modification of metal oxide nanoparticles with R–COOH as a modifier (from Adschiri's work)

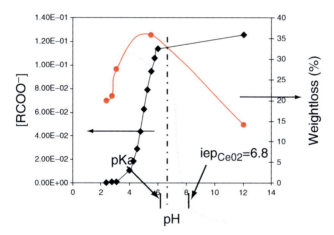

Fig. 11.18. Concentration of [RCOO$^-$] and amount of surface modifier on particles surface vs. pH of the reaction media

reactions, mass balances, charge balance in the actual system, pH, and the modifier can be fixed for most of the systems.

From the literature data, the values of pKa for several surfactants are available. Here, an example of CeO$_2$ nanoparticle surface modification using oleic acid is described. Oleic acid has a pKa value 5.02 [68], and iep of cerium oxide is 6.8 [69]. By calculating [RCOO$^-$] from low pH up to the base conditions, it was found that amount of dissociated part of modifier increases with pH, and the maximum point for surface modification is in the pH range around 5.5 somewhere between pKa and iep, in which the surface charge is negative and chemical reaction occurs between nanoparticles and dissociated part of the modifier as $RCOO^-$. However, it is necessary to consider such calculations at elevated temperature and pressure, or the experimental PT conditions. Figure 11.18 shows the concentration of $[RCOO^-]$ and amount of surface modifier on particles surface vs. pH of the reaction media. The optimum surface modification is seen in between iep and pKa.

11.6.4 Experimental Study of the Supercritical Hydrothermal Synthesis of Organic–Inorganic Hybrid Nanoparticles with Size and Shape Control

The experimental studies on the preparation of some selected the experimental studies on the preparation of some selected hybrid organic–inorganic nanoparticles under supercritical hydrothermal conditions are discussed here for the benefit of the readers. Figure 11.19 shows TEM images of ceria nanoparticles obtained through organic ligand-assisted supercritical hydrothermal synthesis. When decanoic acid (molar ratio to ceria precursor 6:1) was added to the reaction system, the resulting nanocrystals were nanocubes with an average

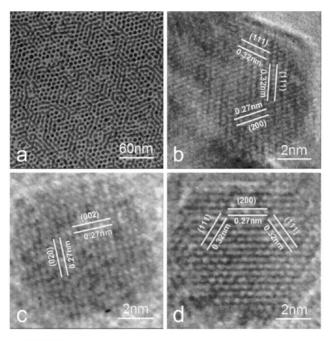

Fig. 11.19. HRTEM images of ceria nanocrystals with varying concentration of decanoic acid (modifier) to ceria [58]

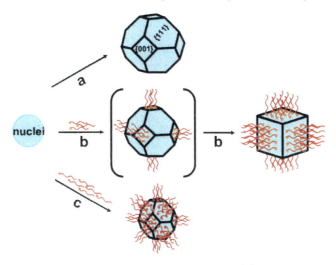

Fig. 11.20. The shape control of ceria nanoparticles: (**a**) truncated octahedron in the case when no organic ligand molecules are used; (**b**) at a low decanoic acid to ceria precursor ratio, the preferential interaction of the ligand molecules with the ceria 0 0 1 planes slows the growth of 0 0 1 faces relative to (1 1 1) faces, which leads to the formation of nanocubes; (**c**) at a high decanoic acid to ceria precursor ratio, organic ligand molecules block growth on both (0 0 1) and (1 1 1) faces, which leads to the formation of truncated octahedral and smaller crystals

size of 6 nm. The ceria nanocrystals unmodified were truncated octahedral, which corresponds with the results of the earlier workers [70]. However, when decanoic acid was introduced into the system under supercritical hydrothermal conditions, the organic ligand molecules become miscible with water, they contribute to the change in morphology of ceria nanocrystals from truncated octahedral to cube, which is mainly because of the suppression of the crystal growth on the (0 0 1) face. As the CeO_2 (0 0 1) surface is less stable than the (1 1 1) surface, the organic ligand molecules were likely to interact preferentially with the (0 0 1) surface thereby reducing the crystal growth along the [0 0 1] direction. If the amount of decanoic acid was further increased (molar ratio to ceria precursor, 24:1), the ceria nanocrystal size decreased to about 5 nm, and the shape of the nanocrystals also changed drastically, because of the growth inhibition not only along the [0 0 1] direction, but also along the [1 1 1] direction. Figure 11.20 shows the schematic diagram of the shape control of ceria nanoparticles under supercritical hydrothermal conditions in the presence of organic ligand molecules [58].

Figure 11.21 shows Co_3O_4 nanoparticles synthesized under supercritical hydrothermal conditions in the presence of C_6NH_2 as a modifier. The morphology of the nanoparticles was varied by changing the concentration of C_6NH_2. In the presence of smaller amount of the modifier, slightly small cubic shaped nanocrystals were formed. But with the addition of higher concentration of the modifier, the size of the nanoparticles further reduced, and ultimately

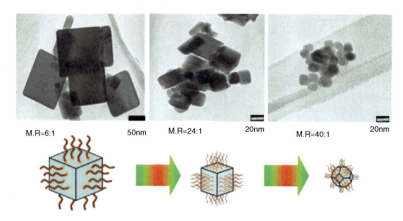

Fig. 11.21. Mechanism for size control in Co_3O_4 nanoparticle fabrication under SCF conditions ($T = 300°C$, $P = 20$ MPa, surfactant is hexaldehyde) with molar ratio (MR) of starting material to the modifier

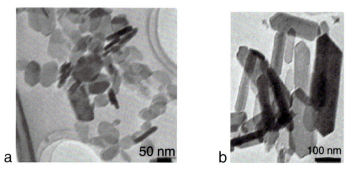

Fig. 11.22. TEM images of the boehmite nanoparticles obtained under supercritical hydrothermal conditions in the presence of $CH_3(CH_2)_5NH_2$ and $CH_3(CH_2)_4CHO$ as organic ligand molecules

ended up with very fine almost truncated morphology. It means that as the nucleation takes place, by increasing the amount of modifier, the number of the attached organic molecules onto the surface of the nuclei rises blocking the active sites for further crystal growth.

The synthesis of boehmite nanoparticles has been carried out under supercritical hydrothermal conditions in the presence of $CH_3(CH_2)_5NH_2$ and $CH_3(CH_2)_4CHO$ as organic ligand molecules and it is found that the size of the particles vary with the concentration and the type of modifier molecules (Fig. 11.22). Usually boehmite nanoparticles crystallize as plate like crystals. But the amines can be more efficient in reducing the size of the particles and also changing their morphology than aldehyde. Figure 11.23 shows a schematic representation of boehmite nanoparticles on a TEM grid, which is hydrophobic, while unmodified boehmite particles are hydrophilic. The

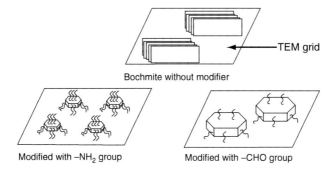

Fig. 11.23. Arrangement of boehmite nanoparticles on the TEM grid depending on the type of reagents

plate-like particles are likely to be stacked on each other and stand perpendicular to the TEM grid. However, for the modified particles which are hydrophobic, the particles show affinity for the TEM grid. Thus, well-distributed nanoparticles lying on the TEM grid can be seen, which suggest successful surface modification of boehmite particles [62].

Figure 11.24 shows very interesting results on the iron oxide nanoparticles modification with different organic ligand molecules. By adding different lengths of the same functional groups as modifier reagent effects on the particle size, arrangement and morphology of the particles obtained under supercritical hydrothermal conditions (400°C and 30 MPa). By adding decanoic and oleic acids to the starting precursors, cubic and spherical magnetite particles were obtained with a mean size about 25 and 5 nm, respectively. Without modifier it yielded hematite particles. However, under subcritical conditions, the addition of oleic acid to the precursor resulted in the formation of nanowires up to 85 nm length.

11.6.5 Dispersibility of Organic–Inorganic Hybrid Nanoparticles

There are several other works on the synthesis of hybrid organic–inorganic nanoparticles under supercritical conditions like TiO_2, ZnO, $CoAl_2O_4$, etc. [59, 70, 71].

It is to be noted that the surface-modified organic–inorganic hybrid nanoparticles synthesized under supercritical hydrothermal conditions can be easily recovered by extraction with organic solvent from the water suspension phase, and well dispersed in organic solvents. Normally, transparent solutions are obtained for the well crystallized and very small hybrid nanoparticles when dispersed. Through the dynamic light scattering (DLS) studies also, it can be confirmed that the particles are not aggregated, but suspended as individual nanoparticles. Figure 11.25 shows the perfect dispersion of various nanoparticles like Ni, $CoAl_2O_4$, Fe_3O_4, TiO_2, and CeO_2 in the organic solvents.

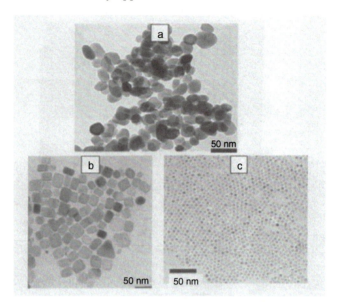

Fig. 11.24. TEM images of iron oxide nanoparticles synthesized under supercritical hydrothermal conditions (400°C and 30 MPa) without and with organic ligand molecules: (**a**) without modifier (hematite); (**b**) modified with decanoic acid (magnetite); and (**c**) modified with oleic acid (magnetite) (from the works of Adschiri)

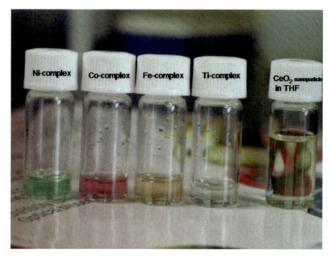

Fig. 11.25. Perfect dispersion of hybrid organic–inorganic nanoparticles of Ni, $CoAl_2O_4$, Fe_3O_4, TiO_2, and CeO_2 in organic solvents

11.6.6 Evaluation Techniques for Organic Ligand Molecules in Organic–Inorganic Hybrid Nanoparticles

Several methods are commonly employed to understand the surface bonding or interaction between the inorganic and organic molecules. The most commonly used techniques are FT-IR and thermal analysis in addition to the specific characteristic properties like luminescence, magnetic, etc. Depending upon the type of inorganic particle and its properties appropriate technique is to be selected. Figure 11.26 shows the FT-IR spectrum of ceria nanoparticles modified with decanoic acid under supercritical conditions. Bands in the 2,800–2,960 cm^{-1} region were attributed to the C–H stretching mode of methyl and methylene groups. The bands at 1,532 and 1,445 cm^{-1} corresponded to the stretching frequency of the carboxylate group, which suggests that the carboxylate group from decanoic acid was chemically bonded to the surface of the ceria nanocrystals and the other hydrocarbon groups were oriented outward [72]. This is probably the result of reactions forming chemical bonds between the nanocrystal surface and the organic ligand molecule in the unique reaction conditions of supercritical water, which are essential for the perfect dispersion of nanocrystals in the organic solvents.

Similarly, the thermogravimetric studies assist in estimating more or less precisely the quantity of organic molecules surrounding the inorganic metal oxide surfaces. Figure 11.27 shows that for the modified iron oxide nanoparticles, by increasing the temperature to 100°C and keeping it constant for several minutes, all the water molecules and hydroxyl groups on the surface will be evaporated and removed. By increasing the temperature until boiling temperature of the modifier, the weight loss corresponds to the

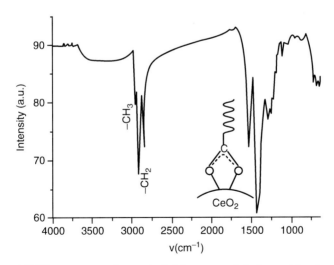

Fig. 11.26. FT-IR spectrum of ceria hybrid nanocrystals formed by decanoic acid-assisted supercritical hydrothermal synthesis

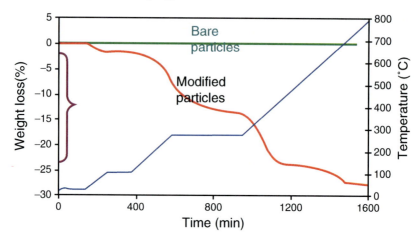

Fig. 11.27. Weight loss of Fe$_2$O$_3$ hybrid nanoparticles modified with decanoic acid and bare particles without modification and water molecules

decomposition of modifier from surface of iron oxide nanoparticles. Keeping temperature constant at 270°C indicates that weight loss is continuing at this temperature.

Such studies are available for all other hybrid organic–inorganic nanoparticles in the literature [59–62, 70, 71].

11.7 Self-Assembly of Hybrid Organic–Inorganic Nanoparticle

Self-assembly is a generic term used to describe a process leading to the ordered arrangement of molecules and small components such as small particles like nanoparticles occurring spontaneously under the influence of certain forces such as chemical reactions, capillary forces, electrostatic attraction, etc. Self-assembly and more generally self-organization of particles in a solvent is considered to be a powerful process for building patterns up to nanoscopic level through multiple interactions among the components of the system under consideration. Ordered assemblies of nanometer size particles represent an interesting class of nanomaterials that provide exceptional potentials for a wide variety of applications. These structures would be useful for various applications like displays, sensors, data storage, photonic band gap materials, etc., owing to their exceptional physical, chemical, and electronic properties [73–76]. Common approaches to noncovalent assembly strategies employ van der Waals/packing interactions, hydrogen bonding, ion pairing, and host–guest inclusion chemistry [77–80]. These self-assembly methods provide direct access to extended structures from appropriately designed nanoparticle building blocks. Although the assembly of nanoparticles from solution into close-packed monolayers and superlattice structures on solid surfaces has met with a fair degree of success [81–83], and in contrary, the controlled assembly of nanoparticles in the organic solvent has not been understood precisely. To meet the major challenge in nanotechnology in realizing many of the desired technological goals, the self-assembly provides a possible convenient route, but controlling size, size distribution, shape, and surface chemistries of the nanohybrid particles is critical in achieving desired structures. Initial syntheses have yielded nearly spherical shapes due to the thermodynamic driving force of minimizing surface area, and self-assembly has been limited to the close packing of spheres [84, 85]. The are several reviews published on this aspect [86, 87]. However, it is to be noted that the self-assembly of hybrid organic–inorganic nanoparticles synthesized through supercritical hydrothermal routes are seldom found in the literature.

The mechanism of making self-assembly structures involves two steps. First step is the positioning of perfect dispersed particles (in solvent) on the substrate by controlling interactions between substrate and particles surfaces (hydrophilicity and hydrophobicity). In the second step, capillary forces between the particles and the surface laterally displace the particles during drying. Steric repulsion between particles because of capping agent on the surface prevents aggregation of nanoparticles. After complete evaporation in the third stage, an irreversible reorganization of the particle–substrate interface occurs that prevents displacement of nanoparticles [88]. Figure 11.28 shows the HRTEM image of the self-assembly of ceria nanoparticles obtained under supercritical hydrothermal conditions in the presence of decanoic acid as the modifier. It is clearly seen from these images that the organic ligand molecules are bonded to the surface of ceria nanocubes. With increasing in

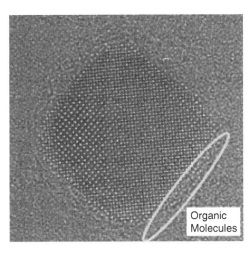

Fig. 11.28. HRTEM image of 7 nm size decanoic acid-modified ceria nanocubes synthesized at 400°C and 30 MPa

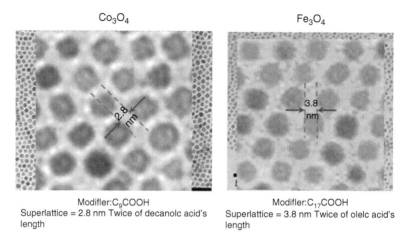

Fig. 11.29. Superlattices of modified Co_3O_4 and Fe_3O_4 synthesized by supercritical hydrothermal method. The gap between the nanoparticles is twice the length of the organic modifier carbon chain

the concentration of organic ligands, particle size not only decreases, but also the morphology changes from cubic to truncated octahedron. The organic layer around the ceria particle is quite thin and it is shown with a yellow loop. Figure 11.29 shows the self-assembly of Co_3O_4 and Fe_3O_4 nanoparticles obtained under supercritical hydrothermal conditions in the presence of organic ligand molecules C_9COOH and $C_{17}COOH$, respectively. The decanoic acid has a carbon chain length of 1.4 nm, and it can be seen that distance between self-assembled nanoparticles is about 2.8 nm. Similarly, oleic acid has a carbon chain length of 1.9 nm, but the distance between the oleic acid-modified magnetite particles is about 3.8 nm. Thus, by choosing different types of modifiers with different carbon chain length, the superlattice of the self-assembled nanoparticles can be monitored precisely.

Similar superlattice structures have been obtained for modified titania, zinc oxide, boehmite, cobalt spinels, etc. [59–62].

This area of self-assembly of the hybrid organic–inorganic nanoparticles is fast growing and it has several advantages in obtaining desired superlattice structures with different packing arrangement for specific applications.

11.8 Conclusion

Supercritical hydrothermal synthesis of advanced nanomaterials like organic–inorganic hybrid nanoparticles has been reviewed in detail starting from the basic principles to the specific features of the methodology. The supercritical fluids are very active in such extreme PT conditions and are highly suitable to synthesize even complex structures, because of the drastic change in the fluid properties like density and dielectric constant, which in turn lower the solubility of the materials. Although supercritical hydrothermal technique gives highly crystalline nanoparticles with homogeneous composition, still there is a problem of larger size, coagulation and poor dispersibility of nanoparticles in aqueous solutions. Hence, there is a trend to use organic ligands, capping agents, surfactants, etc., which generate a new class of nanomaterials, viz., organic–inorganic hybrid nanomaterials. This strategy is based on the miscibility of the organic ligand molecules with supercritical water due to the lower dielectric constant of the water; and the nanocrystal shape control by selective reaction of organic ligand molecules to the specific inorganic crystal surface. A great variety of organic ligand molecules including DNA, proteins, peptides, amino acids, decanoic acid, oleic acid, etc., have been used for this purpose to obtain high quality organic–inorganic nanoparticles, which can be used as sensors, phosphors, data storage, photonic band gap, displays, biomedical, etc. These hybrid nanoparticles can be obtained with such a precise size and shape control. They show excellent dispersibility in the organic solvents and also self-assemble to form superlattice structures. By changing the organic ligands, the particles size and shape can be tailored. Usually the organic layer surrounding the inorganic particle is very thin and it can be clearly observed through HRTEM, FT-IR, and thermal analysis studies. By choosing different types of modifiers with varying carbon chain length, the superlattice of the self-assembled nanoparticles can be monitored precisely. Hence the supercritical hydrothermal synthesis method holds a strategy for the future advanced nanomaterials processing.

Acknowledgments

The authors wish to thank Prof. T. Naka, Prof. S. Ohara, and Prof. M. Umetsu, Institute of Multidisciplinary Research for Advanced Materials, Tohoku University, Japan, for their useful discussion in preparing this chapter. Also, our thanks to the research staff Dr. T. Mousavand, Dr. Y. Hatakeyama, and Dr. J. Zhang, for their cooperation in preparing this chapter.

References

1. N. Taniguichi, *On the Basic Concept of Nanotechnology* (ICPE, Japan, 1974)
2. A.K. Cheetham, P.S.H. Grubstein, *Nanotoday* (16–19 December 2003)
3. R. Rosetti, S. Nakahara, L.E. Brus, J. Chem. Phys. **79**(2), 1086 (1983)
4. A.I. Ekimov, Al.L. Efros, A.A. Onushchenko, Solid State Commun. **56**(11), 921 (1985)
5. A.K. Cheetham, *Hybrid Inorganic–Organic Materials and Their Applications, Plenary Talk – International Conference on Materials for Advanced Technologies (ICMAT-2007)*, 1–6 July 2007, Singapore
6. P.D. Cozzoli, A. Kornowski, H.J. Weller, J. Am. Chem. Soc. **125**, 14548 (2003)

278   T. Adschiri and K. Byrappa

7. E. Katz, I. Willner, in *Nanobiotechnology*, ed. by C.M. Niemeyer, C.A. Mirkin (Wiley-VCH, Weinheim, 2004), pp. 200–226
8. B.M. Novak, Adv. Mater. **5**, 422 (1993)
9. P. Judeinstein, C. Sanchez, J. Mater. Chem. **6**, 511 (1996)
10. D.V. Shevchencko, D.V. Talapin, C.B. Murray, S. O'Brien, J. Am. Chem. Soc. **128**, 3620 (2006)
11. K. Byrappa, M. Yoshimura, *Handbook of Hydrothermal Technology* (Noyes Publications, Park Ridge, NJ, 2001)
12. K. Byrappa, T. Adschiri, Prog. Cryst. Growth Charact. **53**, 117 (2007)
13. K. Byrappa, S. Ohara, T. Adschiri, Adv. Drug Deliv. Rev. **60**, 299 (2008)
14. E. Lester, P. Blood, J. Li, M. Poliakoft, *Reactor Geometry and Supercritical Water Reactions, Joint ISHR & ICSTR 2006*, Sendai, Japan, 5–9 August 2006, p. 62
15. M. Yoshimura, K. Byrappa, J. Mater. Sci. **43**, 2085 (2008)
16. T.M. Seward, E.U. Franck, Phys. Chem. **85**, (1981)
17. K. Sue, Y. Hakuta, R.L. Smitha Jr., T. Adschiri, K. Aria, J. Chem. Eng. Data **44**, 1422 (1999)
18. R.E. Riman, W.L. Suchanek, K. Byrappa, C.W. Chen, P. Shuk, C.S. Oakes, Solid State Ionics **151**, 393 (2002)
19. J.B. Hannay, J. Hogarth, Proc. R. Soc. Lond. **30**, 178 (1880)
20. G. Tammann, *The States of Aggregation* (D. Van Nostrand Company, New York, 1925)
21. J.F. Brennecke, C.A. Eckert, Am. Inst. Chem. Eng. J. **35**, 1409 (1989)
22. K. Sue, S. Muneyuki, A. Kunio, T. Ohashi, H. Ura, K. Matsui, Y. Hakuta, H. Hayashi, M. Watanabe, T. Hiaki, Green Chem. **8**, 634 (2006)
23. T. Adschiri, K. Kanazawa, K. Arai, J. Am. Ceram. Soc. **75**, 2615 (1992)
24. T. Adschiri, K. Kanazawa, K. Arai, J. Am. Ceram. Soc. **75**, 1019 (1992)
25. M. Yoshimura, W.L. Suchanek, K. Byrappa, MRS Bull. **25**(9), 17 (2000)
26. J. Jung, M. Perrut, J. Supercrit. Fluids **20**, 179 (2001)
27. S. Palakodaty, R. Sloan, A. Kordikowski, in *Supercritical Fluid Technology in Materials Science and Engineering*, ed. by Y.-P. Sun (Dekker, New York, 2002), pp. 439–490
28. J.N. Hay, A. Khan, J. Mater. Sci. **37**, 4743 (2002)
29. M. Uematsu, E.U. Franck, J. Phys. Chem. Ref. Data **9**, 1291 (1991)
30. M. Uematsu, in *Supercritical Fluids – Molecular Interaction, Physical Properties, and New Applications*, ed. by Y. Arai, T. Sako, Y. Takebayashi (Springer, Germany, 2002), pp. 71–78
31. H.C. Helgeson, D.H. Kirkham, G.C. Flowers, Am. J. Sci. **281**, 1249 (1981)
32. G.M. Anderson, S. Castet, J. Schoot, R.E. Mesmer, Geochim. Cosmochim. Acta **55**, 1769 (1991)
33. E.U. Franck, Pure Appl. Chem. **53**, 1401 (1981)
34. Y. Hakuta, T. Adschiri, T. Suzuki, T. Chida, K. Seino, K. Arai, J. Am. Ceram. Soc. **81**, 2461 (1998)
35. Y. Hakuta, K. Seino, H. Ura, T. Adschiri, H. Takizawa, K. Arai, J. Mater. Chem. **9**, 2671 (1999)
36. T. Adschiri, Y. Hakuta, K. Arai, Ind. Eng. Chem. Res. **39**, 4901 (2000)
37. T. Adschiri, Y. Hakuta, K. Kanamura, K. Arai, High Pressure Res. **20**, 373 (2001)
38. T. Adschiri, Y. Hakuta, K. Sue, K. Arai, J. Nanoparticle Res. **3**, 227 (2001)

11 Supercritical Hydrothermal Synthesis 279

39. K. Sue, Y. Hakuta, R.L. Smith Jr., T. Adschiri, K. Aria, J. Chem. Eng. Data **44**, 1422 (1999)
40. K. Sue, T. Adschiri, K. Arai, Ind. Eng. Chem. Res. **41**, 3298 (2002)
41. T. Adschiri, K. Arai, in *Supercritical Fluid Technology in Materials Science and Engineering*, ed. by Y.-P. Sun (Dekker, New York, 2002), pp. 311–325
42. A. Cabanas, M. Poliakoff, J. Mater. Chem. **11**, 1408 (2001)
43. E. Reverchon, R. Adami, J. Supercrit. Fluids **37**, 1 (2006)
44. Y. Hakuta, T. Haganuma, K. Sue, T. Adschiri, K. Arai, Mater. Res. Bull. **38**, 1257 (2003)
45. K. Kanamura, A. Goto, R.Y. Ho, T. Umegaki, K. Toyoshima, K. Okada, Y. Hakuta, T. Adschiri, K. Arai, Electrochem. Solid State Lett. **3**, 256 (2000)
46. K. Kanamura, T. Umegaki, K. Toyoshima, K. Okada, Y. Hakuta, T. Adschiri, K. Arai, Electrocerem. Jpn. III Key Eng. Mater. **181**, 147 (2001)
47. M. Le Cleroq, T. Adschiri, K. Arai, Biomass Energy **21**, 73 (2001)
48. K. Sue, N. Kakinuma, T. Adschiri, K. Arai, Ind. Eng. Chem. Res. **43**, 2073 (2004)
49. T. Adschiri, R. Shibata, T. Sato, M. Watanabe, K. Aria, Ind. Eng. Chem. Res. **37**, 2634 (1998)
50. T. Adschiri, S Okazaki, M. Mochiduki, S. Kurosawa, K. Arai, Int. J. Soc. Mater. Eng. Resour. **7**, 273 (1999)
51. H.O. Finklen, R. Vithanage, J. Am. Chem. Soc. 185 (1982)
52. J.D. Miller, H. Ishida, Surf. Sci. **148**, 601 (1984)
53. J.D. Miller, H. Ishida, J. Chem. Phys. **86**, 1593 (1987)
54. Y. Kera, M. Kamada, Y. Hanada, H. Kominami, Compos. Interfaces **8**, 109 (2001)
55. C.A. Mirkin, R.L. Letsinger, R.C. Mucic, J.J. Storhoff, Nature **383**, 607 (1996)
56. J.J. Storhoff, C.A. Mirkin, Chem. Rev. **99**, 1849 (1999)
57. A.P. Alivisatos, K.P. Johnsson, X. Peng, T.E. Wilson, C.J. Loweth, M.P. Bruchez Jr., P.G. Schultz, Nature **382**, 609 (1996)
58. J. Zhang, S. Ohara, M. Umetsu, T. Naka, Y. Hatakeyama, T. Adschiri, Adv. Mater. **19**, 203 (2007)
59. T. Mousavand, S. Takami, M. Umetsu, S. Ohara, T. Adschiri, J. Mater. Sci. **41**, 1445 (2006)
60. T. Mousavand, J. Zhang, S. Ohara, M. Umetsu, T. Naka, T. Adschiri, J. Nanoparticle Res. **6**, 9186 (2007)
61. K. Kaneko, K. Inoke, B. Freitag, A.B. Hungria, P.A. Midley, T.W. Hansen, J. Zhang, S. Ohara, T. Adschiri, Nano Lett. **7**, 421 (2007)
62. T. Mousavand, S. Ohara, M. Umetsu, J. Zhang, S. Takami, T. Naka, T. Adschiri, J. Supercrit. Fluids **40**, 397 (2007)
63. Z. Sun, F. Su, W. Forsling, P.J. Samskog, J. Colloid Interface Sci. **197**, 151 (1998)
64. E. Tombacz, C. Csanaky, E. Illes, Colloid Polym. Sci. **279**, 484 (2000)
65. E. Illes, E. Tombacz, J. Colloid Interface Sci. **295**, 115 (2006)
66. A.S. Piers, C.H. Rochester, J. Colloid Interface Sci. **174**, 97 (1995)
67. S. Liufu, H. Xiao, Y. Li, J. Colloid Interface Sci. **281**, 155 (2005)
68. http://www.cas.astate.edu/draganjac/Oleicacid.html (2006)
69. A.J. Lewis, J. Am. Ceram. Soc. **83**, 2342 (2000)
70. T. Mousavand, J. Zhang, S. Ohara, M. Umetsu, T. Naka, T. Adschiri, J. Nanoparticle Res. **9**, 1067 (2006)

280   T. Adschiri and K. Byrappa

71. D. Rangappa, T. Naka, A. Kondo, M. Ishii, T. Kobayashi, T. Adschiri, J. Am. Chem. Soc. **129**, 11061 (2007)
72. J.E. Tackett, Appl. Spectrosc. **43**, 483 (1989)
73. M.P. Pileni, J. Phys. Chem. B **105**, 3358 (2001)
74. J. Dutta, H. Hofmann, Encycl. Nanosci. Nanotechnol. **9**, 617 (2003)
75. A.C. Templeton, W.P. Wuelfing, R.W. Murray, Acc. Chem. Res. **33**, 27 (2000)
76. V.F. Puntes, K.M. Krishnan, A.P. Alivisatos, Science **291**, 2115 (2001)
77. J.E. Martin, J.P. Wilcoxon, J. Odinek, P. Provencio, J. Phys. Chem. B **104**, 9475 (2000)
78. A.K. Boal, V.M. Rotello, Langmuir **16**, 9527 (2000)
79. J. Kolny, A. Kornowski, H. Weller, Nano Lett. **2**, 361 (2002)
80. J. Liu, S. Mendoza, E. Roman, M.J. Lynn, R. Xu, A.E. Kaifer, J. Am. Chem. Soc. **121**, 4304 (1999)
81. M. Brust, M. Walkder, D. Bethell, D.J. Schiffrin, R. Whyman, J. Chem. Soc., Chem. Commun. 801 (1994)
82. J.H. Fendler, F. Meldrum, Adv. Mater. **7**, 607 (1995)
83. S. Yamamuro, D.G. Farrell, S.A. Majetich, Phys. Rev. B **65**, 224431 (2002)
84. C.B. Murray, D.J. Norris, M.G. Bawendi, J. Am. Chem. Soc. **115**, 8706 (1993)
85. C.M. Murray, C.R. Kagan, M.G. Bawendi, Science **270**, 1335 (1995)
86. E.V. Shevchenko, D.V. Talapin, C.B. Murray, S. O'Brien, J. Am. Chem. Soc. **128**, 3620 (2006)
87. J. Park, J. Joo, S.G. Kwon, Y. Jang, T. Hyeon, Angew. Chem. Int. Ed. **46**, 4630 (2007)
88. P.A. Kralchevsky, N.D. Denkov, Curr. Opin. Colloid Interface Sci. **6**, 383 (2001)

Index

absorption coefficient, 209, 210, 281
absorption edge, 281, 283, 284
absorption spectroscopy, 207, 209
absorption spectrum, 12, 29, 207–210
acrylamide, 103–105, 114, 115
action potential, 198
active center, 197, 205
AFM, 14, 119, 130, 134
Ag nanoparticle, 24, 57, 58, 60–62, 95
Ag-core/Pd-shell, 58, 68
Ag-core/Rh-shell, 58, 68
Al_2O_3, 222
ALA, 202
alcothermal, 249
Allyl Precursors, 161
AlMepO, 172
$ALPO_4$, 172, 188
5-aminolevulinic acid (ALA), 202
ammonothermal, 249
amphoteric metal oxides, 265
anemia, 196, 197
angular momentum, 211
anomalous diffraction, 205
anomalous scattering, *see* anomalous
    diffraction
apoprotein, 200
aqueous solvents, 249
ASES, 251
asymmetry, 214
atom transfer radical polymerization
    (ATRP), 114–116
atomic cross section, 210
atomic force microscopy (AFM), 134

Au-core/Pt-interlayer/Rh-shell
    trimetallic nanoparticle, 68, 73, 74
Au/Pt bimetallic nanoparticle, 69,
    72–74
Au/Pt/Rh, 74, 76
Au/Pt/Rh trimetallic nanoparticle,
    68–72, 74–77
Auger electron, 226, 239
average diameter, 7, 9, 28, 61, 63, 65–67,
    74, 75, 236

back-scattering, 209
$BaFe_{12}O_{19}$, 258
*BEA, 171, 172
batch reactor, 251, 252
$BaTiO_3$, 258
$BaZrO_3$, 258
Bijvoet mate, 206, 207
bilirubin, 202
biliverdin, 202
bimetallic, 24, 56, 58, 59, 69, 72, 74–76
bimetallic nanoparticle, 56, 57, 59,
    62–69, 73, 74, 76
binding energy, 72, 73
Biological Molecular Interaction, 125
biological specific interactions, 125
biosynthesis, 200, 202
biprism, 228
bis(triethoxysilyl)methane (BTESM),
    172
blood coagulation, 198
blood plasma, 197
boehmite, 271, 276
boehmite nanoparticle, 270, 271

282 Index

Bohr's magneton, 211
Bragg's law, 206
bridging organic group, 171
bright-field, 221, 224, 236, 238, 239
Brust–Schiffrin method, 6

C_s corrector, 221
c-type cytochrome, 201, 202
capping agent, 5, 26, 30, 261, 275, 277
carbonothermal, 249
catalytic activity, 24, 55, 56, 59, 62,
    64–68, 70, 71, 74–76, 131, 134,
    155, 211
cell division, 198
CeO_2, 258, 265, 266, 272, 273
CeO_2 nanoparticle, 259, 267
ceria nanocrystals, 268, 269, 272
ceruloplasmin, see CP
channel electron, 205
chaperone, 200
characteristic X-ray, 224, 226
chelating agents, 154, 261
Chiral PMOs, 155
chromophores, 211
^{13}C NMR, 174, 185
CO, 61, 70
Co_3O_4, 265, 266, 276
Co_3O_4 nanoparticle, 269, 270
co-condensation, 142–146, 149, 152,
    154, 155, 163
CO-FT-IR, 62–64
$CoAl_2O_4$, 261, 265, 271–273
cobalt spinels, 276
cofactor, 125, 131, 136, 193, 199, 200
$CoFe_2O_4$, 258
colloidal probe AFM, 126, 127, 131
colloidal probe atomic force microscopy,
    125, 136
complementary nucleic acid base
    pairing, 126
conduction band, 22, 95, 96
CoO, 239, 242, 258
coordination, 4, 10, 195, 200, 209–211,
    248
core-shell nanocrystal, 82
Core-Shell Nanoparticles, 114
core-shell structure, 55, 116, 219
core-shell-type nanocrystal, 90

coreduction, 56, 72
coupling constant, 212
CP, 197
critical point, 250, 253–255, 257, 258
crystal, 26, 41, 42, 44–46, 48, 49, 63,
    64, 69, 70, 81, 84–90, 94, 148, 172,
    174, 180, 182–184, 204–207, 209,
    211, 220, 259, 260, 262, 265, 269,
    270, 277
crystal-like pore walls, 141, 157–163
Crystallography, 204–207
Cu_2O, 258
cubic silsesquioxanes, 4, 15, 19
CuO, 257
cytochrome c oxidase, 201
cytochrome b5, 201
cytochrome P450, 201
cytochromes, 202, 203
cytoplasm, 200
cytosol, 198

D-amino acid, 205
D3R, 188
dark-field, 223
DELOS, 251
deoxy, 196, 202
Deoxyribonucleic acid, see DNA
depolarization, 198
diamagnetic, 211
dielectric constant, 93, 98, 99, 253–255,
    258, 262, 277
difference spectroscopy, 211
diffraction, 45, 48, 85, 95, 96, 158, 159,
    161, 162, 174, 183, 204, 205, 209,
    220–223, 226, 230, 238–242
dimple grinder, 231
dipolar dephasing pulse sequence, 185,
    186
dipole–dipole interaction, 97, 185
Direct method, 206
Dispersibility, 261, 263, 271, 277
dispersibility of nanoparticle, 261, 277
Dispersibility of Organic–Inorganic
    Hybrid Nanoparticle, 271
dispersion of the nanoparticle, 55, 263
dissociation constant, 265
DNA, 10, 14, 125, 136, 193, 198, 203,
    204, 262, 263, 277

dose, 205, 209
double-probe piezodriving holder, 243
drug design, 204

EDS, 63, 69, 225, 226, 228, 231, 235, 236, 240
EELS, 62, 219, 225, 226, 228, 231
EF-TEM, 58, 62–64, 76
elastic scattering, 224
electric-dipole, 210
electromagnetic wave, 206
electron density, 56, 72, 200, 206
Electron Energy-Loss Spectroscopy, 224
Electron holography, 227–230, 240, 242
Electron Paramagnetic Resonance, *see* EPR
electron transfer, 12, 26, 68, 72, 104, 200, 202, 203, 211, 214
electron-nuclear double resonance, *see* ENDOR
electronic effect, 55, 59, 74
electrostatic interaction, 14, 19, 21, 24, 25, 31, 33, 104
Elemental Mapping, 219, 226–228
ENDOR, 212
energy dispersive X-ray spectroscopy, 219, 225
energy filter, 226, 227
ensemble effect, 55, 76
Enzyme, 125, 130, 134–136, 197, 200, 202–205
Enzyme–Substrate Complexes, 130
EPR, 211, 212
ESEM, 111, 112
ex-situ surface modification, 263
EXAFS, 209
excitonic absorption (EA), 84, 87, 93
exothermic interaction between particles, 73
extended X-ray absorption fine structure, *see* EXAFS
extinction, 87–89, 93, 97–99, 104, 112, 113, 121

α-Fe_2O_3, 48, 51, 258
α-Fe_2O_3 particle, 48, 49, 51, 52
FAU, 172
Fe_2CoO_4, 258
Fe_2O_3, 24, 52, 261, 265, 266, 274

Fe_3O_4, 52, 115–121, 272, 273, 276
Fe–S, 200
Fe–S cluster, 200, 201
ferredoxin, 200
ferric, 211
ferricytochrome c, 211
Ferrochelatase, 202
ferrous, 202
FIB, 219, 232, 233, 239, 240, 243
flow reactor, 252–254, 260
focused ion beam, 219, 230, 232, 243
Fourier transformation, 206
frataxin, 200
FSM-16, 141
FT-IR, 23, 61–64, 70, 74, 76, 116, 117, 272, 274, 277

g-value, 211
Ga_2O_3, 258
GAS, 251
Gel-Sol method, 44, 45, 48
GeO_2, 258
geographic effect, 56
Glomus cells, 198
glucuronic acid, 202
glycogen, 198
glycothermal, 249
Goldman-Shen pulse sequence, 187
grafting, 11, 12, 114, 116, 141–143, 145, 146, 152

HAADF, 219, 223, 224, 226, 228
Hb, 194–197, 201, 203, 204
Hc, 197, 239
hematite particle, 271
heme, 157, 194, 196, 201, 202, 205, 207, 210–212
heme oxygenase, 205
hemerythrin, *see* Hr
hemocyanin, *see* Hc
hemoglobin, *see* Hb
High-Angle Annular Dark-Field, 223
high-resolution electron microscope image, 221, 222
high-resolution electron microscopy, 221, 224, 234
high-resolution transmission electron microscopy, 219

284   Index

high-spin, 211
HKF model, 255
HmuO, 205
Horse radish peroxidase, *see* HRP
Hr, 202
HR-TEM, 63, 64, 69, 74, 76
HREM, 219, 234–237, 239
HRP, 208, 209
HTSSB, 251
HTSSF, 251
hybrid material, 41, 42, 161, 178
hybrid metal oxide, 248, 263
Hybrid Nanoparticle, 247, 248, 251,
    262, 265, 266, 271, 274, 277
hybrid organic–inorganic nanoparticle,
    252, 255, 262, 273, 276
hydrogen bond, 14, 19, 27, 31, 194, 265,
    266, 275
hydrogenation of methyl acrylate, 64,
    65, 70, 74, 75
hydroperoxo, 202
hydrophilicity, 165, 264, 275
hydrophobic interaction, 117, 201
hydrophobicity, 163, 165, 177, 178, 262,
    264, 275
hyperfine, 212, 215
hypoxia, 198

π-A isotherm, 116–118
in situ reduction, 12, 15, 82, 90, 98
in situ surface modification, 264
in situ surface modification of metal
    oxide nanoparticle, 267
in situ surface modification of
    nanoparticle, 263
In_2O_3, 258
inelastic scattering, 224, 225
infrared, 48, 51, 159, 207, 211
intelligent engineering, 250
Ion Milling, 231, 232
ionic bonds, 194
ionic liquid, 28–31
ionic species, 255, 256
ionization potential, 59, 68, 73, 74
iron-sulfur, 199, 212, 214
iron-sulfur cluster, 199, 203
IS, 213–215
isoelectric point, 265
isomer shift, *see* IS

isomorphous replacement, 206, 207
isothermal titration calorimetry (ITC),
    68
ITC, 73
ITO, 104

Kendrew, 204
kinetics of supercritical hydrothermal
    synthesis, 257

L-amino acid, 205
La_2O_3, 258
Langmuir–Blodgett (LB), 14, 103, 125,
    126
lattice, 43, 51, 70, 88, 95, 96, 158, 161,
    172, 183, 204, 206, 213, 220–222,
    231, 234–236
Laue condition, 206
layer-by-layer (LBL), 21, 103
LB modification method, 125
Lewis acid, 203
Lewis base, 203
$LiCoO_2$, 258
ligand, 5, 8–12, 21, 32, 55, 116, 142,
    145, 155, 207, 210–212, 262, 263,
    265, 269–271, 276, 277
ligand effect, 76
ligand-metal charge-transfer, 210
ligand-stabilized gold nanoparticles, 5
ligand-stabilized metal nanoparticles, 8
$LiMn_2O_4$, 258, 260
$LiMn_2O_4$ nanoparticle, 260
Liquid crystal, 42, 149
localized surface plasmon (LSP), 93,
    104, 121
localized surface plasmon resonance
    (LSPR), 111
LTA, 173, 174, 178, 181, 182
luminescence, 121, 156, 259, 272
lyothermal, 249

M41S, 141
macromolecule, 14, 15, 151, 193, 199,
    204
magnetic, 13, 32, 41, 52, 55, 114, 120,
    121, 157, 211, 212, 214, 219, 230,
    233, 234, 236–240, 242, 247, 258,
    272

magnetic dipole moment, 211
magnetic field, 43, 44, 211, 215, 219, 227, 243
Magnetic Nanoparticle, 105, 114–116, 233
Mb, 196, 201, 204, 207, 208
Mechanism for size control, 270
Mechanism of Formation of Organic–Inorganic Hybrid Nanoparticles, 265
mesoporous silica, 141–146, 148, 150–152, 154–157, 163, 165, 171, 188
Metal Complex: PMO, 154
metal nanoparticle, 3–6, 9–16, 18, 19, 21–24, 27, 28, 30–33, 55, 56, 58, 70, 72, 77, 81, 104, 106, 107, 111, 112, 121, 233
Metal oxide, 25, 233, 239, 248, 250, 253, 255, 257–259, 262, 264–266
Metalloenzymes, 203, 204
metalloprotein, 193, 202–205, 207, 209–211, 213
MFI, 172, 174, 183
microdiffraction, 220
microspectrophotometer, 205, 208, 209
γ-MnO_2, 258
modified Co_3O_4, 276
modified metal oxide particles, 264
modified titania, 276
molar extinction coefficient, 111, 112
molecular periodicity, 157, 158, 165
molecular recognition, 125, 126
molecular replacement method, 206
monofunctionalized gold nanoparticles, 9
monosilylated organic precursor, 141, 143, 144, 146
MOR, 174
Mössbauer, 213–215
Multifunctionalized PMOs, 152
multiple wavelength anomalous diffraction method, 206
muscle contraction, 198
myofibril, 198
myoglobin (Mb), 196

N2 fixation, 203
Nanoparticle, 3–25, 27–33, 41–45, 47, 52, 55–62, 65–77, 82–99, 103, 104, 106–113, 115–122, 157, 234–237, 247–251, 259–266, 268–273, 275–277
neuronal, 198
Neuronal signals, 198
neurotransmitter, 198, 199
Ni, 233, 272, 273
Ni nanoparticle, 260, 261
NiO, 258
nitrogenase, 214
NLO, 91, 94
NMR, 116, 151, 174, 176, 178, 180–183, 185–187, 212, 213
NOE, 213
Nonlinear optical (NLO) materials, 81
Nuclear Magnetic Resonance, 212
Nuclear Overhauser, 212
Nuclear Overhauser Effects, 213
nucleation, 85, 90, 180, 182, 249, 258, 270

OFMSs, 171, 172
ordered benzene-bridged hybrid LCs, 158
organic framework, 171, 180, 183, 187, 188
organic ligand, 16, 199, 248, 261–265, 268–273, 276, 277
organic modifier, 263, 265, 276
organic solvents, 30, 248, 251, 253, 262, 263, 271, 273, 275, 277
organic–inorganic hybrid LCs, 41, 42, 44, 45, 51, 52, 141, 164, 172, 188, 248, 262, 263, 271, 277
Organic–Inorganic Hybrid Liquid Crystal, 41, 44
Organic–Inorganic Hybrid Nanoparticle, 247, 251, 262, 263, 265, 268, 272
Organic–Inorganic Hybrid Zeolites, 171, 188
Organic–inorganic hybridization, 41, 44, 171
organic–inorganic nanoparticle, 273, 275

286   Index

organic-ligand-assisted supercritical hydrothermal technology, 263
oxidation state, 207, 213, 214

PAMAM dendimer, 15–19
paramagnetic, 211–213, 239
PBG, 202
PCA, 251
Pd-core/Ag-shell, 59, 61, 62, 64, 68
Pd-core/Au-interlayer/Ag-shell, 59
Pd-nanoparticle, 24, 57–61
Pd/Ag bimetallic nanoparticle, 59–62, 64
Pd/Ag/Rh Trimetallic Nanoparticle, 59, 63–68, 75–77
Pd/Pt/Rh, 75, 76
PDA Nanocrystal, 82–99
peroxidase, 157, 202, 209
peroxo, 202
Perutz, 204
PGSS, 251
phase, 5–7, 9, 10, 19, 27, 30, 31, 41, 43, 45–52, 55, 81, 82, 89, 116, 129, 130, 144, 147, 148, 157, 174, 175, 181, 207, 226–230, 232, 236–240, 242, 249, 250, 253–256, 261, 264, 271
phase angle, 206, 207, 209
Phase Problems, 205, 206
phasing problem, *see* phase problem
Photoactive PMOs, 156
photocatalytic reduction, 82, 94, 96–98
photoelectron, 205, 209
photon, 205, 209, 213
photopattern, 109
physical mixture method, 58, 59
plasma membrane, 198
plasmon band, 16, 22, 28, 57, 93
PMOs, 142, 146–150, 152–159, 161, 164, 165
poly(N-vinyl-2-pyrrolidone) (PVP), 56
Poly(histidine)-tagged (His-tag) proteins, 127
polydiacetylene (PDA), 81
Polymer Grafted Metal Nanoparticle, 10
polymer monolayer, 103
polypeptide, 43, 193, 194
polyscale, 249

polysilylated organic precursors, 141, 146, 147
porphobilinogen, *see* PBG
porphyrin, 196, 201–203, 207, 210, 212
postsynaptic, 198, 199
presynaptic, 199
presynaptic cell, 198
prosthetic group, 196, 199, 201
protoheme, 202
protoporphyrin, 201, 202
Pt/Pd/Rh, 75
Pt/Pd/Rh trimetallic nanoparticle, 74–77
public database, 281
PVT relation of water, 253

QS, 214, 215
quadrupole splitting, *see* QS
quaternary structure, 196

radiation, 209
radiation damage, 205
reaction center, 202
recoil energy, 213
recombinant, 204
reconstructed phase image, 230, 237, 242
redox, 196, 201–203, 205, 209, 213
redox potential, 97
redshift, 112
Reprecipitation Method, 83, 85, 89
residence time, 258
resolution, 212
Resonance Raman, *see* RR
respiration, 202
respiratory chain, 201
RESS, 251
Rh nanoparticle, 59, 62, 66
ribonucleic acid, *see* RNA
RNA, 193, 203
RR, 210

scH$_2$O, 252, 258, 262
SAA, 251
sacrificial hydrogen reduction method, 56, 62
SAS, 251

Index   287

SAS-EM, 251
scaffold proteins, 200
scalar couplings, 213
scanning transmission electron microscopy (STEM), 219
scattering factor, 207
SCFD, 251
SEDS, 251
seed deposition, 82, 90, 91
seed growth, 82
selected area diffraction, 220, 238
self-assembled nanoparticle structure, 248
Self-assembly, 151, 273, 274
Self-Assembly of Hybrid Organic-Inorganic Nanoparticle, 273
self-assembly of the organic-inorganic hybrid nanoparticle, 276
self-assembly structures, 275
self-organization method, 56, 62, 69
Semi-pilot plant flow reactor, 254
sequential electron transfer, 68
sequential electronic charge transfer, 73
SFEE, 251
^{29}Si-NMR, 147, 152, 174
Si–CH$_2$–Si linkage, 182
signal transmission, 198
Silane coupling agent, 262, 263
silsesquioxane precursor, 149
simultaneous reduction, 56
SiO$_2$, 258
site-directed mutagenesis, 204
Size and Shape Control, 51, 268
size distribution histogram, 69
skeletal muscle, 198
solid-state polymerization, 81, 84, 93
solubility, 253, 257
solvothermal, 249
Soret band, 207
space group, 205
spherical aberration constant, 221
spin, 207, 214
spin diffusion, 188
spin–lattice relaxation (T1c), 184
spin–orbit interaction, 211
SSF, 251
STEM, 223, 228
structure factor, 206

structure-directing agent, 174
subunit, 196, 202
succinyl-CoA, 202
sugar, 198
sulfuric acid, 163
supercritical fluid technique, 250
supercritical hydrothermal, 249
supercritical hydrothermal flow reactor, 252
supercritical hydrothermal synthesis, 257, 259, 274
Supercritical Hydrothermal Synthesis of Hybrid Organic–Inorganic Nanoparticles, 262
Supercritical Hydrothermal Synthesis of Metal Oxides, 258
supercritical hydrothermal synthesis of nanoparticle, 277
Supercritical Hydrothermal Technique, 249
supercritical water, 253
superhyperfine, 212
Superlattice, 276
superlattice of the self-assembled nanoparticles, 276
superlattice structure, 275, 277
superparamagnetic, 120
surface area, 66, 71
surface chemistries, 275
Surface Forces Measurement, 125
surface modification, 249, 262, 263, 265
surface modifiers, 263
surface plasmon coupling, 112
surface plasmon resonance (SPR), 106
surface-catalyzed reducing agent, 92
Surfactant, 88, 90, 92, 267, 277
symmetry, 205
synapse, 198
synaptic cleft, 198
synchrotron, 209, 210

T1c decay, 184
TEM, 63
tetrathiafulvalene, 26
thermal analysis, 272
thermal stability, 178
thermochemical computation, 250
thermogravimetric, 273

288   Index

three-dimensional ordered structures or
   superlattice structures, 248
three-layered core/shell structure, 59
TiO_2, 266, 272, 273
TiO_2 nanoparticles, 44
Torchia's pulse sequence, 185
torsion, 212
Transcription Factor, 136
transferrin, 197
trimetallic, 58
Trimetallic Nanoparticles, 59
tropomyosin, 198
troponins, 198
two-dimensional ordered structures or
   super-lattice structure, 248

Ultramicrotomy, 230, 231, 239
uroporphyrinogen III, 202

V_2O_5, 258
valence band, 95
van der Waals attraction, 194
Vibrational Spectroscopy, 210

Vitamin, 198
VPI-5, 172

X-ray, 207, 209, 212, 213, 226
X-ray absorption, 209
X-ray absorption fine structure, *see*
   XAFS
X-ray absorption spectroscopy, *see* XAS
XAFS, 210
XAS, 209
XRD, 69

YAG:Tb nanoparticles, 259
YAG:Tb phosphors nanoparticle, 259

Z-contrast, 223
zinc finger, 203
zinc oxide, 276
$ZnAl_2O_4$, 258
$ZnFe_2O_4$, 258
ZnO, 258, 263, 271
ZOL, 174
ZrO_2, 258